United Nations,
Economic Commission for Europe,
Geneva

Food and Agriculture Organization
of the United

W9-AOD-288

European Timber Trends and Prospects to the year 2000 and beyond

Volume II

UNITED NATIONS
New York, 1986

NOTE

The designations employed and the presentation of material in this publication do not imply the expression of any opinion whatsoever on the part of the Secretariat of the United Nations concerning the legal status of any country, territory, city or area or of its authorities, or concerning the delimitation of its frontiers or boundaries.

ECE/TIM/30
Volume II

UNITED NATIONS PUBLICATION

Sales No. E.86.II.E.19

ISBN 92-1-116368-4
(complete set of two volumes)
ISBN 92-1-116370-6 (Vol. II)

12000P, Vols. I and II
Not to be sold separately

Preface

This volume contains the annexes, mostly of a statistical nature, which are referred to in volume I of the study.

Annex tables, annex figures and annexes consisting mostly of text are numbered consecutively. The numbers of the annexes refer to the chapters in volume I, e.g. annex table 12.8 concerns subjects treated in chapter 12.

Abbreviations, symbols, country groups and data sources used are presented in the preface to volume I.

Almost all tables and figures are bilingual. For technical reasons, a few annexes are presented in English only.

Préface

Ce volume contient les annexes, essentiellement statistiques, auxquelles se réfère le volume I de l'étude.

Les tableaux, les figures, les textes de l'annexe sont numérotés en fonction des chapitres du volume I, par exemple le tableau annexe 12.8 se rapporte aux sujets traités dans le chapitre 12.

Les abréviations, les symboles, les groupes de pays et les sources de données sont décrits dans la préface au volume I.

La plupart des tableaux et des figures sont bilingues. Pour des raisons techniques, quelques annexes ne sont présentées qu'en anglais.

Table of contents

Table des matières

Annex table 1.1
Europe: population 1950-2025

Tableau annexe 1.1
Europe: population 1950-2025

(millions)

Country	1950	1965	1980	2000 a/	2025 a/	Pays
Finland	4.0	4.6	4.8	5.0	4.8	Finlande
Iceland	0.1	0.2	0.2	0.3	0.3	Islande
Norway	3.3	3.7	4.1	4.2	4.3	Norvège
Sweden	7.0	7.7	8.3	8.1	7.5	Suède
NORDIC COUNTRIES	14.4	16.2	17.4	17.6	16.9	PAYS NORDIQUES
Belgium-Luxembourg	8.9	9.8	10.2	10.3	10.2	Belgique-Luxembourg
Denmark	4.3	4.8	5.1	5.1	4.8	Danemark
France	41.7	48.8	53.8	57.1	58.5	France
Germany, Fed.Rep. of	50.0	59.0	61.7	59.8	53.8	Allemagne, Rép.féd. d'
Ireland	3.0	2.9	3.4	4.2	5.2	Irlande
Italy	46.8	51.9	56.2	58.2	56.9	Italie
Netherlands	10.1	12.3	14.2	15.0	14.6	Pays-Bas
United Kingdom	50.6	54.5	55.7	56.2	56.4	Royaume-Uni
EEC(9)	215.4	244.0	260.3	265.9	260.4	CEE(9)
Austria	6.9	7.3	7.5	7.5	7.3	Autriche
Switzerland	4.7	5.9	6.4	5.9	4.9	Suisse
CENTRAL EUROPE	11.6	13.2	13.9	13.4	12.2	EUROPE CENTRALE
Israel	1.3	2.6	3.9	5.4	7.0	Israël
Malta	0.3	0.3	0.4	0.4	0.5	Malte
Greece	7.6	8.6	9.6	10.7	11.8	Grèce
Cyprus	0.5	0.6	0.6	0.8	0.9	Chypre
Portugal	8.4	9.2	9.7	11.0	11.9	Portugal
Spain	27.9	31.9	37.5	43.4	49.2	Espagne
Turkey	20.8	31.2	44.5	68.5	99.3	Turquie
Yugoslavia	16.3	19.4	22.3	25.2	26.6	Yougoslavie
SOUTHERN EUROPE	83.1	103.8	128.5	165.4	207.2	EUROPE MERIDIONALE
Bulgaria	7.3	8.2	9.0	9.7	10.2	Bulgarie
Czechoslovakia	12.4	14.2	15.3	16.8	18.4	Tchécoslovaquie
German Dem. Rep.	18.4	17.0	16.7	16.6	16.1	Rép. dém. allemande
Hungary	9.3	10.1	10.7	10.9	10.9	Hongrie
Poland	24.8	31.5	35.8	41.4	45.9	Pologne
Romania	16.3	19.0	22.2	25.6	29.2	Roumanie
EASTERN EUROPE	88.5	100.1	109.8	121.1	131.2	EUROPE ORIENTALE
EUROPE	413.0	477.3	529.9	583.4	627.9	EUROPE

a/   UN medium variant / Variante moyenne ONU.

Annex table 1.2
Past GDP and scenarios to 2000

Tableau annexe 1.2
PIB dans le passé et scénarios jusqu'à 2000

(billion US$ at 1980 prices and exchange rates)
(millard $US aux prix et aux taux de change)

Country	Actual/Réel			Low/Bas		High/Haut		Pays
	1965	1973	1980	1990	2000	1990	2000	
Finland	25.2	37.0	44.8	59.0	84.0	61.6	97.5	Finlande
Iceland	1.3	2.0	2.3	2.4	3.4	2.5	4.0	Islande
Norway	29.9	41.7	57.3	70.5	106.3	73.6	123.3	Norvège
Sweden	73.4	92.6	112.6	132.6	183.4	139.2	214.1	Suède
NORDIC COUNTRIES	129.8	173.3	217.0	264.5	377.1	276.9	438.9	PAYS NORDIQUES
Belgium-Luxembourg	71.7	105.2	123.9	142.2	193.9	147.1	219.9	Belgique-Luxembourg
Denmark	36.9	51.2	56.2	67.8	91.1	69.2	94.8	Danemark
France	351.6	545.4	652.9	795.0	1154.3	818.3	1270.7	France
Germany, Fed.Rep. of	504.9	704.3	825.3	961.8	1286.3	999.8	1430.8	Allemagne, Rép.féd. d'
Ireland	8.3	11.9	15.5	19.3	27.8	20.8	37.0	Irlande
Italy	220.3	324.4	395.6	476.9	672.8	490.9	733.7	Italie
Netherlands	91.8	136.8	159.4	174.2	229.6	180.2	251.8	Pays-Bas
United Kingdom	373.9	476.8	515.3	577.6	725.1	603.7	788.0	Royaume-Uni
EEC(9)	1659.4	2356.0	2744.1	3214.8	4380.9	3330.0	4826.7	CEE(9)
Austria	40.9	61.8	77.1	93.5	134.4	100.5	159.1	Autriche
Switzerland	73.2	99.6	101.8	112.5	141.2	118.7	165.8	Suisse
CENTRAL EUROPE	114.1	161.4	178.9	206.0	275.6	219.2	324.9	EUROPE CENTRALE
Israel	8.4	15.7	20.2	28.4	44.2	31.9	58.7	Israël
Malta	0.3	0.6	1.1	1.6	2.6	1.7	3.1	Malte
Greece	16.1	29.0	35.9	44.6	67.9	47.2	82.1	Grèce
Cyprus	1.0	1.6	2.0	2.9	4.3	3.1	5.3	Chypre
Portugal	9.7	16.9	21.9	28.5	43.4	30.3	53.1	Portugal
Spain	99.8	163.8	202.6	246.8	348.1	256.5	394.6	Espagne
Turkey	24.1	39.7	57.7	93.1	159.0	98.0	196.5	Turquie
Yugoslavia	25.5	38.5	56.9	76.7	128.6	78.9	141.3	Yougoslavie
SOUTHERN EUROPE	184.9	305.8	398.3	522.3	798.1	547.6	934.7	EUROPE MERIDIONALE
Bulgaria	8.0	15.3	24.1	39.4	73.9	42.3	91.2	Bulgarie
Czechoslovakia	44.2	71.8	89.5	106.5	150.2	117.8	184.8	Tchécoslovaquie
German Dem. Rep.	46.9	70.6	95.9	146.9	205.2	152.6	230.2	Rép. dém. allemande
Hungary	11.1	18.8	25.1	34.4	52.4	36.8	61.1	Hongrie
Poland	30.1	50.6	65.0	68.1	99.8	74.2	192.2	Pologne
Romania	12.0	24.2	42.4	68.1	124.9	73.1	161.5	Roumanie
EASTERN EUROPE	152.3	251.3	342.0	463.4	706.4	496.8	921.0	EUROPE ORIENTALE
EUROPE	2240.5	3247.8	3880.3	4671.0	6538.1	4870.5	7446.2	EUROPE

Annex table 1.3
Past and projected rates
of GDP growth

Tableau annexe 1.3
Taux de croissance du PIB,
passé et avenir

(Average annual percent change)
(Taux moyens annuels)
(%)

Country	1965-1973	1973-1980	1980-2000 Low/Bas	1980-2000 High/Haut	Pays
Finland	4.9	2.8	3.2	4.0	Finlande
Iceland	5.5	2.1	2.0	2.8	Islande
Norway	4.2	4.6	3.1	3.9	Norvège
Sweden	3.0	2.8	2.5	3.3	Suède
NORDIC COUNTRIES	3.7	3.3	2.8	3.6	PAYS NORDIQUES
Belgium-Luxembourg	3.9	2.4	2.3	2.9	Belgique-Luxembourg
Denmark	4.2	1.3	2.4	2.6	Danemark
France	5.7	2.6	2.9	3.4	France
Germany, Fed.Rep. of	4.2	2.3	2.2	2.8	Allemagne, Rép.féd. d'
Ireland	4.6	3.9	2.9	4.4	Irlande
Italy	5.0	2.9	2.7	3.1	Italie
Netherlands	5.1	2.2	1.8	2.3	Pays-Bas
United Kingdom	3.1	1.1	1.7	2.2	Royaume-Uni
EEC(9)	4.5	2.2	2.4	2.9	CEE(9)
Austria	5.3	3.2	2.8	3.7	Autriche
Switzerland	3.9	0.3	1.6	2.5	Suisse
CENTRAL EUROPE	4.4	1.5	2.2	3.0	EUROPE CENTRALE
Israel	8.1	3.7	4.0	5.5	Israël
Malta	7.1	10.3	4.3	5.2	Malte
Greece	7.6	3.1	3.2	4.2	Grèce
Cyprus	6.4	3.1	3.9	5.1	Chypre
Portugal	7.2	3.8	3.5	4.5	Portugal
Spain	6.4	3.1	2.7	3.4	Espagne
Turkey	6.4	5.5	5.2	6.3	Turquie
Yugoslavia	5.3	5.7	4.2	4.7	Yougoslavie
SOUTHERN EUROPE	6.5	3.9	3.5	4.3	EUROPE MERIDIONALE
Bulgaria	8.4	6.7	5.8	6.9	Bulgarie
Czechoslovakia	6.2	3.2	2.6	3.7	Tchécoslovaquie
German Dem. Rep.	5.2	4.5	3.9	4.5	Rép. dém. allemande
Hungary	6.8	4.2	3.7	4.5	Hongrie
Poland	6.7	3.7	2.2	3.5	Pologne
Romania	9.2	8.4	5.6	6.9	Roumanie
EASTERN EUROPE	6.5	4.5	3.7	5.1	EUROPE ORIENTALE
EUROPE	4.7	2.6	2.6	3.3	EUROPE

Annex table 1.4
Residential investment in
selected countries

Tableau annexe 1.4
Investissement résidentiel
dans certains pays

(billions of national currency units, constant prices)
(milliards d'unités de monnaie nationale, prix constants)

Country	1965	1970	1975	1980	Pays
Finland	4.5	6.2	8.0	7.6	Finlande
Iceland	21.8	18.3	30.4	31.2	Islande
Norway	4.3	6.1	8.1	9.6	Norvège
Sweden	12.5	13.4	12.1	12.2	Suède
Belgium	128.6	126.5	144.8	156.5	Belgique
Denmark	13.0	16.9	14.5	12.3	Danemark
France	..	52.4	66.5	62.5	France
Germany, Fed.Rep. of	44.8	44.8	42.8	53.4	Allemagne, Rép.féd. d'
Ireland a/	..	125.7	209.7	294.0c/	Irlande
Italy	..	4026.0	3365.0	3560.0	Italie
Netherlands	8.0	10.3	10.6	15.7d/	Pays-Bas
United Kingdom	3.7	3.8	4.1	3.1	Royaume-Uni
Austria b/	56.4	73.7	99.5	98.2	Autriche b/
Switzerland b/	13.4	15.5	15.9	17.1	Suisse
Malta					Malte
Greece	15.5	19.7	20.5	27.3	Grèce
Portugal	3.8	3.6	5.1	..	Portugal
Spain	105.2	143.8	164.4	..	Espagne
Turkey	2.5	4.7	..	..	Turquie

a/ Millions of national currency units / Millions d'unités de monnaire nationale.

b/ Includes all construction / Comprend toute la construction.

c/ 1979.

d/ 1977 prices. Other years are at 1975 prices / Prix de 1977. Les autres années sont aux prix de 1975.

Source: OECD / OCDE.

Annex table 1.5
Scenarios for GDP/manufacturing
production elasticities

Tableau annexe 1.5
Scénarios pour l'élasticité de la
production industrielle
par rapport au PIB

Country	Low/Bas a/		High/Haut b/		Pays
	1980–1990	1990–2000	1980–1990	1990–2000	
Finland	1.40	1.20	1.40	1.20	Finlande
Norway	0.80	0.80	1.00	1.00	Norvège
Sweden	0.95	0.90	1.00	1.00	Suède
Belgium-Luxembourg	0.95	0.90	1.00	1.00	Belgique-Luxembourg
Denmark	0.95	0.90	1.00	1.00	Danemark
France	0.95	0.90	1.00	1.00	France
Germany, Fed. Rep. of	0.95	0.90	1.00	1.00	Allemagne, Rép. féd. d'
Ireland	0.95	0.90	1.00	1.00	Irlande
Italy	1.0	0.95	1.10	1.05	Italie
Netherlands	1.0	0.95	1.10	1.05	Pays-Bas
United Kingdom	1.0	0.95	1.10	1.05	Royaume-Uni
Austria	0.95	0.90	1.00	1.00	Autriche
Switzerland	0.95	0.90	1.00	1.00	Suisse
Greece	1.45	1.30	1.50	1.40	Grèce
Portugal	1.35	1.20	1.40	1.30	Portugal
Spain	1.20	1.05	1.30	1.15	Espagne
Turkey	1.10	1.10	1.20	1.20	Turquie
Yugoslavia	1.50	1.35	1.60	1.40	Yougoslavie

a/   Elasticities relative to low GDP scenario.
a/   Elasticités par rapport au scénario bas du PIB.

b/   Elasticities relative to high GDP scenario.
b/   Elasticités par rapport au scénario haut du PIB.

Annex table 1.6
Scenarios for residential investment
used in end-use elasticity model

Tableau annexe 1.6
Scénarios pour l'investissement
dans le logement, utilisés pour le
modèle "élasticité par rapport aux
utilisations finales"

Country	Actual/Réel		Low/Bas			High/Haut			Pays
	1965	1979–81	1990	1995	2000	1990	1995	2000	
			(1975=100)						
Finland	55.8	92.2	92.2	92.2	92.2	108.5	124.3	140.1	Finlande
Norway	52.6	117.0	117.0	117.0	117.0	133.8	143.7	153.5	Norvège
Sweden	104.0	100.4	100.4	100.4	100.4	116.0	131.6	147.2	Suède
Belgium–Luxembourg a/	..	100.0	100.0	100.0	100.0	109.4	122.9	136.4	Belgique–Luxembourg a/
Denmark a/	..	100.0	100.0	100.0	100.0	111.6	121.9	132.2	Danemark a/
France	47.6	95.2	95.2	95.2	95.2	106.5	121.2	135.9	France
Germany, Fed. Rep. of	94.8	111.8	111.8	111.8	111.8	123.8	137.1	150.5	Allemagne, Rép. féd. d'
Ireland a/	..	100.0	100.0	100.0	100.0	117.0	136.5	159.1	Irlande a/
Italy	80.5	104.5	104.5	104.5	104.5	118.4	133.0	147.7	Italie
Netherlands	74.9	112.4	112.4	112.4	112.4	120.0	131.9	143.8	Pays–Bas
United Kingdom	88.0	78.8	78.8	78.8	78.8	86.5	93.1	99.7	Royaume–Uni
Austria	56.7	94.0	94.0	94.0	94.0	110.2	126.3	142.3	Autriche
Switzerland	84.5	105.8	105.8	105.8	105.8	115.1	126.5	137.9	Suisse
Greece a/	..	100.0	100.0	100.0	100.0	115.7	134.2	155.6	Grèce a/
Portugal	73.8	75.5	75.5	75.5	75.5	93.8	111.4	128.9	Portugal
Spain	64.0	84.4	84.4	84.4	84.4	97.2	110.3	123.3	Espagne
Turkey a/	..	100.0	100.0	100.0	100.0	134.9	163.0	196.9	Turquie a/
Yugoslavia a/	..	100.0	100.0	100.0	100.0	118.8	139.5	163.1	Yougoslavie a/

a/ Base period = 100. See section 12.3.
a/ Année de base = 100. Se référer à la section 12.3.

Annex table 1.7
Scenarios for manufacturing production
used in end-use elasticity model

Tableau annexe 1.7
Scénarios pour la production industrielle,
utilisés pour le modèle "élasticité par
rapport aux utilisations finales"

Country	Actual/Réel		Low/Bas			High/Haut			Pays
	1965	1979–81	1990	1995	2000	1990	1995	2000	
			(1975=100)						
Finland	56.6	124.3	175.7	220.4	265.2	185.7	250.7	315.7	Finlande
Norway	71.0	100.0	118.7	142.8	167.0	128.8	147.8	166.7	Norvège
Sweden	68.0	97.3	120.3	141.1	161.9	127.6	162.0	196.3	Suède
Belgium–Luxembourg a/	..	100.0	114.1	132.7	151.3	118.8	144.5	177.5	Belgique–Luxembourg a/
Denmark a/	..	100.0	119.7	138.1	156.5	123.1	144.0	168.6	Danemark a/
France	62.3	115.5	137.7	165.7	193.7	143.0	182.5	222.0	France
Germany, Fed. Rep. of	74.8	118.8	137.7	158.6	179.5	144.3	175.4	206.4	Allemagne, Rép. féd. d'
Ireland a/	..	100.0	123.0	147.2	171.5	134.1	178.6	237.9	Irlande a/
Italy	64.8	127.5	156.8	187.4	218.0	164.8	207.6	250.3	Italie
Netherlands	63.0	116.0	127.3	146.5	165.7	131.7	166.7	201.7	Pays–Bas
United Kingdom	86.3	96.1	107.1	118.1	129.0	114.8	132.3	149.8	Royaume–Uni
Austria	62.0	123.3	152.7	182.8	212.9	165.8	214.2	262.2	Autriche
Switzerland	77.7	112.2	124.5	138.8	153.0	132.0	158.2	184.4	Suisse
Greece a/	..	100.0	134.9	180.9	226.8	144.0	208.4	301.6	Grèce
Portugal	51.0	139.0	212.9	279.8	346.8	233.1	346.6	460.2	Portugal
Spain	42.0	113.3	147.7	179.5	211.4	157.9	206.7	255.6	Espagne
Turkey a/	..	100.0	167.4	232.6	297.8	183.9	275.6	413.1	Turquie a/
Yugoslavia a/	..	100.0	152.2	221.7	291.2	160.2	238.4	351.3	Yougoslavie a/

a/   Base period = 100. See section 12.3.
a/   Année de base = 100. Se référer à la section 12.3.

Annex table 1.8
Scenarios for furniture production
used in end-use elasticity model

Tableau annexe 1.8
Scénarios pour la production de meubles
utilisés pour le modèle "élasticité par
rapport aux utilisations finales"

Country	Actual/Réel		Low/Bas			High/Haut			Pays
	1965	1979–81	1990	1995	2000	1990	1995	2000	
				(1975=100)					
Finland	42.8	92.7	131.0	164.4	197.7	138.5	187.0	235.4	Finlande
Norway	60.0	105.7	125.5	151.0	176.6	136.1	156.1	176.2	Norvège
Sweden	59.0	96.3	119.1	139.7	160.2	126.3	160.3	194.3	Suède
Belgium-Luxembourg a/	..	100.0	114.1	132.7	151.3	118.8	144.5	177.5	Belgique-Luxembourg a/
Denmark a/	..	100.0	119.7	138.1	156.5	123.1	144.0	168.6	Danemark a/
France	75.8	96.0	114.4	137.6	160.9	118.9	151.8	184.6	France
Germany, Fed. Rep. of	63.8	110.3	127.9	147.3	166.7	133.9	162.7	191.6	Allemagne, Rép. féd. d'
Ireland a/	..	100.0	123.0	147.2	171.5	134.1	178.6	237.9	Irlande a/
Italy	53.1	178.9	220.0	262.9	305.9	231.2	291.2	351.2	Italie
Netherlands b/	..	..	..	..	..	..	..	..	Pays-Bas b/
United Kingdom	90.1	89.5	99.7	109.9	120.1	106.9	123.2	139.5	Royaume-Uni
Austria b/	..	..	..	..	..	..	..	..	Autriche b/
Switzerland	101.8	113.7	126.1	140.6	155.0	133.8	160.4	186.9	Suisse
Greece a/	..	100.0	134.9	180.9	226.8	144.0	208.4	301.6	Grèce a/
Portugal b/	..	..	..	..	..	..	..	..	Portugal b/
Spain	58.0	119.3	155.5	189.0	222.5	166.3	217.7	269.2	Espagne
Turkey a/	..	100.0	167.4	232.6	297.8	183.9	275.6	413.1	Turquie a/
Yugoslavia a/	..	100.0	152.2	221.7	291.2	160.2	238.4	351.3	Yougoslavie a/

a/   Base period = 100. See section 12.3.
a/   Année de base = 100. Se référer à la section 12.3.

b/   As data for furniture production, were not available, manufacturing production was used.
b/   Les données pour la production de meubles n'étant pas disponibles, celles de la production
      industrielles ont été utilisées.

Annex figure 1.9 Figure annexe 1.9

Indices of GDP, residential investment and manufacturing production

Indices du PIB, de l'investissement dans le logement et de la production des industries manufacturières

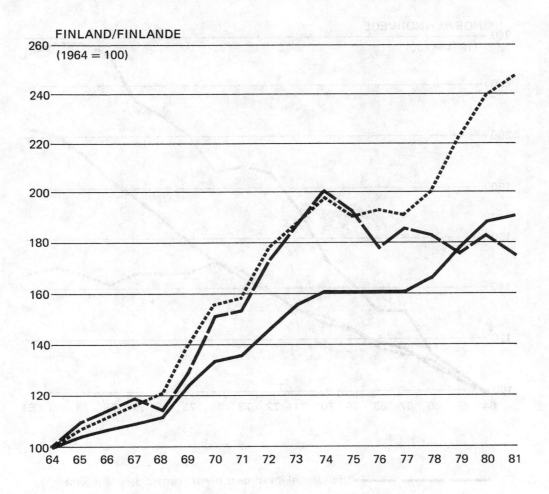

FINLAND/FINLANDE
(1964 = 100)

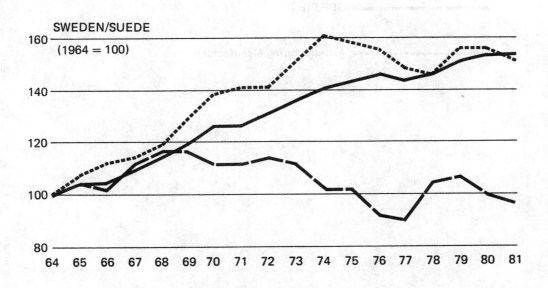

SWEDEN/SUEDE
(1964 = 100)

Fig.1.9 (continued/suite)

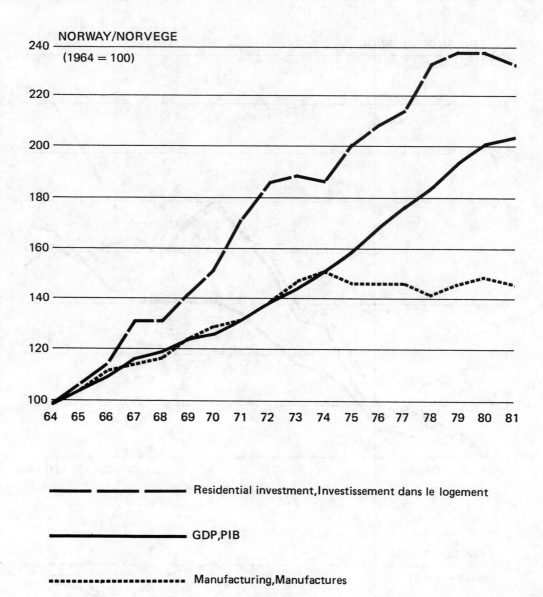

NORWAY/NORVEGE
(1964 = 100)

— — — — Residential investment,Investissement dans le logement

———————— GDP,PIB

•••••••••••••••••• Manufacturing,Manufactures

Fig.1.9 (continued/suite)

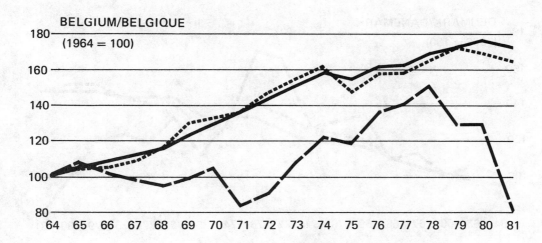

BELGIUM/BELGIQUE
(1964 = 100)

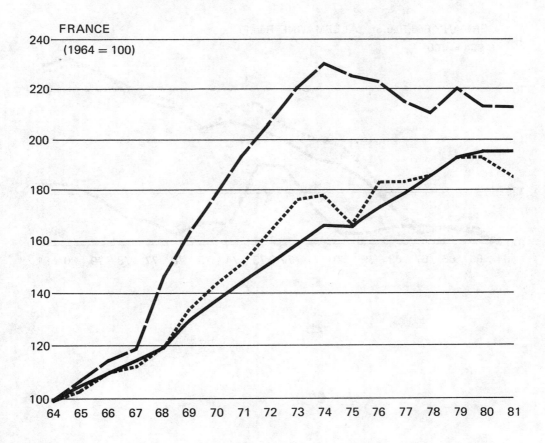

FRANCE
(1964 = 100)

Fig.1.9 (continued/suite)

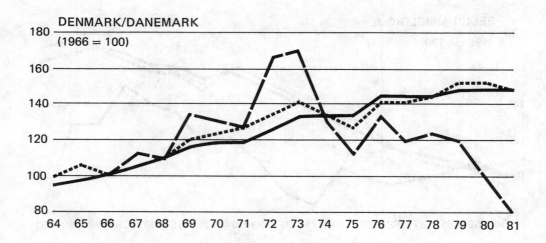

DENMARK/DANEMARK
(1966 = 100)

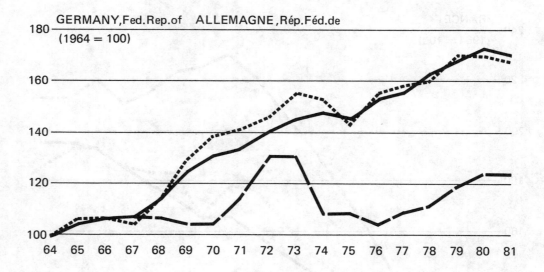

GERMANY,Fed.Rep.of   ALLEMAGNE,Rép.Féd.de
(1964 = 100)

Fig.1.9 (continued/suite)

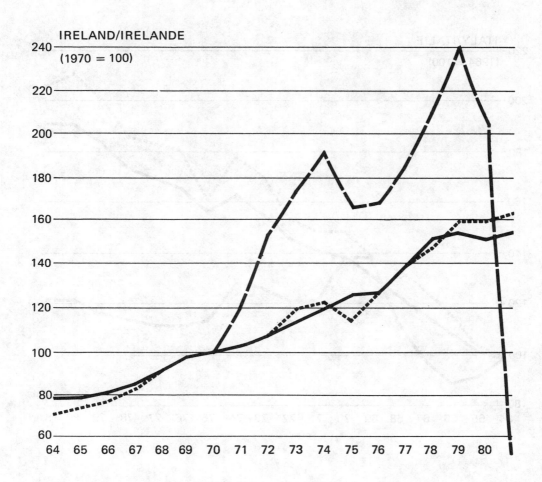

IRELAND/IRELANDE
(1970 = 100)

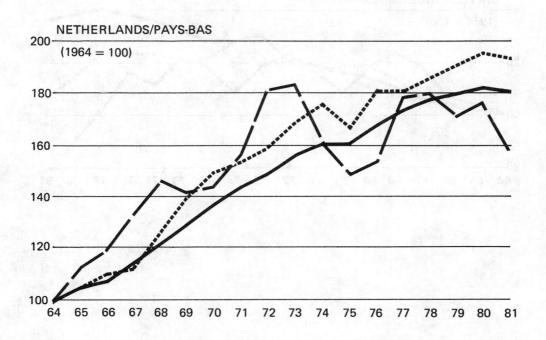

NETHERLANDS/PAYS-BAS
(1964 = 100)

Fig.1.9 (continued/suite)

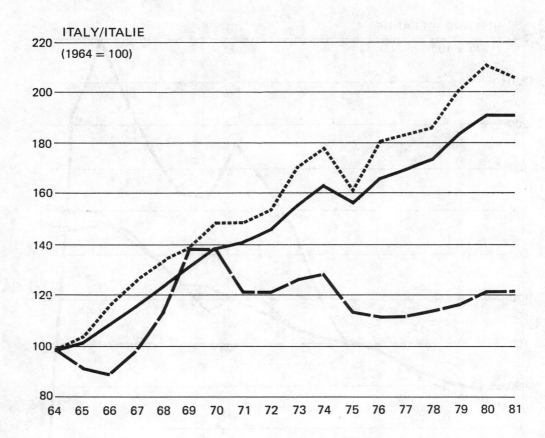

ITALY/ITALIE
(1964 = 100)

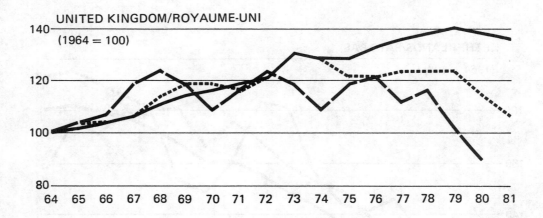

UNITED KINGDOM/ROYAUME-UNI
(1964 = 100)

Fig.1.9 (continued/suite)

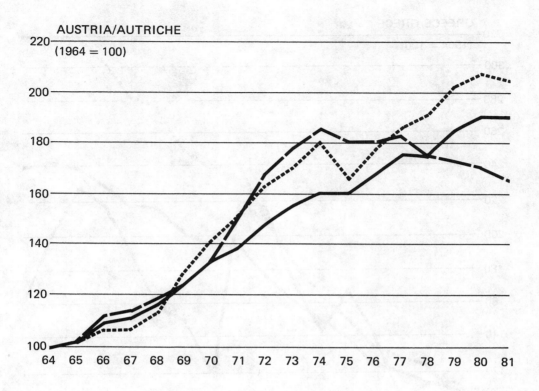

AUSTRIA/AUTRICHE
(1964 = 100)

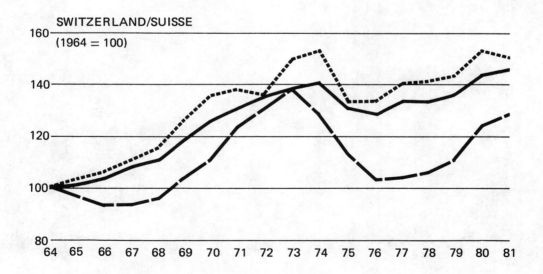

SWITZERLAND/SUISSE
(1964 = 100)

Fig.1.9 (continued/suite)

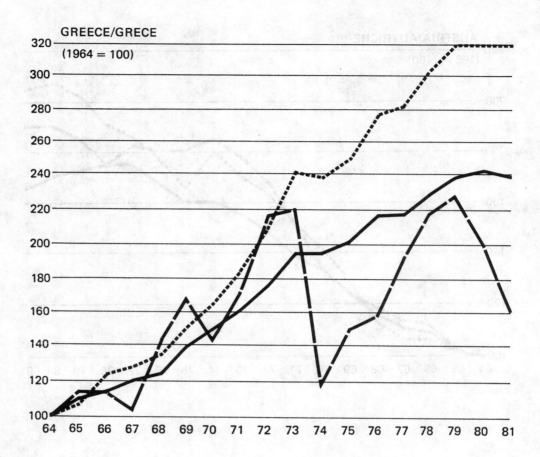

GREECE/GRECE

(1964 = 100)

Fig.1.9 (continued/suite)

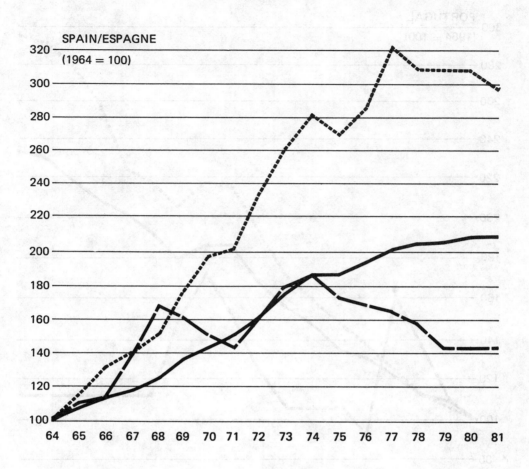

SPAIN/ESPAGNE
(1964 = 100)

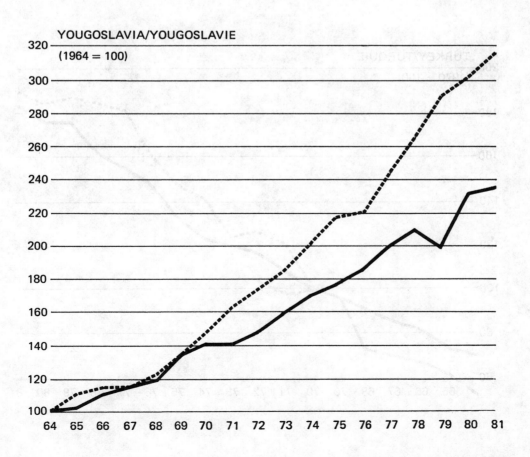

YOUGOSLAVIA/YOUGOSLAVIE
(1964 = 100)

Fig.1.9 (concluded/fin)

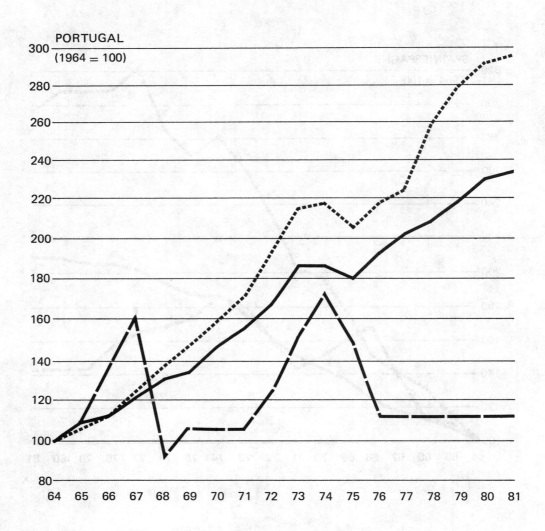

PORTUGAL
(1964 = 100)

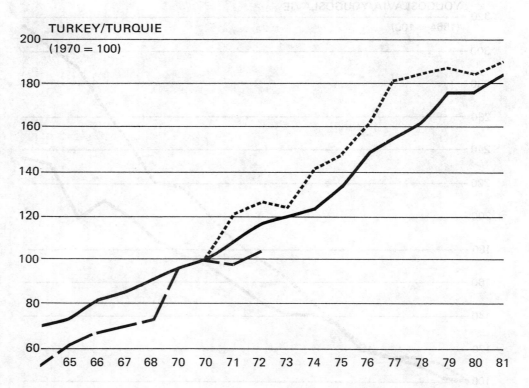

TURKEY/TURQUIE
(1970 = 100)

Annex 2.1

Extracts from summary of "Effects of taxation on forest management and roundwood supply" 1/

8. Summary of effects on forest management

8.1 <u>Forest management</u> will have been seen from the above to be affected by the various taxes according to their provisions, which vary among countries. Even within individual taxes there are variations of incidence between countries. In some instances the incidence of individual taxes constitutes a relief, in others a burden, and in others is neutral - respectively encouraging, discouraging or neutral to sustained and/or proper management.

A clear pattern of tax incidence is difficult to discern. Sometimes the encouragement rests precisely upon comparison of the incidence on forest with that on other property. Furthermore, the effect depends usually on:

- the personal/family circumstances of the taxpayer

- other taxable income being available for set-off against (i.e. recouped from) forest losses

- (as to inheritance taxes) the relationship of the testator (donor) to the beneficiary.

Another factor, from the forest owner's point of view, is the extent to which the over-all support or burden, as the case may be, is affected by monetary aid. A discouraging fiscal incidence may be cushioned by an encouraging direct subsidy; and vice versa.

8.2 <u>Continuity of ownership</u>: It is important to consider the effect of the foregoing taxes on the continuity of ownership. When positive it generally leads to sustained and/or proper forest management. (However, encouragement of continuity could be construed as negative if the objective of grouping of properties into larger units would lead to more economic management). Where income tax is modest (e.g. Finland) this generally encourages continuity. Where it is high (e.g. on actual income) it may be a disincentive to continuity. As to wealth tax and inheritance taxes, where the assessment basis is relatively low (e.g. France, Finland, Norway) this is generally an encouragement to continuity.

In Denmark the inheritance tax sometimes has an effect on forests. The tax is levied on the beneficiary; but rarely does it lead to fragmentation. Though under normal circumstances the total tax burden on forests in Denmark is not high as compared with other forms of land use (e.g. agriculture), when a forest is inherited, especially if inheritance takes

---

1/ Supplement 4 to volume XXXIII of the <u>Timber Bulletin for Europe</u>, ECE/FAO, Geneva, 1980.

place two or more times with short intervals, the tax burden can be disastrous to the ownership. In Denmark again, there are several regulations making it virtually impossible to raise funds to pay tax by selling part of the forest. Furthermore, some 90 per cent are "obligatory forests" which are not permitted to be converted to another use.2/

In Italy and Luxembourg the inheritance tax is relatively high and the provisions are generally negative to continuity of ownership, although in the latter country the direct heir is exempt from the tax.

In almost all countries much depends on the relationship of the beneficiary to the testator (donor): the closer the relationship, the less is the tax rate.

Some other countries have a positive approach towards continuity, and ostensibly towards management. In the Federal Republic of Germany the inheritance tax, being based on income value, is modest - unless the beneficiary sells his forest. France reduces income tax and land tax on young stands for 10-30 years; and reduces inheritance tax and registration tax if the beneficiary undertakes to practice proper management for 30 years. The Netherlands, through the benefits of its Natural Beauty Law, encourages continuity. (If outside that law, the effect is a heavy burden.) Portugal would seem to be neutral in effect, but positive when within its National Forest of Pendena-Gêres. In Greece the effect is negative.

8.3   On rotation: The fiscal provisions have varying effects on length of rotations:

In Denmark, the site value tax is closely related to rotations as well as to species, although the rotation choice for tax assessment is made by the taxation expert and not by the silviculturist.

Public access may be a factor which affects rotation length. In Denmark's site value tax there is a 50 per cent reduction for forests which are almost exclusively managed as recreational areas, and this may have the effect of lengthening rotations and possibly reducing removals, thereby affecting both management and roundwood supply. The same effect may be evident in areas under the Natural Beauty Law in the Netherlands.

Where in some countries, as noted earlier, there is fiscal (though unintentional) encouragement to over-cut - either to pay the tax or because the yield from over-cutting is tax free or tax-reduced - it may have an effect on rotation length. To the contrary, where over-cutting is taxed at progressive rates (e.g. Norway) rotations tend to lengthen.

In Denmark, in order to obtain a reduced assessment for annual wealth tax, theremay be a tendency to shorten rotations; the opposite is the case where the intention is to keep the growing stock high to meet future inheritance tax.

In the United Kingdom, where the Capital Transfer Tax on trees following a disposal on death can be deferred until sold, there may be a tendency to lengthen rotations.

2/   Denmark for over 150 years has had a law forbidding "obligatory forests" to be split up into holdings of less than 50 hectares; and for most of the largest private forest estates the restrictions are even much greater.

In Greece, the fact that afforestation is exempt from the forest State tax for the whole rotation may lead to a lengthened rotation.

8.4 <u>On species</u>: Some of the fiscal provisions have an effect on species:

In Italy coppice is exempt from income tax for 15 years, high forest for 40 years, and the conversion to high forest for 25 years. In France, the exemptions are for 10 years for poplars, 20 years for conifers, and 30 years for other broadleaves. In Greece, all afforestations are exempt from the forest state tax for the whole rotation.

Where growing stock is intentionally retained, even to over-maturity (as is sometimes the case in Denmark to meet future inheritance tax), the tendency is to choose oak and beech, particularly because there is a long period between the timber being physically saleable and physically unsaleable - through decay.

11. Summary of effects on roundwood supply

11.1 <u>Roundwood supply</u> will have been seen to be affected by the various taxes according to their provisions, which vary among countries. The factors noted in section 8.1 above relating to forest management are generally pertinent also to roundwood supply.

11.2 <u>Over-cutting</u> (i.e. in excess of sustained yield or of allowable annual cut): over-cuttings other than those caused by calamities (see below) are generally encouraged, tax being treated not as income but as a reduction in capital. As to over-cuttings caused by calamities (e.g. windthrow):

- in the Federal Republic of Germany and Luxembourg, for cuttings <u>within</u> the allowable cut the normal income tax rate is reduced by one half. For cuttings <u>above</u> the allowable cut in the Federal Republic of Germany the rate is reduced to one quarter to one eighth of the normal rate according to the extent. In Austria the reduction is one half in both cases.

- in Denmark, in cases of exceptional windthrow one half of the resulting extra income is considered as a reduction in capital and therefore exempt from income tax.

- in Norway, assessment of income from calamities can be spread over several years.

In Austria, most large forest owners plan their annual cut to provide income sufficient to cover their fixed costs but to minimize taxable income. Thinnings provide little income but are generally considered silviculturally desirable. Therefore, taxes are unlikely to have much influence on limiting volume derived from thinnings. However, if owners wish to avoid high progressive rates in times of rising prices for roundwood, they reduce cuttings, and prefer thinning to clear cuttings; and vice versa in times of decreasing prices.

11.3 <u>Afforestation</u>: In the long term, afforestation will increase wood production and, eventually, wood supply. The time-scale involved for coppice may be some 10-20 years, for conifers 20-70 years, for poplars 15-30 years, and for other broadleaves 50-200 years.

Some countries encourage afforestation by wholly or partly exempting young stands from income tax and/or land tax (Belgium for 21 years, France for 10-30 years, Italy for 15-40 years, Portugal for 20 years and Poland for 30 years. Italy permanently exempts from income tax stands in mountainous regions and those lying 700 m or more above sea level). In the United Kingdom, afforestation is substantially encouraged by tax relief (under Schedule D) supplemented by grant-aid. However, the Capital Gains Tax on land sometimes discourages farmers from releasing land for afforestation. In Greece, afforestation is relieved of the forest State tax for the whole rotation.

Countries where the annual tax assessment is on an income/outgoings calculation, virtually discourage afforestation in so far as they do not tax-relieve the expenditure. (The countries include the Federal Republic of Germany, Luxembourg, Ireland, the Netherlands and Denmark).

11.4 <u>Roads</u>: In countries where the annual tax assessment is on an income/outgoings calculation, road maintenance is tax relieved, but road construction is not.

Where road construction, etc., is not tax-relieved, such expenditure is discouraged (e.g. the Federal Republic of Germany, Portugal, Denmark).

Norway tax-relieves both construction and maintenance of forest roads, while other roads are not tax-relieved.

In the United Kingdom under Schedule D, construction is tax-relieved over eight years, whereas maintenance is tax-relieved in one year.

In Greece, road construction is encouraged under the provisions of the forest State tax.

11.5 <u>Machinery and equipment</u>: Depreciation is tax-relieved in Norway; also in the United Kingdom under Schedule D.

11.6 <u>Taxes on wood sales</u>: The information stated in 5.9 above should be noted.

11.7 <u>Wood as fuel</u>: Taxes linked with the use of fuels could affect the competitive position between the use of wood for fuel and for other purposes, and between wood and other fuels. This aspect is likely to become of increasing importance. In Greece, fuelwood bears a lower rate of forest tax than does other roundwood. In the United Kingdom, fuelwood is exempt from turnover tax (V.A.T.).

12.9.2 <u>Specific fiscal incidences</u>:

(a) <u>Valuation of forest assets for tax purposes</u>: Where valuation for capital tax and inheritance taxes and also sometimes for land and other real property taxes is calculated as a market value, it is generally greatly at variance with the capitalized-income value, with the result that the sustained management of the forest is endangered.

(b) <u>Taxation of capital as an addition to income tax</u>: Where capital is taxed other than on a capitalized-income value this tends to lead to the breaking up of properties or to bad husbandry, neither of which is in the public interest. Furthermore, there is need for a definite relationship (as in Denmark) between taxable income and taxable assets, i.e. between income tax and capital tax.

(c) <u>Inheritance and gift taxes</u>: The burden of these taxes tends to cause a breakdown of the principles of sustained forest management or to lead to an agronomically unjustifiable fragmentation of properties. They cannot be considered a suitable means of land reform or redistribution of property on account of their serious repercussions on the state of the forest. On top of a capital tax, the only object of an inheritance tax in the form of a succession or gift tax can be to tax the pecuniary gain obtained by the beneficiary. For forest holdings under sustained management it is the capitalized-income value alone which must be used as the basis of assessment for inheritance tax as the pecuniary gain consists only in the yield of a forest holding. If a forest holding is sold by the inheritors or if its intrinsic value is much modified by actions inimical to its sustained management, the pecuniary gain consists in its market value or in the sum for which it is sold. Only in these cases is it right to use the market value as the basis for taxation.

(d) <u>Capital gain on forest land</u>: Taxation of capital gain in respect of forest land transferred without payment (if this capital gain is not realized by selling forest) is an additional tax burden suffered by forest management on account of an increase in land prices which is unrelated to an increase in forestry yields.

(e) <u>Land and other real property taxes</u>: Where these taxes are normally allotted to the local authorities, they are based either on the average annual yield (cadastral value current value) or on a basis of assessment derived from the net asset value. Often the bases of these taxes are out-dated. No matter what method is used it should be related to recent yield figures in respect of forest holdings. As local authorities generally have the power to fix their own tax rates there is a danger of a very high and, above all, a very uneven tax burden on forest holdings. This danger must be met either by placing an upper limit on tax rates or by reducing the tax assessment basis by an amount corresponding to the excessive burden. Yield taxes are not consistent with the principle of taxation according to the ability to pay: this defect should be remedied by linking these taxes in some way with income tax.

(f) <u>Assessment of forestry income for income tax purposes</u>: One of the various methods employed to ascertain the income from forestry for purposes of income tax is the use of average rates or average yields, which only very seldom corresponds to the yield situation of the forest holding. In forests with mainly young stands it is extremely difficult for forest holdings which have no other sources of income to pay the tax without drawing on capital.

Total exemption of forestry from income tax to avoid the problem of assessing income would not be a satisfactory solution since losses incurred by the forest holding could not then be offset against other taxable income.

(g) <u>Taxes on turnover</u>: Generally the incidence of these taxes on forestry is neutral, because in most European countries (and in all member States of the EEC) it is possible to opt for a procedure which makes the tax merely a self-balancing item in the forest holding's accounts. It may occasionally appear to forest owners as an encouragement compared with other enterprises. However, differences between the VAT rates imposed on agriculture and forest products could affect farmers' decisions regarding the use of their forest holdings. Changes towards simplicity of administration and accountancy appear desirable.

(h) <u>Consideration of changes in forest taxation systems</u>: Forest conditions vary as between European countries, and forest management is at varying stages of development, with the result that forest policy is reflected in widely differing regulations. In considering the making of changes in forest taxation systems (such as the introduction of forest valuations based on yield) individual countries would appraise the effects on their tax revenue. Consideration would need to be supported by reliable statistics, and take into account the increasing claims made on forests for public purposes, as well as parafiscal charges – in particular those relating to social security (connected with wage costs and other labour costs of a social nature). Furthermore, consideration would need to extend over a long reference period so that current developments in tax laws may be taken into account."

Annex table 3.1

Comparable data on area of forest and other wooded land between around 1970 and around 1980

	Do definitions agree with those of categories shown in headings?	Around 1970					Around 1980				
		Precise year or period (Period I)	Forest and other wooded land (1000 ha)				Precise year or period (Period II)	Forest and other wooded land (1000 ha)			
			Total (TFOWL)	Exploitable closed forest (ECF)	Non-exploitable closed forest (UCF)	Other wooded land (OWL)		Total (TFOWL)	Exploitable closed forest (ECF)	Non-exploitable closed forest (UCF)	Other wooded land (OWL)
Finland	Yes	1964-70	22500	18700	140	3660	1975-81	23225	19445	440	3340
Iceland			..	..	..	..		..	..	-	..
Norway	Yes	1970	8700*	6500	975*	1225*	1980	8700*	6600	975*	1125*
Sweden	No	1968-72	27842*	23459	(------	4383*)	1978-82	27842	23501	(------	4341)
NORDIC COUNTRIES			59042	48659	(------	10383)		59767	49546	(------	10221)
Denmark	No	1965	472	373	31	68	1980	505	365	42	98
France	Yes	1970	14800	13090	440	1270	1981	15075	13340	535	1200
Ireland	Yes	1970	301	268	-	33	1980	380	347	-	33
Netherlands	No	1964-68	360*	296	-	64*	1980-83	397	331	-	66
United Kingdom	Yes	1970	1882	1521	-	361	1980	2178	2017	10	151
Austria	No	1961-70	3691	3230	461	-	1971-80	3754	3165	589	-
Switzerland	Yes	1970	1100	(970 --------	)	130	1985	1186	(968 --------	)	218
Cyprus	Yes	1970	173	120*	33	20	1982	173	100*	53	20
Spain	Yes	1965-74	25622	5931	2179	17512	1980-84	25622	6506	1882	17234
Czechoslovakia	Yes	1970	4456	4190	204	62	1980	4535	4126	309	100
Hungary	Yes	1970	1529	1466	-	63	1980	1635	1596	-	39
Poland	Yes	1967	8638	8371	129	138*	1978	8726	8410	178	138
USA	No	1970	305000	201000	8000	96000	1977	298000	195000	10000	93000

Annex table 3.2

Changes in area of forest and other wooded land between around 1970 and around 1980

	Do definitions agree with those of categories shown in headings ?	Area change adjusted to a 10-year period a/ (1000 ha)				% change adjusted to a 10-year period a/ (%)			
		Total (TFOWL)	Exploitable closed forest (ECF)	Non-exploitable closed forest (NCF)	Other wooded land (OWL)	Total (TFOWL)	Exploitable closed forest (ECF)	Non-exploitable closed forest (NCF)	Other wooded land (OWL)
Finland	Yes	+ 659	+ 677	+ 273	− 291	+ 2.9	+ 3.6	+ 195.0	− 8.0
Norway	Yes	−*	+ 100	−*	− 100*	−*	+ 1.5	−*	− 8.2*
Sweden	No	−*	+ 42	(-----------	− 42*)	−*	+ 0.2	(-----------	− 1.0)
NORDIC COUNTRIES		+ 659	+ 819	(-----------	− 160)	+ 0.1	+ 1.6	(-----------	− 1.5)
Denmark	No	+ 22	− 5	+ 7	+ 20	+ 4.7	− 1.3	+ 22.6	+ 29.4
France	Yes	+ 250	+ 227	+ 86	− 63	+ 1.7	+ 1.7	+ 19.5	− 5.0
Ireland	Yes	+ 79	+ 79	−	−	+ 26.2	+ 29.5	−	−
Netherlands	No	+ 24*	+ 23	−	+ 1*	+ 6.7*	+ 7.8	−	+ 1.6*
United Kingdom	Yes	+ 296	+ 496	+ 14	− 210	+ 15.7	+ 32.6	−	− 58.2
Austria	No	+ 63	− 65	+ 128	−	+ 1.7	− 2.0	+ 28.8	−
Switzerland	Yes	+ 57	(− 1 -----------)		+ 59	+ 5.2	(− -----------)		+ 45
Cyprus	Yes	−	−	+ 17	−	−	− 14.2*	+ 51.5	−
Spain	Yes	−	+ 548	− 283	− 265	−	+ 9.2	− 13.0	− 1.5
Czechoslovakia	Yes	+ 79	− 64	+ 105	+ 38	+ 1.8	− 1.5	+ 51.4	+ 61.3
Hungary	Yes	+ 106	+ 130	−	− 24	+ 6.9	+ 8.9	−	− 38.1
Poland	Yes	+ 80	+ 35	+ 45	−	+ 0.9	+ 0.4	+ 34.9	−
USA	No	− 10000	− 8580	+ 2860	− 4280	− 3.3	− 4.3	+ 35.8	− 4.5

a/ Not all countries could provide comparable data precisely for 1970 and 1980. Consequently, to achieve consistency between countries, changes in this table and annex tables 3.3 and 3.4 are adjusted to a 10-year period.

Annex table 3.3

Comparable data on growing stock around 1970 and around 1980

Country	Do definitions agree with those shown in headings?	Around 1970 Precise year or period (Period I)	Around 1970 Growing stock on ECF Total (million m3 o.b.)	Coniferous	Non-coniferous	Around 1980 Precise year or period (Period II)	Around 1980 Growing stock on ECF Total (million m3 o.b.)	Coniferous	Non-coniferous	Volume change adjusted to 10-year period Growing stock on ECF Total (million m3 o.b.)	Coniferous	Non-coniferous	Percent change adjusted to 10-year period Growing stock on ECF Total (%)	Coniferous	Non-coniferous	Growing stock per ha on ECF Period I (m3 o.b./ha)	Period II
Finland	Yes	1964-70	1445	1188	257	1975-81	1568	1290	278	+ 112	+ 93	+ 19	+ 7.8	+ 7.8	+ 7.4	77	81
Norway	Yes	1970	516	427	89	1980	575	459	116	+ 59	+ 32	+ 27	+ 11.4	+ 7.5	+ 30.3	79	87
Sweden	No	1968-72	2289	1966	323	1978-82	2405	2037	368	+ 116	+ 71	+ 45	+ 5.1	+ 3.6	+ 13.9	98	102
NORDIC COUNTRIES			4250	3581	669		4548	3786	762	+ 287	+ 196	+ 91	+ 6.8	+ 5.5	+ 13.6	87	92
Denmark	No	1985	42	22	20	1980	47	29	18	+ 3	+ 5	- 1	+ 7.9	+ 21.4	- 7.0	113	129
France	Yes	1970	1174	515	659	1981	1550	605	945	+ 342	+ 82	+ 260	+ 29.1	+ 15.9	+ 39.5	90	116
Ireland	Yes	1970	15	10	5	1980	27	20	7	+ 12	+ 10	+ 2	+ 80.0	+100.0	+ 40.0	56	78
Netherlands *	No	1964-68	20	16	4	1980-83	23	15	8	+ 2	- 1	+ 3	+ 10.0	- 3.8	+ 65.0	68	69
United Kingdom	Yes	1970	135	70	65	1980	203	111	92	+ 68	+ 41	+ 27	+ 50.4	+ 58.6	+ 41.5	89	101
Austria	No	1961-70	752	642	110	1971-80	798	675	123	+ 46	+ 33	+ 13	+ 6.1	+ 5.1	+ 11.8	233	252
Switzerland	Yes	1970	270	..	..	1985	344	251	93	+ 50	..	..	+ 18.4	..	..	a/ 278	a/ 356
Cyprus	Yes	1970	3	3	-	1982	3	3	-	-	-	-	+ 1.3	+ 0.8	+ 11.1	27*	33*
Spain	Yes	1970	427	254	173	1980	453	280	173	+ 26	+ 26	-	+ 6.1	+ 10.2	-	72*	70
Czechoslovakia	Yes	1970	801	598	203	1980	923	686	237	+ 122	+ 88	+ 34	+ 15.3	+ 14.7	+ 16.7	191	224
Hungary	Yes	1970	215	16	199	1980	253	29	224	+ 38	+ 13	+ 25	+ 17.7	+ 81.2	+ 12.6	147	158
Poland	Yes	1967	1049	857	192	1978	1162	897	265	+ 103	+ 36	+ 67	+ 9.8	+ 4.2	+ 34.9	125	138
USA	No	1970	22410	14990	7420	1977	23390	15180	8210	+1400	+ 270	+1130	+ 6.2	+ 1.8	+ 15.2	111	120

a/ Includes non-exploitable closed forest.

Annex table 3.4

Comparable data net annual increment around 1970 and around 1980

Country	Do definitions agree with those shown in headings?	Around 1970 Precise year or period (Period I)	Around 1970 Net annual increment on ECF Total	Conif-erous	Non-conif-erous	Precise year or period (Period II)	Around 1980 Net annual increment on ECF Total	Conif-erous	Non-conif-erous	Volume change adjusted to 10-year period Net annual increment on ECF Total	Conif-erous	Non-conif-erous	Percent change adjusted to 10-year period Total	Conif-erous	Non-conif-erous	Net annual increment per ha on ECF Period I	Period II
			(million m3 o.b.)				(million m3 o.b.)			(million m3 o.b.)			(%)			(m3 o.b./ha)	
Finland	Yes	1964-70	55.8	44.2	11.6	1975-81	61.9	48.1	13.8	+ 5.5	+ 3.5	+ 2.0	+ 9.9	+ 7.9	+ 17.2	2.98	3.18
Norway	Yes	1970	15.8	12.6	3.2	1980	17.3	13.7	3.6	+ 1.5	+ 1.1	+ 0.4	+ 9.5	+ 8.7	+ 12.5	2.43	2.62
Sweden	No	1968-72	69.6	60.0	9.6	1978-82	78.7	64.1	14.6	+ 9.1	+ 4.1	+ 5.0	+ 13.1	+ 6.8	+ 52.1	2.97	3.35
NORDIC COUNTRIES			141.2	116.8	24.4		157.9	125.9	32.0	+ 16.1	+ 8.7	+ 7.4	+ 11.4	+ 7.4	+ 30.3	2.90	3.19
Denmark	No	1965	2.3	1.5	0.8	1980	2.8	2.0	0.8	+ 0.3	+ 0.3	-	+ 13.0	+ 20.0	-	6.17	7.67
France	Yes	1970	46.5	22.4	24.1	1981	54.0	23.5	30.5	+ 6.8	+ 1.0	+ 5.8	+ 14.6	+ 4.5	+ 24.1	3.55	4.05
Ireland	Yes	1970	1.8	1.7	0.1	1980	2.5	2.4	0.1	+ 0.7	+ 0.7	-	+ 38.9	+ 41.2	-	6.72	7.20
Netherlands *	No	1964-68	1.2	1.0	0.2	1980-83	1.2	0.9	0.3	-	- 0.1	+ 0.1	+ 2.2	- 4.6	+ 32.2	4.05	3.75
United Kingdom	Yes	1970	6.3	5.1	1.2	1980	11.2	8.6	2.6	+ 4.9	+ 3.5	+ 1.4	+ 77.8	+ 68.6	+116.7	4.14	5.55
Austria	No	1961-70	18.5	15.7	2.8	1971-80	19.6	16.7	2.9	+ 1.1	+ 0.1	+ 0.1	+ 5.9	+ 6.4	+ 3.6	5.73	6.19
Switzerland	Yes	1970	5.2	..	..	1985	5.2	..	..	-	..	..	-	-	-	a/5.36	a/5.37
Cyprus	Yes	1970	0.1	..	..	1982	0.1	0.1	-	-	-	-	+ 0.8	+ 0.8	+ 0.8	0.8*	0.8*
Spain	Yes	1970	25.5	17.6	7.9	1980	27.8	19.5	8.3	+ 2.3	+ 1.9	+ 0.4	+ 9.0	+ 10.8	+ 5.1	4.30	4.27
Czechoslovakia	Yes	1970	15.7	11.6	4.1	1980	15.7	11.6	4.1	-	-	-	-	-	-	3.52	3.46
Hungary	Yes	1970	9.0	0.9	8.1	1980	9.7	1.2	8.5	+ 0.7	+ 0.3	+ 0.4	+ 7.9	+36.3	+ 4.7	6.14	6.08
Poland	Yes	1967	34.4	29.0	5.4	1978	28.4	23.4	5.0	- 5.5	- 5.1	- 0.4	-16.0	-17.6	- 7.4	4.11	3.38
USA	No	1970	648.6	374.4	274.2	1976	711.1	409.3	301.8	+104.2	+ 58.2	+ 46.0	+ 16.1	+ 15.5	+ 16.8	3.23	3.65

a/ Includes non-exploitable closed forest.

Notes to annex tables 3.1 to 3.4

The following comments by correspondents accompanied the data set out in these tables.

Finland   Unexploitable closed forest comprises the closed forest in national parks, nature reserves and similar areas where production is not allowed.

Sweden   The data under "exploitable closed forest" refer to "forest land" according to Swedish definitions i.e. exploitable closed forest and approximately 1.3 million ha of forest above a limit for investment. NAI is calculated as gross annual increment minus natural losses (estimated at 7 million m3 in both periods).

France   1980 data are from the National Forest Inventory (NFI), 1971 data are based on NFI data for 30 departments and estimates for other departments where NFI averages for growing stock and increment have been applied to data on forest area from other sources (taking account of the special circumstances of two departments, Gironde and Landes). Poplar plantations are included for both periods.

Netherlands   Estimates were necessary for the following:

- other wooded land around 1970 (only partly inventoried)

- growing stock and NAI for 1964-68 (differing interpretation of exploitable closed forest)

- growing stock and NAI around 1980 (data on these aspects from latest inventory not yet available).

United Kingdom   Some adjustments were made to the 1970 data to allow for missing categories and to bring the private woodland figures to a 1970 base.

Austria   "Protection forests with yield" are included in exploitable closed forest and "protection forests without yield" in non-exploitable closed forest. The former are under real management, but also have a real protection function in the function of soil conservation, especially in steep terrain. The latter are either not accessible, cannot be used for economic reasons or are not productive because of species composition.

Cyprus   The decrease of exploitable closed forest is due to big fires in 1974. The burnt area has been reforested. Although the area of exploitable closed forest decreased between 1970 and 1982, the growing stock shows an increase due to more accurate estimates based on inventory figures.

Spain   Data of growing stock and increment for 1970 in ETTS III referred to all land covered with trees. These data have been adjusted to refer to exploitable closed forest. Wood production from exploitable closed forest is about 85% of total removals.

Poland   For "other wooded land", there were no data for 1967, but this was estimated in 1978. We can assume that this area was more or less the same in 1967.

United States   Data supplied were for underbark volumes (converted to overbark by the secretariat).

Annex table 5.1

Comparison between countries' latest forest inventory data as reported for the Forest Resource Enquiry (FRE) and summarized in chapter 3, and those used as the base period for their forecasts in chapter 5 (ETTS IV)

	Year/period covered		Area of exploitable closed forest (ECF) (1000 ha.)		Growing stock on ECF (million m3 overbark)		Net annual increment on ECF (million m3 overbark)		Fellings on ECF (million m3 overbark)		Total removals (all sources) (million m3 underbark)		
	Forest Resource Enquiry (FRE)	Forestry Forecasts Enquiry (ETTS IV) Base period	FRE	ETTS IV	FRE	ETTS IV	FRE	ETTS IV	FRE	ETTS IV	FRE	ETTS IV	FAO/ECE 1979-81
Finland	1975-81	1975-81	19445	19445	1568	1568	61.93	61.93	55.30	55.69	45.40	45.78	44.99
Norway	1967-79	1967-79	6600	6600	575	575	17.31	17.31	10.89	10.89	9.22	9.52	9.37
Sweden	1973-77	1973-77	22230	22230	2210	2210	66.94	66.94	57.20	57.20	47.54	48.52	48.52
Belgium	1970	1970	600	600	73	73	4.50	4.50	a/2.39	2.78	2.44	2.44	2.44
Denmark	1981	1979-81	400	365	46	47	3.40	2.80	1.94	2.10	..	1.89	2.07
France	1981	1981	13340	13340	1550	1550	54.00	54.00	41.36	41.36	38.48	38.48	31.27
Germany, Fed. Rep. of	1961 and later	..	6838	6960	1062	1500	38.50	39.67	35.62	37.16	32.44	33.03	31.46
Ireland	1980	1980	347	347	32	32	2.53	2.53	0.65	0.65	0.54	0.54	0.38
Italy	1980	..	3868	3868	557	557	11.88	11.88	8.06	8.06	8.96	8.96	8.66
Luxembourg	1983	..	80	80	13	13	0.33	0.33	0.35	0.35	..	0.29	0.29
Netherlands	1982	1982	294	294	23	29	1.24	1.24	1.18	1.18	1.12	1.12	0.88
United Kingdom	1980	1980	2017	2017	203	203	11.20	11.20	4.58	4.58	4.32	4.32	4.13
Austria	1971-80	1971-80	3165	3165	797	803	19.58	19.58	15.17	15.21	12.13	12.17	14.39
Switzerland	1972-78	1972-78	795	b/ 935	312	312	5.20	5.20	4.50	4.80	4.00	4.39	4.39
Cyprus	1980	1983	100	100	3	4	0.10	0.09	0.07	0.07	..	0.07	0.08
Greece	1983	1980	1793	2300	133	159	3.68	4.10	1.50*	2.85	2.62*	2.59	2.59
Israel	1977-82	1981	70	66	3	3	0.18	0.20	0.14	0.09	..	0.14	0.12
Portugal	1980	..	2590	2590	189	189	11.45	11.45	8.50*	10.80	..	8.42	8.42
Spain	1980	1980	6506	6506	453	453	27.83	27.83	13.35	13.34	12.24	12.17	12.38
Turkey	1963-72	1963-72	6642	6642	637	637	19.20	19.21	19.57	19.57	c/24.74	22.38	22.38
Yugoslavia	1979	1979	8500	8500	1084	1135	27.75	28.85	19.94	19.94	14.53	13.79	13.62
Bulgaria	1980	..	3300	3300	298	298	6.00	6.00	6.03	6.03	..	4.44	4.44
Czechoslovakia	1981	..	4435	4185	923	923	16.80	22.50	a/21.46	21.46	19.41	19.32	18.36
German Dem. Rep.	1980-81	..	2590	2590	440	440	15.00	15.00	a/11.30	12.04	10.30	9.99	9.99
Hungary	1981	1981	1563	1596	253	253	9.61	9.71	7.56	7.54	5.85	6.16	6.17
Poland	1978	1978	8410	8410	1162	1162	28.45	28.45	25.30	25.30	21.20	21.20	20.56
Romania	..	d/	5860	5723	1268	1202	26.90	31.59	..	20.70	..	18.42	18.42

a/ Removals overbark (no felling figures provided).

b/ Appears to be figure for closed forest.

c/ 1981.

d/ Forecasts for 1980 in ETTS III for area of ECF, growing stock and net annual increment.

Glossary of terms used in annex tables 5.2 to 5.57
Glossaire des termes utilisés dans les tableaux annexes 5.2 à 5.57

English/Anglais	Français/French
Total removals	Quantités enlevées totales
Sawlogs and veneer logs	Grumes de sciages et de placages
Smallwood (wood in the rough other than sawlogs + veneer logs)	Bois de petites dimensions (bois bruts autres que grumes de sciages et de placages)
Underbark (u.b.)	Sous écorce (u.b.)
Overbark (o.b.)	Sur écorce (o.b.)
Coniferous	Résineux
Non-coniferous	Feuillus
Base period	Période de base
Low (low forecasts)	Faible (prévisions faibles)
High (high forecasts)	Forte (prévisions fortes)
Exploitable closed forest	Forêt dense exploitable
Growing stock (GS)	Matériel sur pied (GS)
Net annual increment (NAI)	Accroissement net annuel (NAI)
Fellings	Abattages
Reported data (3-year average)	Données fournies (moyenne de 3 ans)
ETTS III	Troisième étude
ETTS IV	Quatrième étude
Other removals (e.g. from other wooded land, trees outside the forest)	Autres quantités enlevées (p. ex. d'autres terres boisées et d'arbres hors des forêts)

Explanatory notes to annex tables 5.2 to 5.57 (country forecasts)

General

1.   Based on a framework established by the Joint FAO/ECE Working Party on Forest Economics and Statistics, an enquiry was circulated in 1983 to ETTS IV correspondents on the long-term outlook for the forestry situation and wood raw material supply in their countries up to 2020. The enquiry consisted of:

Part A - Statistical forecasts for exploitable closed forest

Part B - Descriptive questions.

2.   In Part A, the purpose was to obtain estimates of potential wood supplied which spanned the range from the lowest to the highest that were realistic. The following guidelines were given:

"(i)   The lower estimate should be based on modest, but realistic assumptions of biological developments, e.g. for the intensification of silviculture, stand density, species mixture, age-class distribution, length of rotation, and of technical developments, including trends in harvesting or transport technology, labour productivity and methods of road building. For economic development, this estimate should assume low growth in the world economy (as in recent years) with only a small increase in the demand for forest products at best, very competitive conditions in international markets, and no upward pressure on prices of the sort that may be needed to "commercialize" areas at present too costly to exploit;

(ii)   The higher estimate should be based on more expansive (but still realistic) assumptions on developments affecting the biological potential, and more favourable (but again realistic) trends in technical factors which influence the costs of exploitation. For the development of the global economy, this estimate should assume growth rates similar to those achieved in the third quarter of the century, with growing demand for forest products and some upward pressure on the prices of wood and wood products traded internationally. Where increases in relative timber prices (i.e. increases relative to the general level of prices) are important influences on the estimates of removals, countries should state their price assumptions.

Both scenarios should take into account developments in the use of wood for energy and any likely new demands on the forest resource."

3.   Correspondents were asked to ensure that their forecasts for growing stock, net annual increment and fellings were consistent with each other by conforming to the formula:

$$GS_{80} + NAI_{81-90} - Fellings_{81-90} = GS_{90}, \text{ and so on, where}$$

GS_{80} = growing stock in 1980,

NAI_{81-90} = net annual increment aggregated for the years 1981 to 1990 inclusive

$Fellings_{81-90}$ = aggregated fellings for the years 1981 to 1990 inclusive

GS_{90} = growing stock in 1990.

4. In practice, forecasts for each year generally not being available, those for the beginning and end of each 10-year period were worked out and consistency was obtained by means of the following formula:

$$GS_1 = [(NAI_1 + NAI_2) \times \frac{10}{2}] - [(F_1 + F_2) \times \frac{10}{2}] = GS_2 ,$$

where GS_1 = growing stock at beginning of given 10-year period

NAI_1 = net annual increment at beginning of period

F_1 = fellings at beginning of period

GS_2 = growing stock at end of period

NAI_2 = net annual increment at end of period

F_2 = fellings at end of period.

5. The replies of country correspondents to Part A of the enquiry are set out, country by country, in annex tables 5.4 to 5.57. Where replies were not received from countries, estimates were prepared by the secretariat and indicated by a double asterisk - ** - against the country name. In a few cases, partial estimates were provided by countries which were completed by the secretariat, indicated by an asterisk in brackets - (*).

6. In Part B, correspondents were invited to reply to the following questions:

"(1) Area of exploitable closed forest

If in Table 1 (Part A of the enquiry) you have indicated changes of this area between 1980 and 2020, what are the factors causing these changes? Changes might occur as a result of, amongst other things, afforestation, conversion of other wooded land to exploitable forest, decrease of forest area due to road and building expansion, changes in economic accessibility, etc.

(2) Growing stock

(a) What are the basic factors likely to cause a change in the growing stock between 1980 and 2020? Factors might include, amongst others, an expansion of exploitable forest area, fellings less than potential cut due to private owners' reluctance to sell timber, silvicultural measures leading to increased increment rate, impact of pollution or other calamities.

(b) Explain the main arguments of your forest policy measures for improving the quality and composition of the growing stock, e.g. more artificial regeneration, better age class distribution, changes in rotation length, change of species, bringing into production of virgin stands, roundwood market measures etc.

(3) Fellings

Explain your main arguments and assumptions for:

(a) The changes in the volume of fellings between 1980 and 2020;

(b) the difference between the "lower" and the "higher" forecasts of fellings (or removals).

(4)   Wood supply other than from exploitable closed forest

Table 1 and questions (1) to (3) above dealt with exploitable closed forest.   Indicate the volume of _other_ fellings or removals (on other wooded land, trees outside the forest) in 1980 and the expected trend to 2020."

7.   The country tables show data for a 70-year period, 1950 to 2020.   It is important to bear in mind that the data shown for 1950, 1960 and 1970 are _not_ comparable with those for the base period (around 1980) for the forecasts or with the forecasts themselves;   and that they may not refer to exactly the year indicated.   Thus, the 1950 data on area, growing stock and increment are taken from the FAO World Forest Resources, 1955;   those for 1960 from its World Forest Inventory, 1963;   and for 1970 from the FAO/ECE European Timber Trends and Prospects, 1950 to 2000 (ETTS III).   Some ETTS III data for 1970 have, however, been revised in the light of the special enquiry on charges in the forest resource (see section 3.7 of ETTS IV).

8.   Changes in definitions, coverage, etc. often make it difficult to compare one period with another.   For example, in the tables of "Forestry forecasts relating to exploitable closed forest", data were taken from the above-mentioned sources which were under the following headings:

-   for 1950   Forests in use
-   for 1960   Forests in use
-   for 1970   Exploitable forest.

The base period data in those tables are those relating to "exploitable closed forest" taken from country replies to the latest forest resource enquiry.   They are not necessarily comparable to those shown for earlier periods for reasons explained above;   on the other hand, they form the starting point for, and should be comparable with, the forecasts for 1990 to 2020.

9.   It is repeated, therefore, that the information for the base period around 1980 and forecasts for 1990 to 2020 relating to area of exploitable closed forest and growing stock, increment, fellings and removals on exploitable closed forest are not necessarily comparable with those shown for 1950, 1960 and 1970.

10.   The figures shown for removals underbark for 1949-51, 1959-61 and 1969-71 are three-year averages and have been taken from the official returns to FAO/ECE questionnaires.   They should, therefore, be total removals from all sources, not just from exploitable closed forest.   Removals overbark and fellings are in consequence also total figures, and have been derived from removals underbark by using the bark percentages and felling loss estimates applied to the 1980 data. This should be borne in mind when observing the ratio of fellings to net annual increment for some countries for these years.

11.   The figures shown in the tables of forestry forecasts relating to exploitable closed forest for the base period mostly correspond with those in the latest Forest Resource Enquiry.   They may not relate exactly to the year 1980 but to a year or period around 1980, depending on when the inventory was carried out in the country concerned.   In the text of chapter 5, the term "base period around 1980" is used to indicate this situation and the fact that the data for this period formed the starting point for the forecasts for 1990 to 2020.

12.   To sum up, the forecasts for 1990 to 2020 presented in the annex tables to chapter 5 of ETTS IV should be comparable with the data for the "base period around 1980" but not necessarily with those shown for earlier periods.

Annex table 5.2

Forecasts of removals of sawlogs and veneer logs by country and country group, 1990 and 2000
(million m3 underbark)

	Total					Coniferous					Non-coniferous				
	Base period b/	1990 Lower	1990 Higher	2000 Lower	2000 Higher	Base period b/	1990 Lower	1990 Higher	2000 Lower	2000 Higher	Base period b/	1990 Lower	1990 Higher	2000 Lower	2000 Higher
Finland	20.46	20.88	21.60	20.89	22.53	19.02	19.41	19.67	19.22	20.69	1.44	1.47	1.93	1.67	1.84
Norway	4.77	4.39	5.14	4.55	5.11	4.76	4.37	5.13	4.54	5.10	0.01	0.02	0.01	0.01	0.01
Sweden	22.06	27.94	29.91	28.24	30.84	21.67	27.54	29.51	27.84	30.44	0.39	0.40	0.40	0.40	0.40
NORDIC COUNTRIES	47.29	53.21	56.65	53.68	58.48	45.45	51.32	54.31	51.60	56.23	1.84	1.89	2.34	2.09	2.25
Belgium*	1.21	1.50	1.59	1.62	1.81	0.80	0.86	0.93	0.90	1.04	0.41	0.64	0.66	0.72	0.77
Denmark	1.05	1.10	1.09	1.05	1.12	0.64	0.68	0.70	0.67	0.73	0.41	0.42	0.39	0.38	0.39
France	18.49	21.50	22.70	23.87	26.89	10.31	11.85	12.22	13.31	14.24	8.18	9.65	10.48	10.56	12.65
Germany, Fed. Rep. of*	17.70	17.00	19.00	17.00	19.40	13.99	13.65	16.00	13.65	16.30	3.71	3.35	3.00	3.35	3.10
Ireland	0.34	0.87	0.87	2.05	2.05	0.32	0.85	0.85	2.03	2.03	0.02	0.02	0.02	0.02	0.02
Italy*	2.31	2.61	2.72	2.76	3.08	0.90	1.15	1.23	1.32	1.54	1.41	1.46	1.49	1.44	1.54
Luxembourg*	0.14	0.14	0.14	0.13	0.15	0.06	0.06	0.06	0.05	0.07	0.08	0.08	0.08	0.08	0.08
Netherlands*	0.25	0.23	0.25	0.27	0.33	0.13	0.14	0.14	0.16	0.16	0.12	0.09	0.11	0.11	0.17
United Kingdom	2.65	3.30	3.68	4.06	5.16	1.68	2.48	2.63	3.24	4.11	0.97	0.82	1.05	0.82	1.05
EUROPEAN ECONOMIC COMMUNITY (9)	44.14	48.25	52.04	52.81	59.99	28.83	31.72	34.76	35.33	40.22	15.31	16.53	17.28	17.48	19.77
Austria	7.47	8.55	8.84	8.84	9.42	7.08	7.84	8.10	8.12	8.66	0.39	0.71	0.74	0.72	0.76
Switzerland	2.61	2.34	2.67	2.48	2.91	2.18	1.94	2.24	2.06	2.42	0.43	0.40	0.43	0.42	0.49
CENTRAL EUROPE	10.08	10.89	11.51	11.32	12.33	9.26	9.78	10.34	10.18	11.08	0.82	1.11	1.17	1.14	1.25
Cyprus	0.04	0.03	0.03	0.03	0.03	0.04	0.03	0.03	0.03	0.03	-	-	-	-	-
Greece	0.48	0.55	0.58	0.65	0.73	0.30	0.42	0.44	0.49	0.56	0.18	0.13	0.14	0.16	0.17
Israel	0.03	0.04	0.04	0.05	0.05	0.02	0.03	0.03	0.04	0.04	0.01	0.01	0.01	0.01	0.01
Portugal	4.55	4.97	5.18	5.56	5.76	4.15	4.57	4.74	5.08	5.23	0.40	0.40	0.44	0.48	0.53
Spain	3.55	4.35	4.62	5.61	5.97	2.80	3.29	3.50	4.24	4.53	0.75	1.06	1.12	1.37	1.44
Turkey	4.94	5.30	5.57	5.60	6.14	4.08	4.40	4.65	4.60	5.10	0.86	0.90	0.92	1.00	1.04
Yugoslavia	7.63	9.45	9.45	10.84	10.84	3.58	4.65	4.65	5.21	5.21	4.05	4.80	4.80	5.63	5.63
SOUTHERN EUROPE	21.22	24.69	25.477	28.34	29.52	14.97	17.39	18.04	19.69	20.70	6.25	7.30	7.43	8.65	8.82
Bulgaria	1.48	1.66	1.89	1.71	2.18	0.65	0.64	0.73	0.66	0.86	0.83	1.02	1.16	1.05	1.32
Czechoslovakia	10.57	9.27	9.91	9.14	9.99	8.60	7.21	7.85	7.08	7.81	1.98	2.06	2.06	2.06	2.18
German Dem. Rep.	4.05	4.42	4.24	4.40	4.47	3.24	3.54	3.33	3.48	3.53	0.81	0.88	0.91	0.92	0.94
Hungary	1.84	1.79	2.09	1.70	2.02	0.17	0.18	0.25	0.26	0.32	1.67	1.61	1.84	1.44	1.70
Poland	10.79	10.34	10.71	11.03	11.37	9.22	8.67	9.03	9.23	9.57	1.57	1.67	1.68	1.80	1.80
Romania	8.58	9.09	9.37	9.57	10.08	3.34	3.47	3.72	3.63	4.02	5.24	5.62	5.65	5.94	6.06
EASTERN EUROPE	37.31	36.57	38.21	37.55	40.11	25.22	23.71	24.91	24.34	26.11	12.10	12.86	13.30	13.21	14.00
TOTAL EUROPE	160.04	173.61	183.88	183.70	200.43	123.73	133.92	142.36	141.14	154.34	36.32	39.69	41.52	42.57	46.09

a/ Around 1980, but varies from country to country according to period covered by latest inventory.

Annex table 5.3

Forecasts of removals of smallwood a/ by country and country group, 1990 and 2000
(million m3 underbark)

	Total					Coniferous					Non-coniferous				
	Base period b/	1990 Lower	1990 Higher	2000 Lower	2000 Higher	Base period b/	1990 Lower	1990 Higher	2000 Lower	2000 Higher	Base period b/	1990 Lower	1990 Higher	2000 Lower	2000 Higher
Finland	25.32	25.06	28.61	27.72	30.55	18.97	18.48	19.45	19.78	21.24	6.35	6.58	9.16	7.94	9.31
Norway	4.75	4.81	5.56	4.95	6.19	3.91	4.01	4.65	4.14	5.18	0.84	0.80	0.91	0.81	1.01
Sweden	26.46	21.26	36.09	23.46	34.86	20.94	15.41	26.89	17.51	26.66	5.52	5.85	9.20	5.95	8.20
NORDIC COUNTRIES	56.54	51.13	70.26	56.13	71.60	43.82	37.90	50.99	41.43	53.08	12.72	13.23	19.27	14.70	18.52
Belgium*	1.23	1.42	1.49	1.52	1.67	0.68	0.70	0.75	0.74	0.85	0.55	0.72	0.74	0.78	0.82
Denmark	0.84	1.15	1.25	1.29	1.31	0.52	0.75	0.82	0.85	0.88	0.32	0.40	0.43	0.44	0.43
France	19.99	20.35	22.74	20.12	23.99	4.39	5.83	6.57	6.26	7.34	15.60	14.52	16.17	13.86	16.65
Germany, Fed. Rep. of*	15.33	13.15	14.15	13.15	14.25	8.60	7.37	8.77	7.37	8.84	6.73	5.78	5.38	5.78	5.40
Ireland	0.19	0.73	0.73	0.99	0.99	0.17	0.70	0.70	0.95	0.95	0.02	0.03	0.03	0.04	0.04
Italy*	6.65	6.93	7.08	7.01	7.49	0.81	0.94	1.00	1.08	1.26	5.84	5.99	6.08	5.93	6.23
Luxembourg*	0.15	0.14	0.15	0.14	0.15	0.05	0.04	0.05	0.04	0.05	0.10	0.10	0.10	0.10	0.10
Netherlands*	0.87	0.79	0.81	0.86	0.93	0.45	0.42	0.42	0.44	0.43	0.42	0.37	0.39	0.42	0.50
United Kingdom	1.67	2.35	2.47	3.19	3.89	1.34	2.07	2.12	2.91	3.54	0.33	0.28	0.35	0.38	0.35
EUROPEAN ECONOMIC COMMUNITY (9)	46.92	47.01	50.87	48.27	54.67	17.01	18.82	21.20	20.64	24.14	29.91	28.19	29.67	27.63	30.52
Austria	4.70	6.35	6.56	6.51	6.93	2.93	4.61	4.75	4.77	5.09	1.77	1.74	1.81	1.74	1.84
Switzerland	1.78	1.95	2.32	2.11	2.58	0.78	0.92	1.12	1.00	1.24	1.00	1.03	1.20	1.11	1.34
CENTRAL EUROPE	6.48	8.30	8.88	8.62	9.51	3.71	5.53	5.87	5.77	6.33	2.77	2.77	3.01	2.85	3.18
Cyprus	0.03	0.03	0.03	0.03	0.03	0.03	0.03	0.03	0.03	0.03	-	-	-	-	-
Greece	2.11	2.46	2.56	2.70	2.93	0.50	0.61	0.67	0.77	0.87	1.61	1.85	1.89	1.93	2.06
Israel	0.11	0.13	0.16	0.14	0.15	0.04	0.05	0.06	0.06	0.05	0.07	0.08	0.10	0.08	0.10
Portugal	3.87	3.79	4.18	4.42	4.98	1.39	1.53	1.66	1.70	1.93	2.48	2.26	2.52	2.72	3.05
Spain	8.62	11.79	13.02	14.92	16.60	5.33	7.37	8.17	9.37	10.43	3.29	4.42	4.85	5.55	6.17
Turkey	17.44	17.54	18.28	18.13	19.60	9.93	10.28	10.85	10.77	11.94	7.51	7.26	7.43	7.36	7.66
Yugoslavia	6.16	7.85	7.85	8.46	8.46	0.34	0.35	0.35	0.39	0.39	5.82	7.50	7.50	8.07	8.07
SOUTHERN EUROPE	38.34	43.59	46.08	48.80	52.75	17.56	20.22	21.79	23.09	25.64	20.78	23.37	24.29	25.71	27.11
Bulgaria	2.96	2.74	2.85	2.81	2.97	0.55	0.46	0.49	0.48	0.53	2.41	2.28	2.36	2.33	2.44
Czechoslovakia	8.74	8.28	8.54	8.41	8.91	6.23	5.66	5.92	5.79	6.14	2.52	2.62	2.62	2.62	2.77
German Dem. Rep.	5.94	6.63	6.36	6.60	6.71	4.69	5.31	5.01	5.21	5.31	1.25	1.32	1.35	1.39	1.40
Hungary	4.32	4.35	5.26	4.40	5.46	0.21	0.23	0.31	0.31	0.38	4.11	4.12	4.95	4.09	5.08
Poland	10.41	11.11	11.74	12.17	13.59	7.64	7.84	7.84	8.67	8.90	2.77	3.27	3.90	3.50	4.69
Romania	9.84	10.33	10.68	10.92	11.48	3.60	3.75	4.04	3.94	4.36	6.24	6.60	6.64	6.98	7.12
EASTERN EUROPE	42.21	43.44	45.43	45.31	49.12	22.92	23.25	23.61	24.40	25.62	19.30	20.21	21.82	20.91	23.50
TOTAL EUROPE	190.49	193.47	221.52	207.13	237.65	105.02	105.72	123.46	115.33	134.81	85.48	87.77	98.06	91.80	102.83

a/ All wood other than sawlogs and veneer logs (i.e. includes pulpwood, pitprops, other industrial wood and wood for energy).
b/ Around 1980, but varies from country to country according to period covered by latest inventory.

Annex table 5.4
Forestry forecasts relating to exploitable closed forest

Country: FINLAND

	Unit	1950	1960	1970	Base period	1990 Lower	1990 Higher	2000 Lower	2000 Higher	2010 Lower	2010 Higher	2020 Lower	2020 Higher
Exploitable closed forest	1000 ha	18200	18000	18734	19445	20000	20000	20300	20300	20400	20400	20500	20500
Growing stock (GS)	Mill.m3 o.b	1456	1430	1445	1568	1651	1620	1741	1650	1830	1680	1913	1711
- Coniferous (volume)	"	1165	1144	1188	1290	1354	1337	1424	1370	1493	1403	1557	1437
- Non-coniferous (volume)	"	291	286	257	278	297	283	317	280	337	277	356	274
- Coniferous (percent)	%	80	80	82	82	82	83	82	83	82	84	81	84
Growing stock/ha	m3o.b./ha	80	79	77	81	83	81	86	81	90	82	93	83
Net annual increment (NAI)	1000 m3 o.b.	53300	52200	55780	61930	64600	64200	67900	68000	71100	71800	71700	73900
- Coniferous (volume)	"	40500	39672	44240	48120	50030	50380	52440	53710	54760	57080	55070	59120
- Non-coniferous (volume)	"	12800	12528	11540	13810	14570	13820	15460	14290	16340	14720	16630	14780
- Coniferous (percent)	%	76	76	79	78	77	78	77	79	77	79	77	80
NAI/ha	m3o.b. ha	2.9	2.9	3.0	3.2	3.2	3.2	3.3	3.4	3.5	3.5	3.5	3.6
NAI as % of GS	%	3.7	3.7	3.9	3.9	3.9	4.0	3.9	4.1	3.9	4.3	3.8	4.3
- Coniferous	%	3.5	3.5	3.7	3.7	3.7	3.8	3.7	3.9	3.7	4.1	3.5	4.1
- Non-coniferous	%	4.4	4.4	4.5	5.0	4.9	4.9	4.9	5.1	4.9	5.3	4.7	5.4
Fellings*	1000 m3 o.b.	48850	54720	54700	55690	55640	61300	58990	64850	62260	68900	63880	70700
- Coniferous*	"	34620	40340	38390	45200	44880	46560	46220	49960	48980	53960	50460	57960
- Non-coniferous*	"	14240	14380	16310	10490	10760	14740	12770	14890	13280	14940	13420	12740
Felling/NAI ratio*	%	92	105	98	90	86	95	87	95	88	96	89	96
- Coniferous*	%	85	102	87	94	90	92	88	93	89	95	92	98
- Non-coniferous*	%	111	115	141	76	74	107	83	104	81	101	81	86
Removals (overbark)*	1000 m3/o.b.	44890	50290	50270	51180	51350	56330	54440	59590	57470	63320	58960	64970
- Coniferous*	"	32500	37880	36050	42440	42320	43720	43580	46910	46190	50670	47580	54420
- Non-coniferous*	"	12390	12410	14220	8740	9030	12610	10860	12680	11280	12650	11380	10550
Removals (underbark)	1000 m3/u.b.	39367	44100	44080	44885	45030	49310	47700	52180	50360	55480	51670	57000
- Coniferous (volume)	"	28653	33400	31783	37432	37330	38560	38440	41370	40740	44690	41970	48000
- Non-coniferous (volume)	"	10714	10700	12297	7453	7700	10750	9260	10810	9620	10790	9700	9000
- Coniferous (percent)	%	73	76	72	83	83	78	81	79	81	81	81	84
Sawlogs as % of total	%	27	30	37	45	46	44	44	43	42	42	40	43
- Coniferous	"	34	36	44	51	52	51	50	50	48	49	46	48
- Non-coniferous	"	10	9	17	18	19	18	18	17	17	16	14	15

Annex table 5.5
Removals forecasts
(1000 m3 underbark)

Country: FINLAND

	1949-51	1959-61	1969-81	1980	1990	2000	2010	2020
1. Reported data (3-yr av.) + ETTS III forecasts				ETTS III FORECASTS				
1.1 Total removals	39367	44100	44080	44500	46200	49100	...	...
1.1.1 – Coniferous	28653	33400	31783			42700	...	...
1.1.2 – Non-coniferous	10714	10700	12297			6400	...	...
1.2 Of which: Sawlogs + veneer logs, total	10728	12990	16173				...	...
1.2.1 – Coniferous	9628	12025	14073				...	...
1.2.2 – Non-coniferous	1100	965	2100				...	...
2. Forecasts for ETTS IV (Lower)				B.p. b/	LOWER FORECASTS			
2.1 From exploitable closed forest, total				44885	45030	47700	50360	51670
2.1.1 – Coniferous				37432	37330	38440	40740	41970
2.1.2 – Non-coniferous				7453	7700	9260	9620	9700
2.2 Other removals a/				900*	-------	-------	-------	↑
2.2.1 – Coniferous				560*	-------	-------	-------	↑
2.2.2 – Non-coniferous				340*	-------	-------	-------	↑
2.3 Total removals	39367	44100	44080	45785	45930	48600	51260	52570
2.3.1 – Coniferous	28653	33400	31783	37992	37890	39000	41300	42530
2.3.2 – Non-coniferous	10714	10700	12297	7793	8040	9600	9960	10040
2.4 Of which: sawlogs + veneer logs, total	10728	12990	16173	20457	20875	20885	21190	20665
2.4.1 – Coniferous	9628	12025	14073	19023	19410	19220	19555	19305
2.4.2 – Non-coniferous	1100	965	2100	1434	1465	1665	1635	1360
3. Forecasts for ETTS IV (Higher)					HIGHER FORECASTS			
3.1 From exploitable closed forest, total				44885	49310	52180	55480	57000
3.1.1 – Coniferous				37432	38560	41370	44690	48000
3.1.2 – Non-coniferous				7453	10750	10810	10790	9000
3.2 Other removals a/				900*	-------	-------	-------	↑
3.2.1 – Coniferous				560*	-------	-------	-------	↑
3.2.2 – Non-coniferous				340*	-------	-------	-------	↑
3.3 Total removals	39367	44100	44080	45785	50210	53080	56380	57900
3.3.1 – Coniferous	28653	33400	31783	37992	39120	41930	45250	48560
3.3.2 – Non-coniferous	10714	10700	12297	7793	11090	11150	11130	9340
3.4 Of which: sawlogs + veneer logs, total	10728	12990	16173	20457	21600	22525	23755	24390
3.4.1 – Coniferous	9628	12025	14073	19023	19665	20685	22030	23040
3.4.2 – Non-coniferous	1100	965	2100	1434	1935	1840	1725	1350

a/ Other removals: e.g. from other wooded land and trees outside the forest.
b/ Base period (see explanatory notes).

Annex table 5.6
Forestry forecasts relating to exploitable closed forest

Country: NORWAY

	Unit	1950	1960	1970	Base period	1990 Lower	1990 Higher	2000 Lower	2000 Higher	2010 Lower	2010 Higher	2020 Lower	2020 Higher
Exploitable closed forest	1000 ha	6400	6450	6500	6600	6600	6600	6600	6600	6600	6600	6600	6600
Growing stock (GS)	Mill.m3 o.b	436	465	516	575	639	639	714	695	794	746	876	784
- Coniferous (volume)	"	363	390	427	459	497	497	541	523	584	540	624	543
- Non-coniferous (volume)	"	73	75	89	116	142	142	173	172	210	206	252	241
- Coniferous (percent)	%	83	84	83	80	78	78	76	75	74	72	71	69
Growing stock/ha	m3o.b./ha	68	72	79	87	97	97	108	105	120	113	133	119
Net annual increment (NAI)	1000 m3 o.b.	14370	14720	15770	17310	18030	17830	18810	18040	19410	18340	20350	18180
- Coniferous (volume)	"	11960	12170	12590	13710	13960	13790	14160	13460	14170	13290	14540	12710
- Non-coniferous (volume)	"	2410	2550	3180	3600	4070	4040	4650	4580	5240	5050	5810	5470
- Coniferous (percent)	%	83	83	80	79	77	77	75	75	73	72	71	70
NAI/ha	m3o.b./ha	2.2	2.3	2.4	2.6	2.7	2.7	2.9	2.7	2.9	2.8	3.1	2.8
NAI as % of GS	%	3.3	3.2	3.1	3.0	2.8	2.8	2.6	2.6	2.4	2.5	2.3	2.3
- Coniferous	%	3.3	3.1	2.9	3.0	2.8	2.8	2.6	2.6	2.4	2.5	2.3	2.3
- Non-coniferous	%	3.3	3.4	3.6	3.1	2.9	2.8	2.7	2.7	2.5	2.5	2.3	2.3
Fellings*	1000 m3 o.b.	11882	10567	10240	10894	10510	12280	10860	13000	11220	14480	11820	15950
- Coniferous*	"	10412	9314	9149	9882	9540	11190	9890	11780	10130	12950	10600	14130
- Non-coniferous*	"	1470	1253	1091	1012	970	1090	970	1220	1090	1530	1220	1820
Felling/NAI ratio*	%	83	72	65	63	58	69	58	72	58	79	58	88
- Coniferous*	%	87	77	73	72	68	81	70	88	71	97	73	111
- Non-coniferous*	%	61	49	34	28	24	27	21	27	21	30	21	33
Removals (overbark)	1000 m3 o.b.	11317	10065	9660	10278	9920	11590	10250	12260	10590	13660	11150	15050
- Coniferous	"	9917	8871	8631	9323	9000	10560	9330	11110	9560	12220	10000	13330
- Non-coniferous	"	1400	1194	1029	955	920	1030	920	1150	1030	1440	1150	1720
Removals (underbark)	1000 m3 u.b.	10152	9023	8664	9222	8900	10400	9200	11000	9500	12250	10000	13500
- Coniferous (volume)	"	8926	7984	7768	8391	8100	9500	8400	10000	8600	11000	9000	12000
- Non-coniferous (volume)	"	1226	1039	896	831	800	900	800	1000	900	1250	1000	1500
- Coniferous (percent)	%	88	88	90	91	90	91	91	91	91	90	90	89
Sawlogs as % of total	%	36	37	49	52	49	49	49	47	49	45	46	44
- Coniferous	"	40	42	54	57	54	54	54	51	54	50	51	49
- Non-coniferous*	"	2	2	2	2	2	2	2	2	1	1	1	1

Annex table 5.7
Removals forecasts
(1000 m3 underbark)

Country: NORWAY

	1949-51	1959-61	1969-71	1980	1990	2000	2010	2020
1. Reported data (3-yr av.) + ETTS III forecasts				ETTS III FORECASTS				
1.1 Total removals	10152	9023	8664	10000	11000	12000		
1.1.1 – Coniferous	8926	7984	7768			10800*		
1.1.2 – Non-coniferous	1226	1039	896			1200*		
1.2 Of which: Sawlogs + veneer logs, total	3611	3383	4225					
1.2.1 – Coniferous	3587	3362	4209					
1.2.2 – Non-coniferous	24	21	16					
2. Forecasts for ETTS IV (Lower)				B.p. b/	LOWER FORECASTS			
2.1 From exploitable closed forest, total				9222	8900	9200	9500	10000
2.1.1 – Coniferous				8391	8100	8400	8600	9000
2.1.2 – Non-coniferous				831	800	800	900	1000
2.2 Other removals a/				300	300*	→	→	→
2.2.1 – Coniferous				280	280*	→	→	→
2.2.2 – Non-coniferous				20	20*	→	→	→
2.3 Total removals	10152	9023	8664	9522	9200	9500	9800	10300
2.3.1 – Coniferous	8926	7984	7768	8671	8380	8680	8880	9280
2.3.2 – Non-coniferous	1226	1039	896	851	820	820	920	1020
2.4 Of which: sawlogs + veneer logs, total	3611	3383	4225	4775	4389	4550	4657	4602
2.4.1 – Coniferous	3587	3362	4209	4760	4374	4536	4644	4590
2.4.2 – Non-coniferous	24	21	16	15	15	14	13	12
3. Forecasts for ETTS IV (Higher)					HIGHER FORECASTS			
3.1 From exploitable closed forest, total				9222	10400	11000	12250	13500
3.1.1 – Coniferous				8391	9500	10000	11000	12000
3.1.2 – Non-coniferous				831	900	1000	1250	1500
3.2 Other removals a/				300	300*	→	→	→
3.2.1 – Coniferous				280	280*	→	→	→
3.2.2 – Non-coniferous				20	20*	→	→	→
3.3 Total removals	10152	9023	8664	9522	10700	11300	12550	13800
3.3.1 – Coniferous	8926	7984	7768	8671	9780	10280	11280	12280
3.3.2 – Non-coniferous	1226	1039	896	851	920	1020	1270	1520
3.4 Of which: sawlogs + veneer logs, total	3611	3383	4225	4775	5145	5115	5515	5895
3.4.1 – Coniferous	3587	3362	4209	4760	5130	5100	5500	5880
3.4.2 – Non-coniferous	24	21	16	15	15	15	15	15

a/ Other removals: e.g. from other wooded land and trees outside the forest. For the forecasts for 1990 to 2020, it has been assumed that other removals will remain at about the 1980 level.
b/ Base period (see explanatory notes).

Annex table 5.8

Forestry forecasts relating to exploitable closed forest

Country: SWEDEN

	Unit	1950	1960	1970	Base period	1990 Lower	1990 Higher	2000 Lower	2000 Higher	2010 Lower	2010 Higher	2020 Lower	2020 Higher
Exploitable closed forest	1000 ha	22940	20009	23459	22230	22230	22230	22230	22230	22230	22230	22230	22230
Growing stock (GS)	Mill.m3 o.b	1820	2089	2288	2210	2475	2366	2632	2356	2759	2352	2866	2358
- Coniferous (volume)	"	1544	1777	1966	1886	2080	2004	2181	1999	2244	1993	2277	2005
- Non-coniferous (volume)	"	276	312	322	324	395	362	451	357	515	359	589	352
- Coniferous (percent)	%	85	85	86	85	84	85	83	85	81	85	79	85
Growing stock/ha	m3o.b./ha	79	104	98	99	111	106	118	106	124	106	129	106
Net annual increment (NAI)	1000 m3 o.b.	57000	65100	63300	66940	74400	76600	74600	76500	75600	78700	77200	83100
- Coniferous (volume)	"	50000		54000	55310	60300	64000	59500	64700	58900	67400	58400	71200
- Non-coniferous (volume)	"	7000		9300	11630	14100	12600	15100	11800	16700	11300	18800	11900
- Coniferous (percent)	%	88		85	83	81	84	80	85	78	86	76	86
NAI/ha	m3o.b./ha	2.5	3.3	2.7	3.0	3.3	3.4	3.3	3.4	3.4	3.5	3.5	3.7
NAI as % of GS	%	3.1	3.1	2.8	3.0	3.0	3.2	2.8	3.2	2.7	3.3	2.7	3.5
- Coniferous	%	3.2		2.7	2.9	2.9	3.2	2.7	3.2	2.6	3.4	2.6	3.6
- Non-coniferous	%	2.5		2.9	3.6	3.6	3.5	3.3	3.3	3.2	3.1	3.2	3.4
Fellings*	1000 m3 o.b.	47150	53020	72220	57200	58700	77600	61900	76900	64900	78100	68000	80900
- Coniferous*	"	41010	47280	64310	50110	50200	64500	53200	65300	55600	66100	57700	67800
- Non-coniferous*	"	6140	5740	7910	7090	8500	13100	8700	11600	9300	12000	10300	13100
Felling/NAI ratio*	%	85	81	114	85	79	101	83	101	86	99	88	97
- Coniferous*	%	82		119	91	83	101	89	101	94	98	99	95
- Non-coniferous*	%	88		85	61	60	104	58	98	56	106	55	110
Removals (overbark)	1000 m3 o.b.	43500	48910	66620	52780	54200	71300	57100	70900	59900	71900	62600	74400
- Coniferous	"	37830	43620	59330	46240	47200	60600	50000	61400	52300	62100	54200	63700
- Non-coniferous	"	5670	5290	7290	6540	7000	10700	7100	9500	7600	9800	8400	10700
Removals (underbark)	1000 m3 u.b.	37830	42530	57930	45890	47200	62000	49700	61700	52100	62600	54500	64700
- Coniferous (volume)	"	32900	37930	51590	40200	41100	52700	43500	53400	45500	54000	47200	55400
- Non-coniferous (volume)	"	4930	4600	6340	5690	6100	9300	6200	8300	6600	8600	7300	9300
- Coniferous (percent)	%	87	89	89	88	87	85	88	87	87	86	87	86
Sawlogs as % of total	%	34	38	39	48	52	45	55	47	52	48	47	47
- Coniferous	%	38	41	43	52	67	56	64	57	61	58	56	57
- Non-coniferous	"	8	9	7	9	7*	4*	7*	5*	4*	5*	5*	4*

Annex table 5.9
Removals forecasts
(1000 m3 underbark)

	1949-51	1959-61	1969-71	1980	1990	2000	2010	2020
1. Reported data (3-yr av.) + ETTS III forecasts				ETTS III FORECASTS				
1.1 Total removals	37830	42530	57930	63000	64000	65000		
1.1.1 – Coniferous	32900	37930	51590	55440 *	56320 *	57200		
1.1.2 – Non-coniferous	4930	4600	6340	7560 *	7680 *	7800		
1.2 Of which: Sawlogs + veneer logs, total	12970	16070	22680					
1.2.1 – Coniferous	12570	15670	22260					
1.2.2 – Non-coniferous	400	400	420					
2. Forecasts for ETTS IV (Lower)				B.p. b/	LOWER FORECASTS			
2.1 From exploitable closed forest, total				45890	47200	49700	52100	54500
2.1.1 – Coniferous				40200	41100	43500	45500	47200
2.1.2 – Non-coniferous				5690	6100	6200	6600	7300
2.2 Other removals a/				2630	2000	- - -	- - -	↑
2.2.1 – Coniferous				2410	1850 *	- - -	- - -	↑
2.2.2 – Non-coniferous				220	150 *	- - -	- - -	↑
2.3 Total removals	37830	42530	57930	48520	49200	51700	54100	56500
2.3.1 – Coniferous	32900	37930	51590	42610	42950	45350	47350	49050
2.3.2 – Non-coniferous	4930	4600	6340	5910	6250	6350	6750	7450
2.4 Of which: sawlogs + veneer logs, total	12970	16070	22680	22060	27940	28240	27960	26830
2.4.1 – Coniferous	12570	15670	22260	21670	27540	27840	27560	26430
2.4.2 – Non-coniferous	400	400	420	390	400 *	400 *	400 *	400 *
3. Forecasts for ETTS IV (Higher)					HIGHER FORECASTS			
3.1 From exploitable closed forest, total				45890	62000	61700	62600	64700
3.1.1 – Coniferous				40200	52700	53400	54000	55400
3.1.2 – Non-coniferous				5690	9300	8300	8600	9300
3.2 Other removals a/				2630	4000	- - -	- - -	↑
3.2.1 – Coniferous				2410	3700 *	- - -	- - -	↑
3.2.2 – Non-coniferous				220	300 *	- - -	- - -	↑
3.3 Total removals	37830	42530	57930	48520	66000	65700	66600	68700
3.3.1 – Coniferous	32900	37930	51590	42610	56400	57100	57700	59100
3.3.2 – Non-coniferous	4930	4600	6340	5910	9600	8600	8900	9600
3.4 Of which: sawlogs + veneer logs, total	12970	16070	22680	22060	29910	30840	31720	31980
3.4.1 – Coniferous	12570	15670	22260	21670	29510	30440	31320	31580
3.4.2 – Non-coniferous	400	400	420	390	400 *	400 *	400 *	400 *

a/ Other removals: for most countries, relates to removals from other wooded land and trees outside the forest. For 1980, calculated as the difference between total removals (1.1) and removals from exploitable closed forest (2.1). In case of Sweden, nearly half of the forecasts of other removals is the recovery of natural losses on exploitable closed forests.

b/ Base period (see explanatory notes).

Tableau annexe 5.10

Prévisions forestières concernant la forêt dense exploitable

Pays: BELGIQUE(*)

	Unité	1950	1960	1970	Période de base	1990 Faible	1990 Fort	2000 Faible	2000 Fort	2010 Faible	2010 Fort	2020 Faible	2020 Fort
Forêt dense exploitable	1000 ha	601	588	604	600	600	620	600	635	600	645	600	650
Matériel sur pied (GS)	Mill. m3 o.b.	46	57	70	73	86	86	95	96	102	103	105	107
- Résineux (volume)	"	21	31	40	40	45	46	49	51	52	54	54	56
- Feuillus (volume)	"	25	26	30	33	41	40	46	45	50	49	51	51
- Résineux (pourcentage)	%	46	54	57	55	52	53	52	53	51	52	51	52
Matériel sur pied/ha	m3 o.b./ha	77	97	116	122	143	139	158	151	170	160	175	165
Accroissement annuel net (NAI)	1000 m3 o.b.	2260	2430	2620	4500	4440	4710	4380	4890	4320	5030	4260	5135
- Résineux (volume)	"	1360	1525	1520	2438	2400	2590	2365	2740	2335	2865	2300	2980
- Feuillus (volume)	"	900	905	1100	2062	2040	2120	2015	2150	1985	2165	1960	2155
- Résineux (pourcentage)	%	60	63	58	54	54	55	54	56	54	57	54	58
NAI/ha	m3 o.b./ha	3.8	4.1	4.3	7.5	7.4	7.6	7.3	7.7	7.2	7.8	7.1	7.9
NAI en pourcentage de GS	%	4.9	4.3	3.7	6.2	5.2	5.5	4.6	5.1	4.2	4.9	4.1	4.8
- Résineux	%	6.5	4.9	3.8	6.1	5.3	5.6	4.8	5.4	4.5	5.3	4.3	5.3
- Feuillus	%	3.6	3.5	3.7	6.2	5.0	5.3	4.4	4.8	4.0	4.4	3.8	4.2
Abattages*	1000 m3 o.b.	2660	2910	3150	2775	3350	3555	3620	4050	3885	4530	4045	4875
- Résineux*	"	1440	1540	2000	1815	1920	2070	2010	2330	2100	2580	2185	2830
- Feuillus*	"	1220	1370	1150	960	1430	1485	1610	1720	1785	1950	1860	2045
Rapport abattage/NAI*	%	118	120	120	62	75	75	83	83	90	90	95	95
- Résineux*	%	106	101	132	74	80	80	85	85	90	90	95	95
- Feuillus*	%	136	151	105	47	70	70	80	80	90	90	95	95
Quantités enlevées (sur écorce)	1000 m3 o.b.	2400	2620	2840	2500	3015	3200	3260	3645	3495	4075	3640	4390
- Résineux	"	1300	1390	1800	1635	1730	1865	1810	2095	1890	2320	1965	2545
- Feuillus	"	1100	1230	1040	865	1285	1335	1450	1550	1605	1755	1675	1845
Quantités enlevées (sous écorce)	1000 m3 u.b.	2165	2358	2561	2251	2715	2880	2935	3280	3145	3670	3275	3950
- Résineux (volume)	"	1170	1248	1622	1474	1555	1680	1630	1885	1700	2090	1770	2290
- Feuillus (volume)	"	995	1110	939	777	1160	1200	1305	1395	1445	1580	1505	1660
- Résineux (pourcentage)	%	54	53	63	65	57	58	56	57	54	57	54	58
Quantités enlevées de grumes de sciage en % du total	%												
- Résineux	"	40	50	51	54	55	55	55	55	55	55	55	55
- Feuillus	"	31	43	40	53	55	55	55	55	55	55	55	55

N.B. o.b. = volume sur écorce; u.b. = volume sous écorce.

(*) Prévisions préparées par le secrétariat sur la base de prévisions partielles envoyées par le pays.

Tableau annexe 5.11
Prévisions des quantités enlevées
(1000 m3 sous écorce)

Pays: BELGIQUE (*)

	1949-51	1959-61	1969-71	1980	1990	2000	2010	2020
1. Données fournies (moyenne de 3 ans) + Prévisions de la 3ème étude (ETTS III)				Prévisions du ETTS III				
1.1 Quantités enlevées, total	2165	2358	2561	2800*	2900*	3000*	..	..
1.1.1 – Résineux	1170	1248	1622	..	..	2000*	..	..
1.1.2 – Feuillus	995	1110	939	..	..	1000*	..	..
1.2 Dont: grumes de sciages et de placage, total	863	1169	1295	..	..	..	..	..
1.2.1 – Résineux	368	540	641	..	..	..	..	..
1.2.2 – Feuillus	495	629	654	..	..	..	..	..
2. Prévisions pour la 4ème étude (ETTS IV) (Faibles)				PRÉVISIONS FAIBLES				
2.1 Des forêts denses exploitables, total				B.p. b/ 2251	2715	2935	3145	3275
2.1.1 – Résineux				1474	1555	1630	1700	1770
2.1.2 – Feuillus				777	1160	1305	1445	1505
2.2 Autres quantités enlevées a/				188	200	→	→	→
2.2.1 – Résineux				–	–			
2.2.2 – Feuillus				188	200	→	→	→
2.3 Quantités enlevées, total	2165	2358	2561	2439	2915	3135	3345	3475
2.3.1 – Résineux	1170	1248	1622	1474	1555	1630	1700	1770
2.3.2 – Feuillus	995	1110	939	965	1360	1505	1645	1705
2.4 Dont: grumes de sciages et de placage, total	863	1169	1295	1208	1495	1615	1730	1800
2.4.1 – Résineux	368	540	641	796	855	895	935	975
2.4.2 – Feuillus	495	629	654	412	640	720	795	825
3. Prévisions pour la 4ème étude (ETTS IV) (Fortes)				PRÉVISIONS FORTES				
3.1 Des forêts denses exploitables, total				2251	2880	3280	3670	3950
3.1.1 – Résineux				1474	1680	1885	2090	2290
3.1.2 – Feuillus				777	1200	1395	1580	1660
3.2 Autres quantités enlevées, total a/				188	200	→	→	→
3.2.1 – Résineux				–	–			
3.2.2 – Feuillus				188	200	→	→	→
3.3 Quantités enlevées, total	2165	2358	2561	2439	3080	3480	3870	4150
3.3.1 – Résineux	1170	1248	1622	1474	1680	1885	2090	2290
3.3.2 – Feuillus	995	1110	939	965	1400	1595	1580	1860
3.4 Dont: grumes de sciages et de placage, total	863	1169	1295	1208	1585	1805	2020	2170
3.4.1 – Résineux	368	540	641	796	925	1035	1150	1260
3.4.2 – Feuillus	495	629	654	412	660	770	870	910

a/ Autres quantités enlevées : p. ex. d'autres terres boisées et d'arbres hors des forêts. Pour l'année 1980, elles sont calculées comme différence entre le total des quantités enlevées (1.1) et les quantités enlevées des forêts denses exploitables (2.1). Pour les années 1990 à 2020, on a supposé que le volume se maintiendra au niveau de 1980.

b/ Période de base (voir notes explicatives).

(*) Prévisions préparées par le secrétariat sur le base de prévisions partielles envoyées par le pays.

Annex table 5.12

Forestry forecasts relating to exploitable closed forest

Country: DENMARK

	Unit	1950	1960	1970	Base period	1990 Lower	1990 Higher	2000 Lower	2000 Higher	2010 Lower	2010 Higher	2020 Lower	2020 Higher
Exploitable closed forest	1000 ha	438	374	373	365	364	366	371	383	379	406	388	425
Growing stock (GS)	Mill.m3 o.b	40	44	42	47	51	52	53	55	56	59	60	63
- Coniferous (volume)	"	17	21	22	29	34	35	37	39	41	43	46	47
- Non-coniferous (volume)	"	23	23	20	18	17	17	16	16	15	16	14	16
- Coniferous (percent)	%	43	48	52	62	67	67	70	71	73	73	77	75
Growing stock/ha	m3o.b./ha	91	118	113	129	140	142	143	144	148	145	155	148
Net annual increment (NAI)	1000 m3 o.b.	2500	2530	2300	2800	2700	2900	2800	3100	3000	3400	3300	3900
- Coniferous (volume)	"	1530	1630	1500	2000	1900	2100	2000	2200	2200	2500	2400	3000
- Non-coniferous (volume)	"	970	900	800	800	800	800	800	900	800	900	900	900
- Coniferous (percent)	%	61	64	65	71	70	72	71	71	73	74	73	77
NAI/ha	m3o.b./ha	5.7	6.8	6.2	7.7	7.4	7.9	7.5	8.1	7.9	8.4	8.5	9.2
NAI as % of GS	%	6.3	5.8	5.5	6.0	5.3	5.6	5.3	5.6	5.4	5.8	5.5	6.2
- Coniferous	%	9.0	7.8	6.8	6.9	5.6	6.0	5.4	5.6	5.4	5.8	5.2	6.4
- Non-coniferous	%	4.2	3.9	4.0	4.4	4.7	4.7	5.0	5.6	5.3	5.6	6.4	5.6
Fellings*	1000 m3 o.b.	2180	1970	2470	2100	2500	2600	2600	2700	2700	3000	2800	3500
- Coniferous*	"	1020	1020	1430	1300	1600	1700	1700	1800	1800	2100	1900	2600
- Non-coniferous*	"	1160	950	1040	800	900	900	900	900	900	900	900	900
Felling/NAI ratio*	%	87	78	107	75	93	90	93	87	90	88	85	90
- Coniferous*	%	67	63	95	65	84	81	85	82	82	84	79	87
- Non-coniferous*	%	120	106	130	100	113	113	113	100	113	100	100	100
Removals (overbark)	1000 m3 o.b.	2180	1970	2470	2100	2500	2600	2600	2700	2700	3000	2800	3500
- Coniferous	"	1020	1020	1430	1300	1600	1700	1700	1800	1800	2100	1900	2600
- Non-coniferous	"	1160	950	1040	800	900	900	900	900	900	900	900	900
Removals (underbark)	1000 m3 u.b.	1973	1783	2230	1890	2250	2340	2340	2430	2430	2700	2520	3140
- Coniferous (volume)	"	910	907	1273	1160	1430	1520	1520	1610	1610	1880	1700	2320
- Non-coniferous (volume)	"	1063	876	957	730	820	820	820	820	820	825	825	820
- Coniferous (percent)	%	46	51	57	61	64	65	65	66	66	70	67	74
Sawlogs as % of total	%	51	53	65	56	49	47	45	46	44	43	45	45
- Coniferous	"	60	55	59	55	48	46	44	45	43	42	44	44
- Non-coniferous	"	44	51	73	57	51	48	47	48	45	45	46	47

Annex table 5.13
Removals forecasts
(1000 m3 underbark)

Country: DENMARK

	1949-51	1959-61	1969-71	1980	1990	2000	2010	2020
1. Reported data (3-yr. av.) + ETTS III forecasts				ETTS III FORECASTS				
1.1 Total removals	1973	1783	2230	2300*	2300*	2300*		
1.1.1 – Coniferous	910	907	1273			1500*		
1.1.2 – Non-coniferous	1063	876	957			800*		
1.2 Of which: Sawlogs + veneer logs, total	1013	949	1450					
1.2.1 – Coniferous	548	500	752					
1.2.2 – Non-coniferous	465	449	698					
2. Forecasts for ETTS IV (Lower)				B.p. b/	LOWER FORECASTS			
2.1 From exploitable closed forest, total				1890	2250	2340	2430	2520
2.1.1 – Coniferous				1160	1430	1520	1610	1700
2.1.2 – Non-coniferous				730	820	820	820	820
2.2 Other removals a/				–	–	–	–	–
2.2.1 – Coniferous				–	–	–	–	–
2.2.2 – Non-coniferous				–	–	–	–	–
2.3 Total removals	1973	1783	2230	1890	2250	2340	2430	2520
2.3.1 – Coniferous	910	907	1273	1160	1430	1520	1610	1700
2.3.2 – Non-coniferous	1063	876	957	730	820	820	820	820
2.4 Of which: sawlogs + veneer logs, total	1013	949	1450	1050	1100	1050	1060	1130
2.4.1 – Coniferous	548	500	752	640	680	670	690	750
2.4.2 – Non-coniferous	465	449	698	410	420	380	370	380
3. Forecasts for ETTS IV (Higher)					HIGHER FORECASTS			
3.1 From exploitable closed forest, total				1890	2340	2430	2700	3140
3.1.1 – Coniferous				1160	1520	1610	1880	2320
3.1.2 – Non-coniferous				730	820	820	820	820
3.2 Other removals a/				–	–	–	–	–
3.2.1 – Coniferous				–	–	–	–	–
3.2.2 – Non-coniferous				–	–	–	–	–
3.3 Total removals	1973	1783	2230	1890	2340	2430	2700	3140
3.3.1 – Coniferous	910	907	1273	1160	1520	1610	1880	2320
3.3.2 – Non-coniferous	1063	876	957	730	820	820	820	820
3.4 Of which: sawlogs + veneer logs, total	1013	949	1450	1050	1090	1120	1160	1410
3.4.1 – Coniferous	548	500	752	640	700	730	790	1020
3.4.2 – Non-coniferous	465	449	698	410	390	390	370	390

a/ Other removals: e.g. from other wooded land and trees outside the forest.
b/ Base period (see explanatory notes).

Tableau annexe 5.14

Prévisions forestières concernant la forêt dense exploitable

Pays : FRANCE

	Unité	1950	1960	1970	Période de base	1990 Faible	1990 Fort	2000 Faible	2000 Fort	2010 Faible	2010 Fort	2020 Faible	2020 Fort
Forêt dense exploitable	1000 ha	11307	11000	13090	13340	13410	13460	13490	13590	13570	13720	13650	13840
Matériel sur pied (GS)	Mill.m3 o.b.	805	978	1174	1550	1649	1640	1732	1688	1825	1744	1950	1804
- Résineux (volume)	"	332	453	515	605	635	637	657	663	688	711	749	763
- Feuillus (volume)	"	473	525	659	945	1014	1003	1075	1025	1137	1033	1201	1041
- Résineux (pourcentage)	%	41	46	44	39	39	39	38	39	38	41	38	42
Matériel sur pied/ha	m3 o.b./ha	71	89	90	116	123	122	128	124	134	127	143	130
Accroissement annuel net (NAI)	1000 m3 o.b.	32425		46500	54000	54050	56820	59730	63440	62210	67800	66080	71330
- Résineux (volume)	"	10763		22400	23500	24130	26070	28420	31350	30370	35200	33720	38350
- Feuillus (volume)	"	21662		24100	30500	29920	30750	31310	32090	31840	32600	32360	32980
- Résineux (pourcentage)	%	33		48	44	45	46	47	49	49	52	51	54
NAI/ha	m3 o.b./ha	2.9		3.6	4.0	4.0	4.2	4.4	4.7	4.6	4.9	4.8	5.2
NAI en pourcentage de GS	%	4.0		4.0	3.5	3.3	3.6	3.4	3.8	3.4	3.9	3.4	4.0
- Résineux	"	3.2		4.3	3.9	3.8	4.1	4.3	4.7	4.4	5.0	4.5	5.0
- Feuillus	%	4.6		3.7	3.2	3.0	3.1	2.9	3.1	2.8	3.2	2.7	3.2
Abattages*	1000 m3 o.b.	38830	44190	38170	41360	46800	51400	50520	59300	52700	60900	55700	66000
- Résineux*	"	12890	17060	18080	18790	22880	24310	25320	27930	27060	29040	30030	34100
- Feuillus*	"	25940	27130	20090	22570	23920	27090	25200	31370	25640	31860	25670	31900
Rapport abattage/NAI*	%	120		82	77	87	90	85	93	85	90	84	93
- Résineux*	%	120		81	80	95	93	89	89	89	83	89	89
- Feuillus*	%	120		83	74	80	88	80	98	81	98	79	97
Quantités enlevées (sur écorce)	1000 m3 o.b.	35300	40170	34700	37600	42550	46700	45920	53900	47900	55400	50600	60000
- Résineux	"	11720	15510	16440	17080	20800	22100	23020	25390	24600	26400	27300	31000
- Feuillus	"	23580	24660	18260	20520	21750	24600	22990	28510	23300	29000	23300	29000
Quantités enlevées (sous écorce)	1000 m3 u.b.	30421	34586	29826	32324	36600	40190	39490	46380	41180	47670	43480	51580
- Résineux (volume)	"	9984	13213	14001	14550	17680	18790	19570	21580	20910	22440	23210	26350
- Feuillus (volume)	"	20437	21373	15825	17774	18920	21400	19920	24800	20270	25230	20270	25230
- Résineux (pourcentage)	%	33	38	47	45	48	47	50	47	51	47	53	51
Quantités enlevées de grumes de sciage en % du total	%	27	47	57	59	59	56	60	58	62	57	62	58
- Résineux	"	49	68	67	70	67	65	68	66	70	68	69	68
- Feuillus	"	16	34	47	50	51	49	53	51	54	47	54	47

N.B. o.b. = volume sur écorce; u.b. = volume sous écorce.

Tableau annexe 5.15
Prévisions des quantités enlevées
(1000 m3 sous écorce)

Pays: FRANCE

	1949-51	1959-61	1969-71	1980	1990	2000	2010	2020
1. Données fournies (moyenne de 3 ans) + prévisions de la 3ème étude (ETTS III)				Prévisions du ETTS III				
1.1 Quantités enlevées, total	30421	34586	29826*	32800	41400	46800		
1.1.1 – Résineux	9984	13213	14001*		19700			
1.1.2 – Feuillus	20437	21373	15825*		27100			
1.2 Dont: grumes de sciages et de placage, total	8207	16145	16891					
1.2.1 – Résineux	4879	8971	9448					
1.2.2 – Feuillus	3328	7174	7443					
2. Prévisions pour la 4ème étude (ETTS IV) (Faibles)				Période de base	PREVISIONS FAIBLES			
2.1 Des forêts denses exploitables, total				32324	36600	39490	41180	43480
2.1.1 – Résineux				14550	17680	19570	20910	23210
2.1.2 – Feuillus				17774	18920	19920	20270	20270
2.2 Autres quantités enlevées a/				6156	5250*	4500*	3750*	3000*
2.2.1 – Résineux				150	–	–	–	–
2.2.2 – Feuillus				6006	5250*	4500*	3750*	3000*
2.3 Quantités enlevées, total	30421	34586	29826*	38480	41850	43990	44930	46480
2.3.1 – Résineux	9984	12313	14001*	14700	17680	19570	20910	23210
2.3.2 – Feuillus	20437	21373	15825*	23780	24170	24420	24020	23270
2.4 Dont: grumes de sciages et de placage, total	8207	16145	16891	18485	21495	23870	25580	26960
2.4.1 – Résineux	4879	8971	9448	10306	11845	13310	14635	16015
2.4.2 – Feuillus	3328	7174	7443	8179	9650	10560	10945	10945
3. Prévisions pour la 4ème étude (ETTS IV) (Fortes)					PREVISIONS FORTES			
3.1 Des forêts denses exploitables, total				32324	40190	46380	47670	51580
3.1.1 – Résineux				14550	18790	21580	22440	26350
3.1.2 – Feuillus				17774	21400	24800	25230	25230
3.2 Autres quantités enlevées, total a/				6156	5250*	4500*	3750*	3000*
3.2.1 – Résineux				150	–	–	–	–
3.2.2 – Feuillus				6006	5250*	4500*	3750*	3000*
3.3 Quantités enlevées, total	30421	34586	29826*	38480	45440	50880	51420	54580
3.3.1 – Résineux	9984	13213	14001*	14700	18790	21580	22440	26350
3.3.2 – Feuillus	20437	21373	15825*	23780	26650	29300	28980	28230
3.4 Dont: grumes de sciages et de placage, total	8207	16145	16891	18485	22700	26890	27120	29780
3.4.1 – Résineux	4879	8971	9448	10306	12215	14240	15260	17920
3.4.2 – Feuillus	3328	7174	7443	8179	10485	12650	11860	11860

a/ Autres quantités enlevées : p. ex. d'autres terres boisées et d'arbres hors des forêts. Pour l'année 1980, elles sont calculées comme différence entre le total des quantités enlevées (1.1) et les quantités enlevées des forêts denses exploitables (2.1). Pour les années 1990 à 2020, les prévisions sont fournies par le pays.

Annex table 5.16

Forestry forecasts relating to exploitable closed forest

Country: GERMANY, FED. REP. OF a/

	Unit	1950	1960	1970	Base period	1990 Lower	1990 Higher	2000 Lower	2000 Higher	2010 Lower	2010 Higher	2020 Lower	2020 Higher
Exploitable closed forest	1000 ha	6732	6838	6838	6960	6960	7000	6960	7040	6960	7080	6960	7120
Growing stock (GS)	Mill.m3 o.b.	625	990	1022	1500	1537	1525	1582	1549	1622	1570	1653	1585
- Coniferous (volume)	"	332	656	722	1065	1118	1103	1170	1129	1214	1152	1248	1172
- Non-coniferous (volume)	"	293	334	300	435	419	422	412	420	408	418	405	414
- Coniferous (percent)	%	53	66	71	71	73	72	74	73	75	73	75	74
Growing stock/ha	m3o.b./ha	93	145	149	215	220	217	227	220	233	221	237	222
Net annual increment (NAI)	1000 m3 o.b.	25000	34000	34000	39670	38980	39900	38280	40130	37580	40360	36890	40580
- Coniferous (volume)	"	18400	25800	25800	30150	29230	30320	28330	30500	27430	30670	26560	30840
- Non-coniferous (volume)	"	6600	8200	8200	9520	9750	9580	9950	9630	10150	9690	10330	9740
- Coniferous (percent)	%	74	76	76	76	75	76	74	76	73	76	72	76
NAI/ha	m3o.b./ha	3.7	5.0	5.0	5.7	5.6	5.7	5.5	5.7	5.4	5.7	5.3	5.7
NAI as % of GS	%	4.0	3.4	3.3	2.6	2.5	2.6	2.4	2.7	2.3	2.6	2.2	2.6
- Coniferous	"	5.5	3.9	3.6	2.8*	2.6	2.7	2.4	2.7	2.3	2.7	2.1	2.6
- Non-coniferous	%	2.3	2.5	2.7	2.2	2.3	2.3	2.4	2.3	2.5	2.3	2.6	2.4
Fellings*	1000 m3 o.b.	35225	31344	33689	37164	34000	37300	34000	37850	34000	38400	34000	39500
- Coniferous*	"	22865	19844	21436	25227	23500	27625	23500	28040	23500	28450	23500	29275
- Non-coniferous*	"	12360	11500	12253	11937	10500	9675	10500	9810	10500	9950	10500	10225
Felling/NAI ratio*	%	141	92	99	94	87	93	89	94	90	95	92	97
- Coniferous*	%	124	77	83	84	80	81	83	92	86	93	88	95
- Non-coniferous*	%	187	140	149	125	108	101	106	102	103	103	102	105
Removals (overbark)	1000 m3 o.b.	34225	30344	32689	36164	33000	36300	33000	36850	33000	37400	33000	38500
- Coniferous	"	22465	19444	21036	24827	23100	27225	23100	27640	23100	28050	23500	28875
- Non-coniferous	"	11760	10900	11653	11337	9900	9075	9900	9210	9900	9350	9900	9625
Removals (underbark)	1000 m3 u.b.	31115	27585	29718	32877	30000	33000	30000	33500	30000	34000	30000	35000
- Coniferous (volume)	"	20424	17676	19124	22570	21000	24750	21000	25125	21000	25500	21000	26250
- Non-coniferous (volume)	"	10691	9909	10594	10307	9000	8250	9000	8375	9000	8500	9000	8750
- Coniferous (percent)	%	66	64	64	69	70	75	70	75	70	75	70	75
Sawlogs as % of total	%	49	56	56	54	56	58	57	58	57	58	57	58
- Coniferous	"	62	68	66	62	65	65	65	65	65	65	65	65
- Non-coniferous	"	24	33	37	36	37	37	37	37	37	37	37	37

a/ Unofficial forecasts.

Annex table 5.17
Removals forecasts
(1000 m3 underbark)

Country: GERMANY, FED. REP. OF a/

	1949-51	1959-61	1969-71	1980	1990	2000	2010	2020
1. Reported data (3-yr av.) + ETTS III forecasts				ETTS III FORECASTS				
1.1 Total removals	31115	27585	29718	27000	28500	30000		
1.1.1 – Coniferous	20424	17676	19124			22200		
1.1.2 – Non-coniferous	10691	9909	10594			7800		
1.2 Of which: Sawlogs + veneer logs, total	15150	15392	16565					
1.2.1 – Coniferous	12573	12074	12645					
1.2.2 – Non-coniferous	2577	3318	3920					
2. Forecasts for ETTS IV (Lower)				B. p. c/	LOWER FORECASTS			
2.1 From exploitable closed forest, total				32877	30000	30000	30000	30000
2.1.1 – Coniferous				22570	21000	21000	21000	21000
2.1.2 – Non-coniferous				10307	9000	9000	9000	9000
2.2 Other removals b/				150*	150*⟶			⟶
2.2.1 – Coniferous				20*	20*⟶			⟶
2.2.2 – Non-coniferous				130*	130*⟶			⟶
2.3 Total removals	31115	27585	29718	33027	30150	30150	30150	30150
2.3.1 – Coniferous	20424	17676	19124	22590	21020	21020	21020	21020
2.3.2 – Non-coniferous	10691	9909	10594	10437	9130	9130	9130	9130
2.4 Of which: sawlogs + veneer logs, total	15150	15392	16565	17700	17000	17000	17000	17000
2.4.1 – Coniferous	12573	12074	12645	13990	13650	13650	13650	13650
2.4.2 – Non-coniferous	2577	3318	3920	3710	3350	3350	3350	3350
3. Forecasts for ETTS IV (Higher)					HIGHER FORECASTS			
3.1 From exploitable closed forest, total				32877	33000	33500	34000	35000
3.1.1 – Coniferous				22570	24750	25125	25500	26250
3.1.2 – Non-coniferous				10307	8250	8375	8500	8750
3.2 Other removals b/				150*	150*⟶			⟶
3.2.1 – Coniferous				20*	20*⟶			⟶
3.2.2 – Non-coniferous				130*	130*⟶			⟶
3.3 Total removals	31115	27585	29718	33027	33150	33650	34150	35150
3.3.1 – Coniferous	20424	17676	19124	22590	24770	25145	25520	26270
3.3.2 – Non-coniferous	10691	9909	10594	10437	8380	8505	8630	8780
3.4 Of which: sawlogs + veneer logs, total	15150	15392	16565	17700	19000	19400	19750	20250
3.4.1 – Coniferous	15273	12074	12645	13990	16000	16300	16600	17000
3.4.2 – Non-coniferous	2577	3318	3920	3710	3000	3100	3150	3250

a/ Unofficial forecasts.
b/ Other removals: e.g. from other wooded land and trees outside the forest. For 1980, taken from table 12 of forest resource enquiry. For the forecasts for 1990 to 2020, it has been assumed that other removals will remain at the 1980 level.
c/ Base period (see explanatory notes).

Annex table 5.18

Forestry forecasts relating to exploitable closed forest

Country: IRELAND

	Unit	1950	1960	1970	Base period	1990 Lower	1990 Higher	2000 Lower	2000 Higher	2010 Lower	2010 Higher	2020 Lower	2020 Higher
Exploitable closed forest	1000 ha	124	171	268	347	385	401	423	469	427	538	429	607
Growing stock (GS)	Mill.m3 o.b.	4.8	10.2	15	32	36.1	36.2	48.3	48.8	54.1	57.1	51.0	60.7
- Coniferous (volume)	"	2.7	6.0	10	25	28.4	28.5	39.8	40.3	44.9	47.9	41.2	50.9
- Non-coniferous (volume)	"	2.1	4.2	5	7	7.7	7.7	8.5	8.5	9.2	9.2	9.8	9.8
- Coniferous (percent)	%	56	59	67	78	79	79	82	83	83	84	81	84
Growing stock/ha	m3o.b./ha	39	60	56	92	94	90	114	104	127	106	119	100
Net annual increment (NAI)	1000 m3 o.b.	290	707	1800	2530	2900	2900	4400	4500	4800	5300	4300	5500
- Coniferous (volume)	"	250	581	1700	2450	2800	2800	4200	4300	4600	5100	4100	5300
- Non-coniferous (volume)	"	40	126	100	80	100	100	200	200	200	200	200	200
- Coniferous (percent)	%	86	82	94	97	97	97	95	96	96	96	95	96
NAI/ha	m3o.b./ha	2.3	4.1	6.7	7.3	7.5	7.2	10.4	9.6	11.2	9.9	10.0	9.1
NAI as % of GS	%	6.0	6.9	12.0	7.9	8.0	8.0	9.1	9.2	8.9	9.3	8.4	9.1
- Coniferous	%	9.3	9.7	17.0	9.8	9.9	9.8	10.6	10.7	10.2	10.6	10.0	10.4
- Non-coniferous	%	1.9	3.0	2.0	1.1	1.3	1.3	2.4	2.4	2.2	2.2	2.0	2.0
Fellings*	1000 m3 o.b.	190	390	390	650	1820	1820	3460	3460	4900	5000	4860	5140
- Coniferous*	"	60	275	345	600	1760	1760	3390	3390	4820	4920	4780	5060
- Non-coniferous*	"	130	115	45	50	60	60	70	70	80	80	80	80
Felling/NAI ratio*	%	66	55	22	26	63	63	79	77	102	94	113	93
- Coniferous*	%	24	47	20	24	63	63	81	79	105	96	117	95
- Non-coniferous*	%	325	91	45	62	60	60	35	35	40	40	40	40
Removals (overbark)*	1000 m3 o.b.	190	390	390	650	1820	1820	3460	3460	4900	5000	4860	5140
- Coniferous*	"	60	275	345	600	1760	1760	3390	3390	4820	4920	4780	5060
- Non-coniferous*	"	130	115	45	50	60	60	70	70	80	80	80	80
Removals (underbark)	1000 m3 u.b.	169	346	346	534	1600	1600	3040	3040	4310	4400	4280	4520
- Coniferous (volume)	"	54	243	306	497	1550	1550	2980	2980	4240	4330	4210	4450
- Non-coniferous (volume)	"	115	103	40	37	50	50	60	60	70	70	70	70
- Coniferous (percent)	%	32	70	88	93	97	97	98	98	98	98	98	98
Sawlogs as % of total	%	45	53	65	64	54	54	67	67	74	73	76	75
- Coniferous	"	83	64	68	65	55	55	68	68	75	74	77	76
- Non-coniferous	"	27	29	45	47	35	35	35	35	34	33	33	32

Annex table 5.19
Removals forecasts
(1000 m3 underbark)

	1949-51	1959-61	1969-71	1980	1990	2000	2010	2020
1. Reported data (3-yr av.) + ETTS III forecasts				ETTS III FORECASTS				
1.1 Total removals	169	346	346	700	1500	2200		
1.1.1 – Coniferous	54	243	306			2100		
1.1.2 – Non-coniferous	115	103	40			100		
1.2 Of which: Sawlogs + veneer logs, total	76	185	226					
1.2.1 – Coniferous	45	155	208					
1.2.2 – Non-coniferous	31	30	18					
2. Forecasts for ETTS IV (Lower)				B. P. b/	LOWER FORECASTS			
2.1 From exploitable closed forest, total				534	1600	3040	4310	4280
2.1.1 – Coniferous				497	1550	2980	4240	4210
2.1.2 – Non-coniferous				37	50	60	70	70
2.2 Other removals a/				-	-	-	-	-
2.2.1 – Coniferous				-	-	-	-	-
2.2.2 – Non-coniferous				-	-	-	-	-
2.3 Total removals	169	346	346	534	1600	3040	4310	4280
2.3.1 – Coniferous	54	243	306	497	1550	2980	4240	4210
2.3.2 – Non-coniferous	115	103	40	37	50	60	70	70
2.4 Of which: sawlogs + veneer logs, total	76	185	226	340	865	2045	3205	3265
2.4.1 – Coniferous	45	155	208	325	850	2025	3180	3240
2.4.2 – Non-coniferous	31	30	18	15	15	20	25	25
3. Forecasts for ETTS IV (Higher)					HIGHER FORECASTS			
3.1 From exploitable closed forest, total				534	1600	3040	4400	4520
3.1.1 – Coniferous				497	1550	2980	4330	4450
3.1.2 – Non-coniferous				37	50	60	70	70
3.2 Other removals a/				-	-	-	-	-
3.2.1 – Coniferous				-	-	-	-	-
3.2.2 – Non-coniferous				-	-	-	-	-
3.3 Total removals	169	346	346	534	1600	3040	4400	4520
3.3.1 – Coniferous	54	243	306	497	1550	2980	4330	4450
3.3.2 – Non-coniferous	115	103	40	37	50	60	70	70
3.4 Of which: sawlogs + veneer logs, total	76	185	226	340	865	2045	3230	3400
3.4.1 – Coniferous	45	155	208	325	850	2025	3205	3380
3.4.2 – Non-coniferous	31	30	18	15	15	20	25	20

a/ Other removals: e.g. from other wooded land and trees outside the forest.
b/ Base period (see explanatory notes).

Tableau annexe 5.20
Prévisions forestières concernant la forêt dense exploitable

Pays: ITALIE**

	Unité	1950	1960	1970	Période de base	1990		2000		2010		2020	
						Faible	Fort	Faible	Fort	Faible	Fort	Faible	Fort
Forêt dense exploitable	1000 ha	5648	6029	5342	3868*	3900	3925	3925	3975	3950	4125	3975	4225
Matériel sur pied (GS)	Mill.m3 o.b.	329	296	354	557*	593	594	623	627	648	655	667	678
- Résineux (volume)	"	114	144	128	201*	222	224	239	243	252	258	260	269
- Feuillus (volume)	"	215	152	226	356*	371	370	384	384	396	397	407	409
- Résineux (pourcentage)	%	35	49	36	36	37	38	38	39	39	39	39	40
Matériel sur pied/ha	m3 o.b./ha	58	49	66	144	152	151	159	158	164	159	168	160
Accroissement annuel net (NAI)	1000 m3 o.b.	14810	11716	14000	11880*	12090	12530	11775	13000	11455	13480	11130	13965
- Résineux (volume)	"	2120	2965	2600	4280*	4350	4635	4240	4940	4125	5255	4005	5585
- Feuillus (volume)	"	12690	8751	11400	7600*	7740	7895	7535	8060	7330	8225	7125	8380
- Résineux (pourcentage)	%	14	25	19	36	36	37	36	38	36	39	36	40
NAI/ha	m3 o.b./ha	2.6	1.9	2.6	3.1	3.1	3.2	3.0	3.3	2.9	3.3	2.8	3.3
NAI en pourcentage de GS	%	4.5	4.0	4.0	2.1	2.0	2.1	1.9	2.1	1.8	2.1	1.7	2.1
- Résineux	%	1.9	2.1	2.0	2.1	2.0	2.1	1.8	2.0	1.6	2.0	1.5	2.1
- Feuillus	%	5.9	5.7	5.0	2.1	2.1	2.1	2.0	2.1	1.8	2.1	1.8	2.0
Abattages*	1000 m3 o.b.	15405	20800	111905	8060	8740	9025	9010	9900	9250	10850	9460	11870
- Résineux*	"	2880	1985	1520	1925	2395	2550	2755	3210	3095	3940	3405	4745
- Feuillus*	"	12525	18815	10385	6135	6345	6475	6255	6690	6155	6910	6055	7125
Rapport abattage/NAI*	%	104	178	85	68	72	72	77	76	81	80	85	85
- Résineux*	%	136	67	58	45	55	55	65	65	75	75	85	85
- Feuillus*	%	99	215	91	81	82	82	83	83	84	84	85	85
Quantités enlevées (sur écorce)	1000 m3 o.b.	14740	19905	11390	7710	8365	8635	8625	9475	8850	10385	9055	11360
- Résineux	"	2755	1900	1455	1840	2290	2440	2635	3070	2960	3770	3260	4540
- Feuillus*	"	11985	18005	9935	5870	6075	6195	5990	6405	5890	6615	5795	6820
Quantités enlevées (sous écorce)	1000 m3 u.b.	14052*	19063	10890	7330	7935	8195	8170	8970	8370	9805	8555	10710
- Résineux (volume)	"	2516	1734	1330	1680	2090	2230	2405	2805	2700	3440	2975	4145
- Feuillus (volume)	"	11536	17329	9560	5650	5845	5965	5765	6165	5670	6365	5580	6565
- Résineux (pourcentage)	%	18	9	12	23	26	27	29	31	32	35	35	39
Quantités enlevées de grumes de sciage en % du total	%	18	14	34	30	33	32	34	34	35	36	35	37
- Résineux	"	57	56	61	54	55	55	55	55	55	55	55	55
- Feuillus	"	10	9	30	25	25	25	25	25	25	25	25	25

N.B. o.b. = volume sur écorce; u.b. = volume sous écorce.
** Prévisions préparées par le secrétariat.

Tableau annexe 5.21
Prévisions des quantités enlevées
(1000 m3 sous écorce)

Pays: ITALIE**

	1949-51	1959-61	1969-71	1980	1990	2000	2010	2020
1. Données fournies (moyenne de 3 ans) + prévisions de la 3ème étude (ETTS III)				*Prévisions du ETTS III*				
1.1 Quantités enlevées, total	14052	11951	10885	7000	8000	9000		
1.1.1 – Résineux	2516	5164	1327			3300		
1.1.2 – Feuillus	11536	6787	9558			5700		
1.2 Dont: grumes de sciages et de placage, total	2567	2618	3720					
1.2.1 – Résineux	1436	973	808					
1.2.2 – Feuillus	1131	1645	2912					
2. Prévisions pour la 4ème étude (ETTS IV) (Faibles)				*Période de base*	*PREVISIONS FAIBLES*			
2.1 Des forêts denses exploitables, total				7330	7935	8170	8370	8555
2.1.1 – Résineux				1680	2090	2405	2700	2975
2.1.2 – Feuillus				5650	5845	5765	5670	5580
2.2 Autres quantités enlevées a/				1630	1600	→	→	→
2.2.1 – Résineux				30	-	→	→	→
2.2.2 – Feuillus				1600	1600	→	→	→
2.3 Quantités enlevées, total	14052	11951	10885	8960	9535	9770	9970	10155
2.3.1 – Résineux	2516	5164	1327	1710	2090	2405	2700	2975
2.3.2 – Feuillus	11536	6787	9558	7250	7445	7365	7270	7180
2.4 Dont: grumes de sciages et de placage, total	2567	2618	3720	2315	2610	2760	2900	3030
2.4.1 – Résineux	1436	973	808	905	1150	1320	1485	1635
2.4.2 – Feuillus	1131	1645	2912	1410	1460	1440	1415	1395
3. Prévisions pour la 4ème étude (ETTS IV) (Fortes)					*PREVISIONS FORTES*			
3.1 Des forêts denses exploitables, total				7330	8195	8970	9805	10710
3.1.1 – Résineux				1680	2230	2805	3440	4145
3.1.2 – Feuillus				5650	5965	6165	6365	6565
3.2 Autres quantités enlevées, total a/				1630	1600	→	→	→
3.2.1 – Résineux				30	-	→	→	→
3.2.2 – Feuillus				1600	1600	→	→	→
3.3 Quantités enlevées, total	14052	11951	10885	8960	9795	10570	11405	12310
3.3.1 – Résineux	2516	5164	1327	1710	2230	2805	3440	4145
3.3.2 – Feuillus	11536	6787	9558	7250	7565	7765	7965	8165
3.4 Dont: grumes de sciages et de placage, total	2567	2618	3720	2315	2715	3080	3480	3920
3.4.1 – Résineux	1436	973	808	905	1225	1540	1890	2280
3.4.2 – Feuillus	1131	1645	2912	1410	1490	1540	1590	1640

a/ Autres quantités enlevées : p. ex. d'autres terres boisées et d'arbres hors des forêts. Pour l'année 1980, elles sont calculées comme différence entre le total des quantités enlevées (1.1) et les quantités enlevées des forêts denses exploitables (2.1). Pour les années 1990 à 2020, on a supposé que le volume se maintiendra au niveau de 1980.

** Prévisions préparées par le secrétariat.

Tableau annexe 5.22

Prévisions forestières concernant la forêt dense exploitable

Pays: LUXEMBOURG**

	Unité	1950	1960	1970	Période de base	1990 Faible	1990 Fort	2000 Faible	2000 Fort	2010 Faible	2010 Fort	2020 Faible	2020 Fort
Forêt dense exploitable	1000 ha	81	81	81	80.5	80.5	81.0	80.5	81.5	80.5	82.0	80.5	82.5
Matériel sur pied (GS)	Mill.m3 o.b.	10.1	11.2	13	13.0	12.9	13.0	12.8	13.0	12.8	13.0	12.9	13.0
- Résineux (volume)	"	2.2	2.2	2	2.2	2.2	2.1	2.2	2.1	2.2	2.1	2.3	2.1
- Feuillus (volume)	"	7.9	9.0	11	10.8	10.7	10.9	10.6	10.9	10.6	10.9	10.6	10.9
- Résineux (pourcentage)	%	22	20	15	17	17	16	17	16	17	16	18	16
Matériel sur pied/ha	m3 o.b./ha	125	138	160	161	160	160	159	160	159	159	160	158
Accroissement annuel net (NAI)	1000 m3 o.b.	170	266	400	330 a/	322	340	314	350	306	361	298	371
- Résineux (volume)	"	80	117	100	120 a/	116	129	113	140	110	152	107	163
- Feuillus (volume)	"	90	149	300	210 a/	206	211	201	210	196	209	191	208
- Résineux (pourcentage)	%	47	44	25	36	36	38	36	40	36	42	36	44
NAI/ha	m3 o.b./ha	2.1	3.3	4.9	4.1	4.0	4.2	3.9	4.3	3.8	4.4	3.7	4.5
NAI en pourcentage de GS	%	1.7	2.4	3.1	2.5	2.5	2.6	2.5	2.7	2.4	2.8	2.3	2.9
- Résineux	%	3.6	5.3	5.0	5.4	5.3	6.1	5.1	6.7	5.0	7.2	4.7	7.8
- Feuillus	%	1.1	1.7	3.0	1.9	1.9	1.9	1.9	1.9	1.8	1.9	1.8	1.9
Abattages*	1000 m3 o.b.	225	215	255	348	332	340	324	350	300	361	277	371
- Résineux*	"	120	80	95	134	116	129	113	140	104	152	96	163
- Feuillus*	"	105	135	160	214	216	211	211	210	196	209	181	208
Rapport abattage/NAI*	%	132	81	64	105	103	100	103	100	98	100	93	100
- Résineux*	%	150	68	95	112	103	100	100	100	95	100	90	100
- Feuillus*	%	117	91	53	102	100	100	105	100	100	100	95	100
Quantités enlevées (sur écorce)	1000 m3 o.b.	215	205	245	331	315	323	308	333	285	343	263	352
- Résineux	"	115	75	90	128	110	123	107	133	99	144	91	155
- Feuillus	"	100	130	155	203	205	200	201	200	186	199	172	197
Quantités enlevées (sous écorce)	1000 m3 u.b.	186	183	218	292 a/	278	285	272	294	252	302	232	310
- Résineux (volume)	"	98	66	80	111 a/	96	107	93	116	86	125	79	135
- Feuillus (volume)	"	88	117	138	181 a/	182	178	179	178	166	177	153	175
- Résineux (pourcentage)	%	53	36	37	38 a/	35	38	34	39	34	41	34	43
Quantités enlevées de grumes de sciage en % du total	%	21	36	38	48 a/	49	49	49	50	47	50	46	50
- Résineux	"	23	36	34	57 a/	57	57	57	57	57	57	57	57
- Feuillus	"	19	36	40	43 a/	45	45	45	45	42	45	40	45

N.B. o.b. = volume sur écorce; u.b. = volume sous écorce.

a/ Statistiques Forestières: EEC.

** Prévisions préparées par le secrétariat.

Tableau annexe 5.23
Prévisions des quantités enlevées
(1000 m3 sous écorce)

Pays: LUXEMBOURG**

	1949-51	1959-61	1969-71	1980	1990	2000	2010	2020
1. Données fournies (moyenne de 3 ans) + prévisions de la 3ème étude (ETTS III)				Prévisions du ETTS III				
1.1 Quantités enlevées, total	186	183	218	300*	300*	300*		
1.1.1 – Résineux	98	76	80	100*		100*		
1.1.2 – Feuillus	88	107	138	200*		200*		
1.2 Dont: grumes de sciages et de placage, total	39	65	82					
1.2.1 – Résineux	23	24	27					
1.2.2 – Feuillus	16	41	55					
2. Prévisions pour la 4ème étude (ETTS IV) (Faibles)				Période de base	PREVISIONS FAIBLES			
2.1 Des forêts denses exploitables, total				292	278	272	252	232
2.1.1 – Résineux				111	96	93	86	79
2.1.2 – Feuillus				181	182	179	166	153
2.2 Autres quantités enlevées a/				–	↑	↑	↑	↑
2.2.1 – Résineux				–	↑	↑	↑	↑
2.2.2 – Feuillus				–	↑	↑	↑	↑
2.3 Quantités enlevées, total	186	183	218	292	278	272	252	232
2.3.1 – Résineux	98	76	80	111	96	93	86	79
2.3.2 – Feuillus	88	107	138	181	182	179	166	153
2.4 Dont: grumes de sciages et de placage, total	39	65	82	141	137	134	124	114
2.4.1 – Résineux	23	24	27	63	55	53	49	45
2.4.2 – Feuillus	16	41	55	78	82	81	75	69
3. Prévisions pour la 4ème étude (ETTS IV) (Fortes)					PREVISIONS FORTES			
3.1 Des forêts denses exploitables, total				292	285	294	302	310
3.1.1 – Résineux				111	107	116	125	135
3.1.2 – Feuillus				181	178	178	177	175
3.2 Autres quantités enlevées, total a/				–	↑	↑	↑	↑
3.2.1 – Résineux				–	↑	↑	↑	↑
3.2.2 – Feuillus				–	↑	↑	↑	↑
3.3 Quantités enlevées, total	186	183	218	292	285	294	302	310
3.3.1 – Résineux	98	76	80	111	107	116	125	135
3.3.2 – Feuillus	88	107	138	181	178	178	177	175
3.4 Dont: grumes de sciages et de placage, total	39	65	82	141	141	146	151	156
3.4.1 – Résineux	23	24	27	63	61	66	71	77
3.4.2 – Feuillus	16	41	55	78	80	80	80	79

a/ Autres quantités enlevées : p. ex. d'autres terres boisées et d'arbres hors des forêts.

** Prévisions préparées par le secrétariat.

Annex table 5.24

Forestry forecasts relating to exploitable closed forest

Country: NETHERLANDS

	Unit	1950	1960	1970	Base period	1990 Lower	1990 Higher	2000 Lower	2000 Higher	2010 Lower	2010 Higher	2020 Lower	2020 Higher
Exploitable closed forest	1000 ha	250	260	256	294	305	315	320	335	335	345	335	345
Growing stock (GS)	Mill.m3 o.b.	15	15.5	20	29	31*	31*	33*	33*	35*	34*	36*	35*
- Coniferous (volume)	''	13	13.3	16	21	23*	23*	24*	24*	25*	25*	26*	26*
- Non-coniferous (volume)	''	2	2.2	4	8	8*	8*	9*	9*	10*	9*	10*	9*
- Coniferous (percent)	%	87	86	80	72	74	74	73	73	71	74	72	74
Growing stock/ha	m3o.b./ha	60	60	78	99	102	98	103	99	104	99	107	101
Net annual increment (NAI)	1000 m3 o.b.	647	905	1200	1240*	1320*	1340*	1380*	1450*	1450*	1540*	1560*	1580*
- Coniferous (volume)	''	498	815	980	910*	920*	920*	920*	920*	930*	930*	940*	940*
- Non-coniferous (volume)	''	149	90	220	330*	400*	420*	460*	530*	520*	610*	620*	640*
- Coniferous (percent)	%	77	90	86	73*	70*	69*	67*	63*	64*	60*	60*	59*
NAI/ha	m3o.b./ha	2.6	3.5	4.7	4.2	4.3	4.3	4.3	4.3	4.3	4.5	4.7	4.6
NAI as % of GS	%	4.3	5.8	6.0	4.3	4.3	4.3	4.2	4.4	4.1	4.5	4.3	4.5
- Coniferous	%	3.8	6.1	6.1	4.3	4.0	4.0	3.8	3.8	3.7	3.7	3.6	3.6
- Non-coniferous	%	7.4	4.1	5.5	4.1	5.0*	5.3*	5.1*	5.9*	5.2*	6.8*	6.2*	7.1*
Fellings*	1000 m3 o.b.	780	950	1180	1180	1020	1060	1130	1290	1390	1480	1500	1520
- Coniferous*	''	410	610	790	760	740	740	780	780	855	835	880	880
- Non-coniferous*	''	370	340	390	420	280	320	350	510	535	645	620	640
Felling/NAI ratio*	%	121	105	98	95	77	79	82	89	96	96	96	96
- Coniferous*	%	82	75	81	84	80	80	85	85	92	90	94	94
- Non-coniferous*	%	248	378	177	127	70	76	76	96	103	106	100	100
Removals (overbark)*	1000 m3 o.b.	725	880	1095	1090	950	990	1060	1210	1300	1380	1400	1420
- Coniferous*	''	380	565	730	700	690	690	730	730	800	780	820	880
- Non-coniferous*	''	345	315	365	390	260	300	330	480	500	600	580	600
Removals (underbark)	1000 m3 u.b.	600	728	905	910*	780*	820*	870*	1000*	1080*	1150*	1160*	1180*
- Coniferous (volume)	''	308	460	594	575	560*	560*	590*	590*	650*	630*	660*	660*
- Non-coniferous (volume)	''	292	268	311	335*	220*	260*	280*	410*	430*	520*	500*	520*
- Coniferous (percent)	%	51	63	66	67	72	68	68	59	60	55	57	56
Sawlogs as % of total	%	25	21	28	28	29	30	31	32	33	34	34	35
- Coniferous	''	8	19	15	23	25	25	26	26	27	27	28	28
- Non-coniferous	''	42	24	52	38	40	40	41	41	42	42	43	43

Annex table 5.25
Removals forecasts
(1000 m3 underbark)

Country: NETHERLANDS

	1949-51	1959-61	1969-71	1980	1990	2000	2010	2020
1. Reported data (3-yr av.) + ETTS III forecasts				ETTS III FORECASTS				
1.1 Total removals	600	728	905	900*	900*	1000*	1080*	1160*
1.1.1 – Coniferous	308	460	594			800*	650*	660*
1.1.2 – Non-coniferous	292	268	311			200*	430*	500*
1.2 Of which: Sawlogs + veneer logs, total	149	152	250	214*	240*	260*	280*	300*
1.2.1 – Coniferous	26	88	87	-*	-	-	-	-
1.2.2 – Non-coniferous	123	64	163	214*	240*	260*	280*	300*
2. Forecasts for ETTS IV (Lower)				B. p. b/	LOWER FORECASTS			
2.1 From exploitable closed forest, total				910*	780*	870*	1080*	1160*
2.1.1 – Coniferous				575*	560*	590*	650*	660*
2.1.2 – Non-coniferous				335*	220*	280*	430*	500*
2.2 Other removals a/				214*	240*	260*	280*	300*
2.2.1 – Coniferous				-*	-	-	-	-
2.2.2 – Non-coniferous				214*	240*	260*	280*	300*
2.3 Total removals	600	728	905	1124	1020*	1130*	1360*	1460*
2.3.1 – Coniferous	308	460	594	575	560*	590*	650*	660*
2.3.2 – Non-coniferous	292	268	311	549	460*	540*	710*	800*
2.4 Of which: sawlogs + veneer logs, total	149	152	250	255*	230*	270*	355*	400*
2.4.1 – Coniferous	26	88	87	130*	140*	155*	175*	185*
2.4.2 – Non-coniferous	123	64	163	125*	90*	115*	180*	215*
3. Forecasts for ETTS IV (Higher)					HIGHER FORECASTS			
3.1 From exploitable closed forest, total				910*	820*	1000*	1150*	1180*
3.1.1 – Coniferous				575*	560*	590*	630*	660*
3.1.2 – Non-coniferous				335*	260*	410*	520*	520*
3.2 Other removals a/				214*	240*	260*	280*	300*
3.2.1 – Coniferous				-	-	-	-	-
3.2.2 – Non-coniferous				214	240*	260*	280*	300*
3.3 Total removals	600	728	905	1124	1060*	1260*	1430*	1480*
3.3.1 – Coniferous	308	460	594	575	560*	590*	630*	660*
3.3.2 – Non-coniferous	292	268	311	549	500*	670*	800*	820*
3.4 Of which: sawlogs + veneer logs, total	149	152	250	255*	245*	325*	390*	410*
3.4.1 – Coniferous	26	88	87	130*	140*	155*	170*	185*
3.4.2 – Non-coniferous	123	64	163	125*	105*	170*	220*	225*

a/ Other removals: e.g. from other wooded land and trees outside the forest.
b/ Base period (see explanatory notes).

Annex table 5.26
Forestry forecasts relating to exploitable closed forest

Country: UNITED KINGDOM

	Unit	1950	1960	1970	Base period	1990 Lower	1990 Higher	2000 Lower	2000 Higher	2010 Lower	2010 Higher	2020 Lower	2020 Higher
Exploitable closed forest	1000 ha	1535	1675	1521	2017	2220	2400	2400	2800	2600	3200	2800	3600
Growing stock (GS)	Mill.m3 o.b	95	108	135	203	271	267	341	327	400	389	442	454
- Coniferous (volume)	"	39	52	70	111	164	162	220	211	267	264	298	321
- Non-coniferous (volume)	"	56	56	65	92	107	105	121	116	133	125	144	133
- Coniferous (percent)	%	41	48	52	55	61	61	65	65	67	68	67	71
Growing stock/ha	m3o.b./ha	62	64	89	101	122	111	142	117	154	122	158	126
Net annual increment (NAI)	1000 m3 o.b.	3133	3851	6300	11200	13300	13200	15300	15900	16400	19200	17900	22500
- Coniferous (volume)	"	2120*	2917	5100	8600	11000	10900	13200	13800	14400	17200	16000	20500
- Non-coniferous (volume)	"	1013*	934	1200	2600	2300	2300	2100	2100	2000	2000	1900	2000
- Coniferous (percent)	%	68*	76	81	77	83	83	86	87	88	90	89	91
NAI/ha	m3o.b./ha	2.0	2.3	4.1	5.6	6.0	5.5	6.4	5.7	6.3	6.0	6.4	6.3
NAI as % of GS	%	3.3	3.6	4.7	5.5	4.9	4.9	4.5	4.9	4.1	4.9	4.0	5.0
- Coniferous	%	5.4*	5.6	7.3	7.7	6.7	6.7	6.0	6.5	5.4	6.5	5.4	6.4
- Non-coniferous	%	1.8*	1.7	1.8	2.8	2.1	2.2	1.7	1.8	1.5	1.6	1.3	1.5
Fellings*	1000 m3 o.b.	4245	3930	4230	4580	6200	6900	8300	10400	11400	12200	14500	16200
- Coniferous*	"	1270	1510	2515	3605	5400	5800	7500	9300	10600	11100	13700	15100
- Non-coniferous*	"	2975	2420	1715	975	800	1100	800	1100	800	1100	800	1100
Felling/NAI ratio*	%	135	102	67	41	47	52	54	65	70	64	81	72
- Coniferous*	%	60	52	49	42	49	53	57	67	74	65	86	74
- Non-coniferous*	%	294	259	143	37	35	48	38	52	40	55	42	55
Removals (overbark)	1000 m3 o.b.	3965	3675	3955	4300	5800	6400	7700	9700	10600	11300	13500	15100
- Coniferous	"	1185	1410	2350	3385	5100	5400	7000	8700	9900	10300	12800	14100
- Non-coniferous*	"	2780	2265	1605	915	700	1000	700	1000	700	1000	700	1000
Removals (underbark)	1000 m3 u.b.	3468	3216	3461	3765	5100	5600	6700	8500	9300	9900	11800	13200
- Coniferous (volume)	"	1037	1235	2054	2965	4500	4700	6100	7600	8700	9000	11200	12300
- Non-coniferous (volume)	"	2431	1981	1407	800	600	900	600	900	600	900	600	900
- Coniferous (percent)	%	30	38	59	79	88	84	91	89	94	91	95	93
Sawlogs as % of total	%	76	67	52	66	59	61	56	58	57	59	60	61
- Coniferous	"	52	47	41	63	55	56	53	54	55	56	58	60
- Non-coniferous	"	87	80	67	75	75	75	75	75	75	75	75	75

Annex table 5.27
Removals forecasts
(1000 m3 underbark)

Country: UNITED KINGDOM

	1949-51	1959-61	1969-71	1980	1990	2000	2010	2020
1. Reported data (3-yr av.) + ETTS III forecasts				ETTS III FORECASTS				
1.1 Total removals	3468	3216	3461	4300	6000	7300		
1.1.1 – Coniferous	1037	1235	2054	2900*	4700*	6100		
1.1.2 – Non-coniferous	2431	1981	1407	1400*	1300*	1200		
1.2 Of which: Sawlogs + veneer logs, total	2639	2156	1789					
1.2.1 – Coniferous	535	580	849					
1.2.2 – Non-coniferous	2104	1576	940					
2. Forecasts for ETTS IV (Lower)					LOWER FORECASTS			
2.1 From exploitable closed forest, total				B.p. b/ 3765 c/	5100	6700	9300	11800
2.1.1 – Coniferous				2965	4500	6100	8700	11200
2.1.2 – Non-coniferous				800	600	600	600	600
2.2 Other removals a/				550—	→	→	→	→
2.2.1 – Coniferous				50—	→	→	→	→
2.2.2 – Non-coniferous				500—	→	→	→	→
2.3 Total removals	3468	3216	3461	4315	5650	7250	9850	12350
2.3.1 – Coniferous	1037	1235	2054	3015	4550	6150	8750	11250
2.3.2 – Non-coniferous	2431	1981	1407	1300	1100	1100	1100	1100
2.4 Of which: sawlogs + veneer logs, total	2639	2156	1789	2647	3300	4060	5610	7320
2.4.1 – Coniferous	535	580	849	1678	2475	3235	4785	6495
2.4.2 – Non-coniferous	2104	1576	940	969	825	825	825	825
3. Forecasts for ETTS IV (Higher)					HIGHER FORECASTS			
3.1 From exploitable closed forest, total				3765c/	5600	8500	9900	13200
3.1.1 – Coniferous				2965	4700	7600	9000	12300
3.1.2 – Non-coniferous				800	900	900	900	900
3.2 Other removals a/				550—	→	→	→	→
3.2.1 – Coniferous				50—	→	→	→	→
3.2.2 – Non-coniferous				500—	→	→	→	→
3.3 Total removals	3468	3216	3461	4315	6150	9050	10450	13750
3.3.1 – Coniferous	1037	1235	2054	3015	4750	7650	9050	12350
3.3.2 – Non-coniferous	2431	1981	1407	1300	1400	1400	1400	1400
3.4 Of which: sawlogs + veneer logs, total	2639	2156	1789	2647	3680	5155	6090	8430
3.4.1 – Coniferous	535	580	849	1678	2630	4105	5040	7380
3.4.2 – Non-coniferous	2104	1576	940	969	1050	1050	1050	1050

a/ Other removals: e.g. from other wooded land and trees outside the forest. For 1980, calculated as the difference between total removals (1.1) and removals from exploitable closed forest (2.1). For the forecasts for 1990 to 2020, it has been assumed that other removals will remain at the 1980 level.
b/ Base period (see explanatory notes).
c/ Taken from table 12 of the Forest Resource Enquiry (Part I): data for 1981.

Annex table 5.28
Forestry forecasts relating to exploitable closed forest

Country: AUSTRIA

	Unit	1950	1960	1970	Base period	1990 Lower	1990 Higher	2000 Lower	2000 Higher	2010 Lower	2010 Higher	2020 Lower	2020 Higher
Exploitable closed forest	1000 ha	3139	2991	3230	3165	3170	3185	3175	3210	3175	3210	3175	3210
Growing stock (GS)	Mill.m3 o.b.	348	479	752	803	865	895	925	955	925	955	925	955
- Coniferous (volume)	"	310	432	642	679	736	761	791	817	791	817	791	817
- Non-coniferous (volume)	"	38	47	110	124	129	134	134	138	134	138	134	138
- Coniferous (percent)	%	89	90	85	84	85	85	86	86	86	86	86	86
Growing stock/ha	m3o.b./ha	111	160	233	254	273	281	291	298	291	298	291	298
Net annual increment (NAI)	1000 m3 o.b.	8230	9718	18500	19581	21080	21840	22590	23350	22590	23350	22590	23350
- Coniferous (volume)	"	7032	8474	15700	16657	17940	18580	19220	19870	19220	19870	19220	19870
- Non-coniferous (volume)	"	1198	1244	2800	2924	3140	3260	3370	3480	3370	3480	3370	3480
- Coniferous (percent)	%	85	87	85	85	85	85	85	85	85	85	85	85
NAI/ha	m3o.b./ha	2.6	3.2	5.7	6.2	6.6	6.9	7.1	7.3	7.1	7.3	7.1	7.3
NAI as % of GS	%	2.4	2.0	2.5	2.4	2.4	2.4	2.4	2.4	2.4	2.4	2.4	2.4
- Coniferous	%	2.3	2.0	2.4	2.5	2.4	2.4	2.4	2.4	2.4	2.4	2.4	2.4
- Non-coniferous	%	3.2	2.6	2.5	2.4	2.4	2.4	2.5	2.5	2.5	2.5	2.5	2.5
Fellings*	1000 m3 o.b.	11220	14900	15090	15210	18630	19250	19190	20440	19190	20440	19190	20440
- Coniferous*	"	9805	12775	12795	12510	15555	16070	16115	17170	16115	17170	16115	17170
- Non-coniferous*	"	1415	2125	2295	2700	3075	3180	3075	3270	3075	3270	3075	3270
Felling/NAI ratio*	%	136	153	82	78	88	88	85	88	85	88	85	88
- Coniferous*	%	139	151	81	75	87	86	84	86	84	86	84	86
- Non-coniferous*	%	118	171	82	92	98	98	91	94	91	94	91	94
Removals (overbark)	1000 m3 o.b.	10770	14300	14480	14600	17880	18480	18420	19620	18420	19620	18420	19620
- Coniferous	"	9410	12260	12280	12010	14930	15430	15470	16480	15470	16480	15470	16480
- Non-coniferous	"	1360	2040	2200	2590	2950	3050	2950	3140	2950	3140	2950	3140
Removals (underbark)	1000 m3 u.b.	8976	11915	12070	12168	14900	15400	15350	16350	15350	16350	15350	16350
- Coniferous (volume)	"	7842	10218	10232	10008	12440	12850	12890	13750	12890	13750	12890	13750
- Non-coniferous (volume)	"	1360	2040	2200	2160	2460	2550	2460	2600	2460	2600	2460	2600
- Coniferous (percent)	%	87	86	85	82	83	83	84	84	84	84	84	84
Sawlogs as % of total	%	59	62	66	62	58	58	58	58	58	58	58	58
- Coniferous	"	65	69	74	71	63	63	63	63	63	63	63	63
- Non-coniferous	"	17	24	22	18	29	29	29	29	29	29	29	29

Annex table 5.29
Removals forecasts
(1000 m3 underbark)

Country: AUSTRIA

	1949-51	1959-61	1969-71	1980	1990	2000	2010	2020
1. Reported data (3-yr av.) + ETTS III forecasts				ETTS III FORECASTS				
1.1 Total removals	8976	11915	12070	13330	14730	16270		
1.1.1 – Coniferous	7842	10218	10232			13280*		
1.1.2 – Non-coniferous	1134	1697	1838			2990*		
1.2 Of which: Sawlogs + veneer logs, total	5266	7431	7992					
1.2.1 – Coniferous	5072	7027	7592					
1.2.2 – Non-coniferous	194	404	400					
2. Forecasts for ETTS IV (Lower)				B. p. b/	LOWER FORECASTS			
2.1 From exploitable closed forest, total				12168	14900	15350		
2.1.1 – Coniferous				10008	12440	12890		
2.1.2 – Non-coniferous				2160	2460	2460		
2.2 Other removals a/								
2.2.1 – Coniferous								
2.2.2 – Non-coniferous								
2.3 Total removals	8976	11915	12070	12168	14900	15350		
2.3.1 – Coniferous	7842	10218	10232	10008	12440	12890		
2.3.2 – Non-coniferous	1134	1697	1838	2160	2460	2460		
2.4 Of which: sawlogs + veneer logs, total	5266	7431	7992	7470	8550	8835		
2.4.1 – Coniferous	5072	7027	7592	7080	7835	8120		
2.4.2 – Non-coniferous	194	404	400	390	715	715		
3. Forecasts for ETTS IV (Higher)					HIGHER FORECASTS			
3.1 From exploitable closed forest, total				12168	15400	16350		
3.1.1 – Coniferous				10008	12850	13750		
3.1.2 – Non-coniferous				2160	2550	2600		
3.2 Other removals a/								
3.2.1 – Coniferous								
3.2.2 – Non-coniferous								
3.3 Total removals	8976	11915	12070	12168	15400	16350		
3.3.1 – Coniferous	7842	10218	10232	10008	12850	13750		
3.3.2 – Non-coniferous	1134	1697	1838	2160	2550	2600		
3.4 Of which: sawlogs + veneer logs, total	5266	7431	7992	7470	8835	9415		
3.4.1 – Coniferous	5072	7027	7592	7080	8095	8660		
3.4.2 – Non-coniferous	194	404	400	390	740	755		

a/ Other removals: e.g. from other wooded land and trees outside the forest.
b/ Base period (see explanatory notes).

Tableau annexe 5.30

Prévisions forestières concernant la forêt dense exploitable

Pays : SUISSE

	Unité	1950	1960	1970	Période de base	1990 Faible	1990 Fort	2000 Faible	2000 Fort	2010 Faible	2010 Fort	2020 Faible	2020 Fort
Forêt dense exploitable	1000 ha	982	982	972	935	950	950	950	950	950	950	950	950
Matériel sur pied (GS)	Mill.m3 o.b.	271	281	268	312	313	299	309	275	305	251	302	231
- Résineux (volume)	"	187	194	185	210	211	200	209	184	205	168	203	155
- Feuillus (volume)	"	84	87	83	102	102	99	100	91	100	83	99	76
- Résineux (pourcentage)	%	69	69	69	67	67	67	67	68	67	67	67	67
Matériel sur pied/ha	m3 o.b./ha	276	286	276	334	329	315	325	289	321	264	318	243
Accroissement annuel net (NAI)	1000 m3 o.b.	4820	4820	5250	5200	4900	4400	4700	3800	4800	3600	4900	3700
- Résineux (volume)	"	3326	3326	3623	3600	3400	3100	3300	2700	3400	2500	3400	2600
- Feuillus (volume)	"	1494	1494	1627	1600	1500	1300	1400	1100	1400	1100	1500	1100
- Résineux (pourcentage)	%	69	69	69	69	69	70	70	71	71	69	69	70
NAI/ha	m3 o.b./ha	4.9	4.9	5.4	5.6	5.2	4.6	4.9	4.0	5.1	3.8	5.2	3.9
NAI en pourcentage de GS	%	1.8	1.7	2.0	1.7	1.6	1.5	1.5	1.4	1.6	1.4	1.6	1.6
- Résineux	%	1.8	1.7	2.0	1.7	1.6	1.6	1.6	1.5	1.7	1.5	1.7	1.7
- Feuillus	%	1.8	1.7	2.0	1.6	1.5	1.3	1.4	1.2	1.4	1.3	1.5	1.4
Abattages*	1000 m3 o.b.	4165	4220	4795	4800	4800	5700	5100	6200	5200	6000	5200	5700
- Résineux*	"	3250	3080	3355	3380	3350	4020	3550	4320	3650	4220	3650	4020
- Feuillus*	"	915	1140	1440	1420	1450	1680	1550	1880	1550	1780	1550	1680
Rapport abattage/NAI*	%	86	88	91	92	98	130	109	163	108	167	106	154
- Résineux*	%	98	93	93	94	99	130	108	160	107	169	107	155
- Feuillus*	%	61	76	89	89	97	129	111	171	111	162	103	153
Quantités enlevées (sur écorce)	1000 m3 o.b.	3820	3870	4400	4400	4300	5100	4600	5600	4700	5400	4700	5100
- Résineux 117.4	"	2980	2825	3080	3100	3000	3600	3200	3900	3300	3800	3300	3600
- Feuillus 115.4	"	840	1045	1320	1300	1300	1500	1400	1700	1400	1600	1400	1500
Quantités enlevées (sous écorce)	1000 m3 u.b.	3492	3601	4005	4000	3900	4600	4200	5100	4300	4900	4300	4600
- Résineux (volume)	"	2720	2475	2810	2800	2700	3200	2900	3500	3000	3400	3000	3200
- Feuillus (volume)	"	772	1126	1195	1200	1200	1400	1300	1600	1300	1500	1300	1400
- Résineux (pourcentage)	%	78	69	70	70	69	70	69	69	70	69	70	70
Quantités enlevées de grumes de sciage en % du total	%	39	42	56	62	60	58	59	57	59	57	60	58
- Résineux	"	47	55	70	74	72	70	71	69	71	69	72	70
- Feuillus	"	11	13	22	30	33	31	32	31	31	30	32	31

N.B. o.b. = volume sur écorce; u.b. = volume sous écorce.

Tableau annexe 5.31
Prévisions des quantités enlevées
(1000 m3 sous écorce)

Pays: SUISSE

	1949-51	1959-61	1969-71	1980	1990	2000	2010	2020
1. Données fournies (moyenne de 3 ans) + prévisions de la 3ème étude (ETTS III)					Prévisions du ETTS III			
1.1 Quantités enlevées, total	3490	3600	4000	4500	4800 *	5000		
1.1.1 - Résineux	2720	2470	2810	3300 *	3600 *	3800 *		
1.1.2 - Feuillus	770	1130	1190	1200 *	1200 *	1200 *		
1.2 Dont: grumes de sciages et de placage, total	1370	1500	2240					
1.2.1 - Résineux	1280	1350	1980					
1.2.2 - Feuillus	90	150	260					
2. Prévisions pour la 4ème étude (ETTS IV) (Faibles)				Période de base	PREVISIONS FAIBLES			
2.1 Des forêts denses exploitables, total				4000	3900	4200	4300	4300
2.1.1 - Résineux				2800	2700	2900	3000	3000
2.1.2 - Feuillus				1200	1200	1300	1300	1300
2.2 Autres quantités enlevées a/				390	←	→	→	→
2.2.1 - Résineux				160	←	→	→	→
2.2.2 - Feuillus				230	←	→	→	→
2.3 Quantités enlevées, total	3490	3600	4000	4390	4290	4590	4690	4690
2.3.1 - Résineux	2720	2470	2810	2960	2860	3060	3160	3160
2.3.2 - Feuillus	770	1130	1190	1430	1430	1530	1530	1530
2.4 Dont: grumes de sciages et de placage, total	1370	1500	2240	2610	2340	2480	2540	2580
2.4.1 - Résineux	1280	1350	1980	2180	1940	2060	2130	2160
2.4.2 - Feuillus	90	150	260	430	400	420	410	420
3. Prévisions pour la 4ème étude (ETTS IV) (Fortes)					PREVISIONS FORTES			
3.1 Des forêts denses exploitables, total				4000	4600	5100	4900	4600
3.1.1 - Résineux				2800	3220	3500	3400	3200
3.1.2 - Feuillus				1200	1400	1600	1500	1400
3.2 Autres quantités enlevées, total a/				390	←	→	→	→
3.2.1 - Résineux				160	←	→	→	→
3.2.2 - Feuillus				230	←	→	→	→
3.3 Quantités enlevées, total	3490	3600	4000	4390	4990	5490	5290	4990
3.3.1 - Résineux	2720	2470	2810	2960	3360	3660	3560	3360
3.3.2 - Feuillus	770	1130	1190	1430	1630	1830	1730	1630
3.4 Dont: grumes de sciages et de placage, total	1370	1500	2240	2610	2670	2910	2790	2670
3.4.1 - Résineux	1280	1350	1980	2180	2240	2420	2340	2240
3.4.2 - Feuillus	90	150	260	430	430	490	450	430

a/ Autres quantités enlevées : P. ex. d'autres terres boisées et d'arbres hors des forêts. Pour l'année 1980, elles sont calculées comme différence entre le total des quantités enlevées (1.1) et les quantités enlevées des forêts denses exploitables (2.1).
Pour les années 1990 à 2020, on a supposé que le volume se maintiendra au niveau de 1980.

Annex table 5.32

Forestry forecasts relating to exploitable closed forest

Country: CYPRUS

	Unit	1950	1960	1970	Base period	1990		2000		2010		2020	
						Lower	Higher	Lower	Higher	Lower	Higher	Lower	Higher
Exploitable closed forest	1000 ha	171	147	173	100	100	100	100	100	100	100	100	100
Growing stock (GS)	Mill.m3 o.b	4.5	3.8	3	3.5	3.6	3.6	3.7	3.7	3.8	3.8	4.0	4.0
- Coniferous (volume)	"	4.0	3.3	3	3.5	3.6	3.6	3.7	3.7	3.8	3.8	4.0	4.0
- Non-coniferous (volume)	"	0.5	0.5	-	-	-	-	-	-	-	-	-	-
- Coniferous (percent)	%	89	87	100	100	100	100	100	100	100	100	100	100
Growing stock/ha	m3o.b./ha	26	26	17	35	36	36	37	37	38	38	40	40
Net annual increment (NAI)	1000 m3 o.b.	100	74	60	90	90	90	90	90	100	100	110	110
- Coniferous (volume)	"	83	60	60	90	90	90	90	90	100	100	110	110
- Non-coniferous (volume)	"	17	14	-	-	-	-	-	-	-	-	-	-
- Coniferous (percent)	%	83	81	100	100	100	100	100	100	100	100	100	100
NAI/ha	m3o.b./ha	0.6	0.5	0.3	0.9	0.9	0.9	0.9	0.9	1.0	1.0	1.0	1.0
NAI as % of GS	%	2.2	1.9	2.0	2.6	2.5	2.5	2.4	2.4	2.6	2.6	2.7	2.7
- Coniferous	%	2.1	1.8	2.0	2.6	2.5	2.5	2.4	2.4	2.6	2.6	2.7	2.7
- Non-coniferous	%	3.4	2.3	-									
Fellings*	1000 m3 o.b.	48	58	46	70	70	70	70	70	75	75	80	80
- Coniferous*	"	32	53	42	70	70	70	70	70	75	75	80	80
- Non-coniferous*	"	16	5	4	-	-	-	-	-	-	-	-	-
Felling/NAI ratio*	%	48	78	77	78	78	78	78	78	75	75	73	73
- Coniferous*	%	39	88	70	78	78	78	78	78	75	75	73	73
- Non-coniferous*	%	94	36	91	-	-	-	-	-	-	-	-	-
Removals (overbark)	1000 m3 o.b.	48	58	58	70	70	70	70	70	75	75	80	80
- Coniferous	"	32	53	53	70	70	70	70	70	75	75	80	80
- Non-coniferous	"	16	5	5	-	-	-	-	-	-	-	-	-
Removals (underbark)	1000 m3 u.b.	38	46	46	55	55	55	55	55	60	60	60	60
- Coniferous (volume)	"	25	42	42	55	55	55	55	55	60	60	60	60
- Non-coniferous (volume)	"	13	4	4	-	-	-	-	-	-	-	-	-
- Coniferous (percent)	%	66	91	91	100	100	100	100	100	100	100	100	100
Sawlogs as % of total	%	26	26	18	55	50	50	50	50	50	50	50	50
- Coniferous	"	36	26	17	55	50	50	50	50	50	50	50	50
- Non-coniferous	"	8	25	25	-	-	-	-	-	-	-	-	-

Annex table 5.33
Removals forecasts
(1000 m3 underbark)

Country: CYPRUS

	1949-51	1959-61	1969-71	1980	1990	2000	2010	2020
1. Reported data (3-yr av.) + ETTS III forecasts			ETTS III FORECASTS					
1.1 Total removals	38	46	46					
1.1.1 – Coniferous	25	42	42					
1.1.2 – Non-coniferous	13	4	4					
1.2 Of which: sawlogs + veneer logs, total	10	12	8					
1.2.1 – Coniferous	9	11	7					
1.2.2 – Non-coniferous	1	1	1					
2. Forecasts for ETTS IV (Lower)				B. p. b/	LOWER FORECASTS			
2.1 From exploitable closed forest, total				74	55	55	60	60
2.1.1 – Coniferous				72	55	55	60	60
2.1.2 – Non-coniferous				1	–	–	–	–
2.2 Other removals a/				2	–	–	↑	↑
2.2.1 – Coniferous				1	–	–	↑	↑
2.2.2 – Non-coniferous				1	–	–	↑	↑
2.3 Total removals	38	46	46	76	55	55	60	60
2.3.1 – Coniferous	25	42	42	73	55	55	60	60
2.3.2 – Non-coniferous	13	4	4	3	–	–	–	–
2.4 Of which: sawlogs + veneer logs, total	10	12	8	41	27	27	30	30
2.4.1 – Coniferous	9	11	7	39	27	27	30	30
2.4.2 – Non-coniferous	1	1	1	2	–	–	–	–
3. Forecasts for ETTS IV (Higher)					HIGHER FORECASTS			
3.1 From exploitable closed forest, total				74	55	55	60	60
3.1.1 – Coniferous				72	55	55	60	60
3.1.2 – Non-coniferous				1	–	–	–	–
3.2 Other removals a/				2	–	–	↑	↑
3.2.1 – Coniferous				1	–	–	↑	↑
3.2.2 – Non-coniferous				1	–	–	↑	↑
3.3 Total removals	38	46	46	76	55	55	60	60
3.3.1 – Coniferous	25	42	42	73	55	55	60	60
3.3.2 – Non-coniferous	13	4	4	3	–	–	–	–
3.4 Of which: sawlogs + veneer logs, total	10	12	8	41	27	27	30	30
3.4.1 – Coniferous	9	11	7	39	27	27	30	30
3.4.2 – Non-coniferous	1	1	1	2	–	–	–	–

a/ Other removals: e.g. from other wooded land and trees outside the forest.
b/ Base period (see explanatory notes).

67

Annex table 5.34
Forestry forecasts relating to exploitable closed forest

Country: GREECE (*)

	Unit	1950	1960	1970	Base period	1990 Lower	1990 Higher	2000 Lower	2000 Higher	2010 Lower	2010 Higher	2020 Lower	2020 Highest
Exploitable closed forest	1000 ha	1847	1992	2312	2300	2350	2400	2400	2500	2450	2580	2500	2640
Growing stock (GS)	Mill.m3 o.b	129	148	150	159	170	170	180	180	188	188	194	194
- Coniferous (volume)	"	76	75	75	85	92	93	98	100	102	105	105	109
- Non-coniferous (volume)	"	53	73	75	74	78	77	82	80	86	83	89	85
- Coniferous (percent)	%	59	51	50	53	54	55	54	56	54	56	54	56
Growing stock/ha	m3o.b./ha	70	74	65	69	72	71	75	72	77	73	78	73
Net annual increment (NAI)	1000 m3 o.b.	3750	4145	4000	4100	4350	4560	4560	5000	4780	5420	5000	5810
- Coniferous (volume)	"	2200		1820	1900	2000	2145	2145	2400	2295	2655	2450	2905
- Non-coniferous (volume)	"	1550		2180	2200	2350	2415	2415	2600	2485	2765	2550	2905
- Coniferous (percent)	%	59		45	46	46	47	47	48	48	49	49	50
NAI/ha	m3o.b./ha	2.0	2.1	1.7	1.8	1.85	1.9	1.9	2.0	1.95	2.1	2.0	2.2
NAI as % of GS	%	2.9	2.8	2.7	2.6	2.6	2.7	2.5	2.8	2.5	2.9	2.6	3.0
- Coniferous	%	2.9		2.4	2.2	2.2	2.3	2.2	2.4	2.2	2.5	2.3	2.7
- Non-coniferous	%	2.9		2.9	3.0	3.0	3.1	2.9	3.2	2.9	3.3	2.9	3.4
Fellings*	1000 m3 o.b.	3970	3670	3520	2845	3340	3490	3755	4115	4140	4690	4500	5230
- Coniferous*	"	810	300	900	1045	1340	1435	1630	1825	1930	2230	2205	2615
- Non-coniferous*	"	3160	3370	2620	1800	2000	2055	2125	2290	2210	2460	2295	2615
Felling/NAI ratio*	%	106	89	88	69	77	77	82	82	87	87	90	90
- Coniferous*	%	37		49	55	67	67	76	76	84	84	90	90
- Non-coniferous*	%	204		120	82	85	85	88	88	89	89	90	90
Removals (overbark)	1000 m3 o.b.	3770	3470	3330	2695	3190	3340	3605	3965	3990	4540	4350	5080
- Coniferous	"	770	280	850	995	1290	1385	1580	1775	1880	2180	2155	2565
- Non-coniferous	"	3000	3190	2480	1700	1900	1955	2025	2190	2110	2360	2195	2515
Removals (underbark)	1000 m3 u.b.	3255	3034	2868	2300	2710	2840	3050	3355	3370	3830	3665	4275
- Coniferous (volume)	"	616	223	683	800	1035	1115	1265	1425	1510	1750	1730	2055
- Non-coniferous (volume)	"	2639	2811	2185	1500	1675	1725	1785	1930	1860	2080	1935	2220
- Coniferous (percent)	%	19	7	18	35	38	39	41	42	45	46	47	48
Sawlogs as % of total	%	4	10	15	19	20	21	21	22	23	23	23	23
- Coniferous	"	17	82	34	41	40	40	39	39	38	38	37	37
- Non-coniferous	"	1	4	8	7	8	8	9	9	10	10	11	11

(*) Forecasts prepared by secretariat on basis of partial estimates by country.

Table 5.35
Removals forecasts
(million m3 underbark)

Country: GREECE (*)

	1949-51	1959-61	1969-71	1980	1990	2000	2010	2020
1. Reported data (3-yr av.) + ETTS III forecasts				*ETTS III forecasts*				
1.1 Total removals	3255	3034	2868	3200	3500	3700		
1.1.1 – Coniferous	616	223	683			1700		
1.1.2 – Non-coniferous	2639	2811	2185			2000		
1.2 Of which: Sawlogs + veneer logs, total	129	307	417					
1.2.1 – Coniferous	104	184	232					
1.2.2 – Non-coniferous	25	123	185					
2. Forecasts for ETTS IV (Lower)				*B. p. b/*		*LOWER FORECASTS*		
2.1 From exploitable closed forest, total				2300	2710	3050	3370	3665
2.1.1 – Coniferous				800	1035	1265	1510	1730
2.1.2 – Non-coniferous				1500	1675	1785	1860	1935
2.2 Other removals a/				290	300	→	→	→
2.2.1 – Coniferous				–	–	→	→	→
2.2.2 – Non-coniferous				290	300	→	→	→
2.3 Total removals	3255	3034	2868	2590	3010	3350	3670	3965
2.3.1 – Coniferous	616	223	683	800	1035	1265	1510	1730
2.3.2 – Non-coniferous	2639	2811	2185	1790	1975	2085	2160	2235
2.4 Of which: sawlogs + veneer logs, total	129	307	417	482	550	650	760	850
2.4.1 – Coniferous	104	184	232	300	420	490	580	640
2.4.2 – Non-coniferous	25	123	185	182	130	160	180	210
3. Forecasts for ETTS IV (Higher)						*HIGHER FORECASTS*		
3.1 From exploitable closed forest, total				2300	2840	3355	3830	4275
3.1.1 – Coniferous				800	1115	1425	1750	2055
3.1.2 – Non-coniferous				1500	1725	1930	2080	2220
3.2 Other removals a/				290	300	→	→	→
3.2.1 – Coniferous				–	–	→	→	→
3.2.2 – Non-coniferous				290	300	→	→	→
3.3 Total removals	3255	3034	2868	2590	3140	3655	4130	4575
3.3.1 – Coniferous	616	223	683	800	1115	1425	1750	2055
3.3.2 – Non-coniferous	2639	2811	2185	1790	2025	2230	2380	2520
3.4 Of which: sawlogs + veneer logs, total	129	307	417	482	580	730	870	1000
3.4.1 – Coniferous	104	184	232	300	440	560	660	760
3.4.2 – Non-coniferous	25	123	185	182	140	170	210	240

a/ Other removals: e.g. from other wooded land and trees outside the forest. For 1980, calculated as the difference between total removals (1.1) and removals from exploitable closed forest (2.1). For the forecasts for 1990 to 2020, it has been assumed that other removals will remain at the 1980 level.
b/ Base period (see explanatory notes).
(*) Forecasts prepared by secretariat on basis of partial forecasts by country.

Annex table 5.36

Forestry forecasts relating to exploitable closed forest

Country: ISRAEL

	Unit	1950	1960	1970	Base period	1990 Lower	1990 Higher	2000 Lower	2000 Higher	2010 Lower	2010 Higher	2020 Lower	2020 Higher
Exploitable closed forest	1000 ha	14	66*	43	66	71	71	80	80	90	90	100	100
Growing stock (GS)	Mill.m3 o.b.	0.1	0.5	2	2.9	3.7	3.8	4.5	4.6	5.4	5.5	6.3	6.4
- Coniferous (volume)	"	0.1	0.3	1	1.7	2.2	2.2	2.5	2.6	3.2	3.3	4.0	4.1
- Non-coniferous (volume)	"	-	0.2	1	1.2	1.5	1.6	2.0	2.0	2.2	2.2	2.3	2.3
- Coniferous (percent)	%	100	60	50	59	59	58	56	57	59	60	63	64
Growing stock/ha	m3o.b./ha	71	76*	47	44	52	54	56	58	60	61	63	64
Net annual increment (NAI)	1000 m3 o.b.	37	43	110	200	220	230	210	200	230	220	250	270
- Coniferous (volume)	"	27	24	60	90	110	110	100	90	120	110	140	150
- Non-coniferous (volume)	"	10	19	50	110	110	120	110	110	110	110	110	120
- Coniferous (percent)	%	73	56	55	45	50	48	48	45	52	50	56	56
NAI/ha	m3o.b./ha	2.6	0.7*	2.6	3.0	3.1	3.2	2.6	2.5	2.6	2.4	2.5	2.7
NAI as % of GS	%		8.6	5.5	6.9	5.9	6.1	4.7	4.3	4.3	4.0	4.0	4.2
- Coniferous	%		8.0	6.0	5.3	5.0	5.0	4.0	3.5	3.8	3.3	3.5	3.7
- Non-coniferous	%		9.5	5.0	9.2	7.3	7.5	5.5	5.5	5.0	5.0	4.8	5.2
Fellings*	1000 m3 o.b.	10	40	104	90	100	150	120	160	140	180	160	200
- Coniferous*	"	1	10	36	54	59	64	72	75	82	84	103	106
- Non-coniferous*	"	9	30	68	36	41	86	48	85	58	96	57	94
Felling/NAI ratio*	%	27	93	95	45	45	65	57	80	61	82	64	74
- Coniferous*	%	4	42	60	60	54	58	72	83	68	76	74	71
- Non-coniferous*	%	90	158	136	33	37	72	44	77	53	87	52	78
Removals (overbark)	1000 m3 o.b.	9	36	92	72	98	140	100	150	120	150	140	170
- Coniferous	"	1	9	31	43	58	60	60	70	70	70	90	90
- Non-coniferous	"	8	27	61	29	40	80	40	80	50	80	50	80
Removals (underbark)	1000 m3 u.b.	8	32	81	63	85	123	87	132	105	132	122	149
- Coniferous (volume)	"	1	8	26	37	49	51	51	60	60	60	77	77
- Non-coniferous (volume)	"	7	24	55	26	36	72	36	72	45	72	45	72
- Coniferous (percent)	%	13	25	32	59	58	41	59	45	57	45	63	52
Sawlogs as % of total	%	26		16	18	22	22	25	24	24	24	22	22
- Coniferous	"	36		22	30	37	33	35	39	30	32	25	26
- Non-coniferous	"	8		8	8	8	13	13	12	16	15	18	16

Annex table 5.37
Removals forecasts
(1000 m3 underbark)

Country: ISRAEL

	1949-51	1959-61	1969-71	1980	1990	2000	2010	2020
1. Reported data (3-yr av.) + ETTS III forecasts				ETTS III FORECASTS				
1.1 Total removals	8	32	81		98	100	105	122
1.1.1 – Coniferous	1	8	26		58	60	60	77
1.1.2 – Non-coniferous	7	24	55		40	40	45	45
1.2 Of which: sawlogs + veneer logs, total	–	..	13					
1.2.1 – Coniferous	–	..	6					
1.2.2 – Non-coniferous	–	..	7					
2. Forecasts for ETTS IV (Lower)				B. p. b/	LOWER FORECASTS			
2.1 From exploitable closed forest, total				63	85	87	105	122
2.1.1 – Coniferous				37	49	51	60	77
2.1.2 – Non-coniferous				26	36	36	45	45
2.2 Other removals a/				76	75	103	115	128
2.2.1 – Coniferous				26	26	49	60	73
2.2.2 – Non-coniferous				50	49	54	55	55
2.3 Total removals	8	32	81	139	160	190	220	250
2.3.1 – Coniferous	1	8	26	63	75	100	120	150
2.3.2 – Non-coniferous	7	24	55	76	85	90	100	100
2.4 Of which: sawlogs + veneer logs, total	–	..	13	25	35	47	52	56
2.4.1 – Coniferous	–	..	6	19	28	35	36	38
2.4.2 – Non-coniferous	–	..	7	6	7	12	16	18
3. Forecasts for ETTS IV (Higher)					HIGHER FORECASTS			
3.1 From exploitable closed forest, total				63	123	132	132	149
3.1.1 – Coniferous				37	51	60	60	77
3.1.2 – Non-coniferous				26	72	72	72	72
3.2 Other removals a/				76	77	68	98	111
3.2.1 – Coniferous				26	39	30	60	73
3.2.2 – Non-coniferous				50	38	38	38	38
3.3 Total removals	8	32	81	139	200	200	230	260
3.3.1 – Coniferous	1	8	26	63	90	90	120	150
3.3.2 – Non-coniferous	7	24	55	76	110	110	110	110
3.4 Of which: sawlogs + veneer logs, total	–	..	13	25	44	48	55	57
3.4.1 – Coniferous	–	..	6	19	30	35	38	39
3.4.2 – Non-coniferous	–	..	7	6	14	13	17	18

a/ Other removals: e.g. from other wooded land and trees outside the forest.
b/ Base period (see explanatory notes).

Tableau annexe 5.38
Prévisions forestières concernant la forêt dense exploitable

Pays: PORTUGAL(*)

	Unit	1950	1960	1970	Période de base	1990 Faible	1990 Fort	2000 Faible	2000 Fort	2010 Faible	2010 Fort	2020 Faible	2020 Fort
Forêt dense exploitable	1000 ha	2467	3080	2848	2590	2740	3140	2890	3690	3040	4100	3190	4400
Matériel sur pied (GS)	Mill.m3 o.b.		224	168	189	200	193	218	209	237	226	257	242
- Résineux (volume)	"		128	85	116	129	128	146	144	165	161	185	177
- Feuillus (volume)	"		96	83	73	71	65	72	65	72	65	72	65
- Résineux (pourcentage)	%		57	51	61	65	66	67	69	70	71	72	73
Matériel sur pied/ha	m3o.b./ha		73	59	73	73	68	75	65	78	63	81	55
Accroissement annuel net (NAI)	1000 m3 o.b.	5370	8900	8200	11450	13150	13820	14970	15690	15810	17630	16590	18920
- Résineux (volume)	"	2840	5700	6600	8330	9600	9950	10770	10990	11380	12160	11940	12860
- Feuillus (volume)	"	2530	3200	1600	3120	3550	3870	4200	4700	4430	5470	4650	6060
- Résineux (pourcentage)	%	53	64	80	73	73	72	72	70	72	69	72	68
NAI/ha	m3 o.b./ha	2.2	2.9	2.9	4.4	4.8	4.4	5.2	4.3	5.2	4.3	5.2	4.3
NAI en pourcentage de GS	%		4.0	4.9	6.1	6.6	7.2	6.9	7.5	6.7	7.8	6.5	7.8
- Résineux	%		5.9	7.8	7.2	7.4	7.8	7.4	7.6	6.9	7.6	6.5	7.3
- Feuillus	%		3.3	1.9	4.3	5.0	6.0	5.8	7.2	6.2	8.4	6.6	9.3
Abattages*	1000 m3 o.b.	5355	6940	7455	10800	11500	12300	13100	14100	13800	15900	14500	17500
- Résineux*	"	3150	4735	5445	7300	8000	8400	8900	9400	9400	10400	9900	11400
- Feuillus*	"	2205	2205	2010	3500	3500	3900	4200	4700	4400	5500	4600	6100
Rapport abattage/NAI*	%	100	78	91	94	87	89	88	90	87	90	87	92
- Résineux*	%	111	83	83	88	83	84	83	86	83	86	83	89
- Feuillus*	%	87	69	126	112	99	101	100	100	99	101	99	101
Quantités enlevées (sur écorce)	1000 m3 o.b.	5100	6610	7100	10290	10950	11700	12470	13420	13140	15140	13800	16660
- Résineux*	"	3000	4510	5185	6930	7620	8000	8470	8950	8950	9900	9420	10850
- Feuillus*	"	2100	2100	1915	3360	3330	3700	4000	4470	4190	5240	4380	5810
Quantités enlevées (sous écorce)	1000 m3 u.b.	4517	5828	6250	8420	8760	9360	9980	10740	10510	12110	11040	13330
- Résineux (volume)	"	2610	3920	4510	5544	6100	6400	6780	7160	7160	7920	7540	8680
- Feuillus (volume)	"	1907	1908	1740	2876	2660	2960	3200	3580	3350	4190	3500	4650
- Résineux (pourcentage)	%	58	67	72	66	70	68	68	67	68	65	68	65
Quantités enlevées de grumes de sciage en % du total	%	14	32	50	54	57	55	56	54	56	52	56	51
- Résineux	"	23	43	63	75	75	74	75	73	75	72	75	71
- Feuillus	"	3	10	16	14	15	15	15	15	15	15	15	15

(*) Prévisions préparées par le secrétariat sur la base de prévisions partielles envoyées par le pays.

Tableau annexe 5.39
Prévisions des quantités enlevées
(1000 m3 sous écorce)

Pays: PORTUGAL(*)

	1949-51	1959-61	1969-71	1980	1990	2000	2010	2020
1. Données fournies (moyenne de 3 ans) + prévisions de la 3ème étude (ETTS III)				Prévisions du ETTS III				
1.1 Quantités enlevées, total	4517	5828	6247	6200	6500	6700		
1.1.1 – Résineux	2610	3920	4507			4800		
1.1.2 – Feuillus	1907	1908	1740			1900		
1.2 Dont: grumes de sciages et de placage, total	646	1867	3117					
1.2.1 – Résineux	597	1667	2833					
1.2.2 – Feuillus	49	200	284					
2. Prévisions pour la 4ème étude (ETTS IV) (Faibles)				Période de base	PREVISIONS FAIBLES			
2.1 Des forêts denses exploitables, total				8420	8760	9980	10510	11040
2.1.1 – Résineux				5544	6100	6780	7160	7540
2.1.2 – Feuillus				2876	2660	3200	3350	3500
2.2 Autres quantités enlevées a/					–	–	–	–
2.2.1 – Résineux					–	–	–	–
2.2.2 – Feuillus					–	–	–	–
2.3 Quantités enlevées, total	4517	5828	6247	8420	8760	9980	10510	11040
2.3.1 – Résineux	2610	3920	4507	5544	6100	6780	7160	7540
2.3.2 – Feuillus	1907	1908	1740	2876	2660	3200	3350	3500
2.4 Dont: grumes de sciages et de placage, total	646	1867	3117	4550	4970	5560	5870	6180
2.4.1 – Résineux	597	1667	2833	4150	4570	5080	5370	5650
2.4.2 – Feuillus	49	200	284	400	400	480	500	530
3. Prévisions pour la 4ème étude (ETTS IV) (Fortes)					PREVISIONS FORTES			
3.1 Des forêts denses exploitables, total				8420	9360	10740	12110	13330
3.1.1 – Résineux				5544	6400	7160	7920	8680
3.1.2 – Feuillus				2876	2960	3580	4190	4650
3.2 Autres quantités enlevées a/					–	–	–	–
3.2.1 – Résineux					–	–	–	–
3.2.2 – Feuillus					–	–	–	–
3.3 Quantités enlevées, total	4517	5828	6247	8420	9360	10740	12110	13330
3.3.1 – Résineux	2610	3920	4507	5544	6400	7160	7920	8680
3.3.2 – Feuillus	1907	1908	1740	2876	2960	3580	4190	4650
3.4 Dont: grumes de sciages et de placage, total	646	1867	3117	4550	5180	5760	6330	6860
3.4.1 – Résineux	597	1667	2833	4150	4740	5230	5700	6160
3.4.2 – Feuillus	49	200	284	400	440	530	630	700

a/ Autres quantités enlevées : p. ex. d'autres terres boisées et d'arbres hors des forêts.
(*) Prévisions préparées par le secrétariat, sur la base de prévisions partielles envoyées par le pays.

Annex table 5.40

Forestry forecasts relating to exploitable closed forest

Country: SPAIN

	Unit	1950	1960	1970	Base period	1990 Lower	1990 Higher	2000 Lower	2000 Higher	2010 Lower	2010 Higher	2020 Lower	2020 Higher
Exploitable closed forest	1000 ha			5931	6506	7150	7400	7800	8300	8450	9200	9100	10100
Growing stock (GS)	Mill.m3 o.b.	97	210	427	453	581	574	660	638	704	660	710	660
- Coniferous (volume)	"	73	150	254	280	359	354	408	394	435	408	438	408
- Non-coniferous (volume)	"	24	60	173	173	222	220	252	244	269	252	272	252
- Coniferous (percent)	%	75	71	59	62	62	62	62	62	62	62	62	62
Growing stock/ha	m3o.b./ha			72	70	81	78	85	77	83	72	78	65
Net annual increment (NAI)	1000 m3 o.b.	2600	5270	25500	27830	27950	28400	29350	29820	30810	31310	32350	32880
- Coniferous (volume)	"	2000*	3300	17600	19530	19620	19940	20600	20930	21630	21980	22710	23080
- Non-coniferous (volume)	"	600*	1970	7900	8300	8330	8460	8750	8890	9180	9330	9640	9800
- Coniferous (percent)	%	77	63	69	70	70	70	70	70	70	70	70	70
NAI/ha	m3o.b./ha			4.3	4.3	3.9	3.8	3.8	3.6	3.6	3.4	3.6	3.3
NAI as % of GS	%	2.7	2.5	6.0	6.1	4.8	4.9	4.4	4.7	4.4	4.7	4.6	5.0
- Coniferous	%	2.7*	2.2	6.9	7.0	5.5	5.6	5.0	5.3	5.0	5.4	5.2	5.7
- Non-coniferous	%	2.5*	3.3	4.6	4.8	3.8	3.8	3.5	3.6	3.4	3.7	3.5	3.9
Fellings*	1000 m3 o.b.	14200	16990	16810	13340	18060	19860	23250	25600	28380	31260	33440	32880
- Coniferous*	"	5180	5830	8360	9180	12220	13450	15740	17330	19210	21170	22640	22260
- Non-coniferous*	"	9020	11160	8450	4160	5840	6410	7510	8270	9170	10090	10800	10620
Felling/NAI ratio*	%			66	48	65	70	79	86	92	100	103	100
- Coniferous*	%			48	47	62	67	76	83	89	96	100	96
- Non-coniferous*	%			107	50	70	76	86	93	100	108	112	108
Removals (overbark)	1000 m3 o.b.	14060	16820	16640	13210	17860	19610	23040	25330	28160	30970	33210	32570
- Coniferous	"	5130	5770	8280	9090	12090	13280	15600	17150	19060	20970	22480	22050
- Non-coniferous	"	8930	11050	8360	4120	5770	6330	7440	8180	9100	10000	10730	10520
Removals (underbark)	1000 m3 u.b.	11250	13453	13308	10570	14290	15690	18430	20270	22530	24780	26570	26060
- Coniferous (volume)	"	4106	4618	6621	7270	9670	10620	12480	13720	15250	16780	17990	17640
- Non-coniferous (volume)	"	7144	8835	6687	3300	4620	5070	5950	6550	7280	8000	8580	8420
- Coniferous (percent)	%	36	34	50	69	68	68	68	68	68	68	68	68
Sawlogs as % of total	%	9	17	30	34	30	28	30	28	28	26	28	26
- Coniferous	"	18	31	45	39	34	33	34	33	33	32	33	32
- Non-coniferous	"	4	9	14	23	23	22	23	22	22	21	22	21

Annex table 5.41
Removals forecasts
(1000 m3 underbark)

Country: SPAIN

		1949-51	1959-61	1969-71	1980	1990	2000	2010	2020
1.	Reported data (3-yr av.) + ETTS III forecasts				ETTS III FORECASTS				
1.1	Total removals	11250*	13453	13308	16400	18000	18900	22530	26570
1.1.1	- Coniferous	4106*	4618	6621			9000	15250	17990
1.1.2	- Non-coniferous	7144*	8835	6687			9900	7280	8580
1.2	Of which: Sawlogs + veneer logs, total	1036*	2236	3943					
1.2.1	- Coniferous	727*	1424	2997					
1.2.2	- Non-coniferous	309*	812	946					
2.	Forecasts for ETTS IV (Lower)					LOWER FORECASTS			
					B. p. a/				
2.1	From exploitable closed forest, total				10570	14290	18430	22530	26570
2.1.1	- Coniferous				7270	9670	12480	15250	17990
2.1.2	- Non-coniferous				3300	4620	5950	7280	8580
2.2	Other removals				1600	1850*	2100	———→	———→
2.2.1	- Coniferous				860*	990*	1130*	———→	———→
2.2.2	- Non-coniferous				740*	860*	970*	———→	———→
2.3	Total removals	11250*	13453	13308	12170	16140*	20530	24630	28670
2.3.1	- Coniferous	4106*	4618	6621	8130*	10660*	13610*	16380	19120
2.3.2	- Non-coniferous	7144*	8835	6687	4040*	5480*	6920*	8250	9550
2.4	Of which: sawlogs + veneer logs, total	1036*	2236	3943	3550	4350	5610	6630	7830
2.4.1	- Coniferous	727*	1424	2997	2800	3290	4240	5030	5940
2.4.2	- Non-coniferous	309*	812	946	750	1060	1370	1600	1890
3.	Forecasts for ETTS IV (Higher)					HIGHER FORECASTS			
3.1	From exploitable closed forest, total				10570	15690	20270	24780	26060
3.1.1	- Coniferous				7270	10620	13720	16780	17640
3.1.2	- Non-coniferous				3300	5070	6550	8000	8420
3.2	Other removals				1600	1950*	2300	———→	———→
3.2.1	- Coniferous				860*	1050*	1240*	———→	———→
3.2.2	- Non-coniferous				740*	900*	1060*	———→	———→
3.3	Total removals	11250*	13453	13308	12170	17640*	22570	27080	28360
3.3.1	- Coniferous	4106*	4618	6621	8130*	11670*	14960*	18020	18880
3.3.2	- Non-coniferous	7144*	8835	6687	4040*	5970*	7610*	9060	9480
3.4	Of which: sawlogs + veneer logs, total	1036*	2236	3943	3550	4620	5970	7050	7410
3.4.1	- Coniferous	727*	1424	2997	2800	3500	4530	5370	5640
3.4.2	- Non-coniferous	309*	812	946	750	1120	1440	1680	1770

a/ Base period (see explanatory notes).

Annex table 5.42

Forestry forecasts relating to exploitable closed forest

Country: TURKEY**

	Unit	1950	1960	1970	Base period	1990 Lower	1990 Higher	2000 Lower	2000 Higher	2010 Lower	2010 Higher	2020 Lower	2020 Higher
Exploitable closed forest	1000 ha	10284	......	6642	6642	7040	7440	7440	8240	7840	9040	8240	9840
Growing stock (GS)	Mill.m3 o.b.	524	692	637	637	636	637	639	640	644	646	651	655
- Coniferous (volume)	"	245	499	412	412	414	414	417	418	422	423	428	431
- Non-coniferous (volume)	"	279	193	225	225	222	223	222	222	222	223	223	224
- Coniferous (percent)	%	47	72	65	65	65	65	65	65	66	65	66	66
Growing stock/ha	m3o.b./ha	51	...	96	96	90	86	86	78	82	71	79	67
Net annual increment (NAI)	1000 m3 o.b.	...	12673	19210	19210	20415	21575	21575	23895	22735	26215	23895	28535
- Coniferous (volume)	"	...	6411	11700	11700	12655	13590	13590	15530	14550	17565	15530	19690
- Non-coniferous (volume)	"	...	6262	7510	7510	7760	7985	7985	8365	8185	8650	8365	8845
- Coniferous (percent)	%	...	51	61	61	62	63	63	65	64	67	65	69
NAI/ha	m3o.b./ha	...	...	2.9	2.9	2.9	2.9	2.9	2.9	2.9	2.9	2.9	2.9
NAI as % of GS	%	...	1.8	3.0	3.0	3.2	3.4	3.4	3.7	3.5	4.1	3.7	4.4
- Coniferous	%	...	1.3	4.1	4.1	3.1	3.4	3.3	3.7	3.4	4.2	3.6	4.6
- Non-coniferous	%	...	3.2	3.3	3.3	3.5	3.6	3.6	3.8	3.7	3.9	3.8	3.9
Fellings*	1000 m3 o.b.	7240	10940	19900	19570	20160	21305	21165	23430	22155	25510	23120	27550
- Coniferous*	"	1030	5910	12020	11570	12400	13320	13180	15065	14085	16980	14935	18705
- Non-coniferous*	"	6210	5030	7880	8000	7760	7985	7985	8365	8070	8530	8185	8845
Felling/NAI ratio*	%	...	86	104	102	99	99	98	98	97	97	97	97
- Coniferous*	%	...	92	103	99	98	98	97	97	97	97	96	95
- Non-coniferous*	%	...	80	105	107	100	100	100	100	99	99	98	100
Removals (overbark)	1000 m3 o.b.	7140	10790	19630	19300	19880	21005	20870	23100	21845	25155	22790	27165
- Coniferous	"	1020	5830	11850	11410	12225	13135	12995	14855	13890	16745	14720	18445
- Non-coniferous	"	6120	4960	7780	7890	7655	7870	7875	8245	7955	8410	8070	8720
Removals (underbark)	1000 m3/u.b.	6556	9803	17823	17520	18035	19045	18925	20935	19805	22790	20660	24600
- Coniferous (volume)	"	919	5235	10651	10250	10980	11795	11670	13340	12475	15040	13220	16565
- Non-coniferous (volume)	"	5637	4568	7172	7270	7055	7250	7255	7595	7330	7750	7440	8035
- Coniferous (percent)	%	14	53	60	59	61	62	62	64	63	66	64	67
Sawlogs as % of total	%	9	15	20	22	23	23	23	23	24	24	24	25
- Coniferous	"	49	24	27	29	30	30	30	30	30	30	30	30
- Non-coniferous	"	2	6	11	10	11	11	12	12	13	13	14	14

** Forecasts prepared by the secretariat.

Annex table 5.43
Removals forecasts
(million m3 underbark)

Country: TURKEY**

	1949-51	1959-61	1969-71	1980	1990	2000	2010	2020
1. Reported data (3-yr av.) + ETTS III forecasts				ETTS III forecasts				
1.1 Total removals	6556	9803	17823	21000	22500	22500		
1.1.1 – Coniferous	919	5235	10651			15000		
1.1.2 – Non-coniferous	5637	4568	7172			7500		
1.2 Of which: Sawlogs + veneer logs, total	573	1506	3620					
1.2.1 – Coniferous	446	1234	2863					
1.2.2 – Non-coniferous	127	272	757					
2. Forecasts for ETTS IV (Lower)				B. p. b/		LOWER FORECASTS		
2.1 From exploitable closed forest, total				17520	18035	18925	19805	20660
2.1.1 – Coniferous				10250	10980	11670	12475	13220
2.1.2 – Non-coniferous				7270	7055	7255	7330	7440
2.2 Other removals a/				4856	4800	→	→	→
2.2.1 – Coniferous				3767	3700	→	→	→
2.2.2 – Non-coniferous				1089	1100	→	→	→
2.3 Total removals	6556	9803	17823	22376	22835	23725	24605	25460
2.3.1 – Coniferous	919	5235	10651	14017	14680	15370	16175	16920
2.3.2 – Non-coniferous	5637	4568	7172	8359	8155	8355	8430	8540
2.4 Of which: sawlogs + veneer logs, total	573	1506	3620	4944	5300	5600	5950	6250
2.4.1 – Coniferous	446	1234	2863	4083	4400	4600	4850	5050
2.4.2 – Non-coniferous	127	272	757	861	900	1000	1100	1200
3. Forecasts for ETTS IV (Higher)						HIGHER FORECASTS		
3.1 From exploitable closed forest, total				17520	19045	20935	22790	24600
3.1.1 – Coniferous				10250	11795	13340	15040	16565
3.1.2 – Non-coniferous				7270	7250	7595	7750	8035
3.2 Other removals a/				4856	4800	→	→	→
3.2.1 – Coniferous				3767	3700	→	→	→
3.2.2 – Non-coniferous				1089	1100	→	→	→
3.3 Total removals	6556	9803	17823	22376	23845	25735	27590	29400
3.3.1 – Coniferous	919	5235	10651	14017	15495	17040	18740	20265
3.3.2 – Non-coniferous	5637	4568	7172	8359	8350	8695	8850	9135
3.4 Of which: sawlogs + veneer logs, total	573	1506	3620	4944	5570	6140	6750	7380
3.4.1 – Coniferous	446	1234	2863	4083	4650	5100	5600	6100
3.4.2 – Non-coniferous	127	272	757	861	920	1040	1150	1280

a/ Other removals: e.g. from other wooded land and trees outside the forest. For 1980, calculated as the difference between total removals (1.1) and removals from exploitable closed forest (2.1). For the forecasts for 1990 to 2020, it has been assumed that other removals will remain at the 1980 level.
b/ Base period (see explanatory notes).
** Forecasts prepared by the secretariat.

Annex table 5.44
Forestry forecasts relating to exploitable closed forest

Country: YUGOSLAVIA

	Unit	1950	1960	1970	Base period	1990		2000		2010		2020	
						Lower	Higher	Lower	Higher	Lower	Higher	Lower	Higher
Exploitable closed forest	1000 ha	7345	6833	7045	8500	8550	8550	8700	8700	8900	8900	9100	9100
Growing stock (GS)	Mill.m3 o.b.	718	984	913	1135	1210	1210	1280	1280	1360	1360	1450	1450
- Coniferous (volume)	"	210	271	269	311	330	330	350	350	370	370	400	400
- Non-coniferous (volume)	"	508	713	644	824	880	880	930	930	990	990	1050	1050
- Coniferous (percent)	%	29	28	29	27	27	27	27	27	27	27	28	28
Growing stock/ha	m3o.b./ha	98	144	130	134	142	142	147	147	153	153	159	159
Net annual increment (NAI)	1000 m3 o.b.	14715	20775	22440	28850	29600	29600	30550	30550	31550	31550	32500	32500
- Coniferous (volume)	"	4000*	5309	5730	6150	6500	6500	7000	7000	7350	7350	7800	7800
- Non-coniferous (volume)	"	10715*	15466	16710	22700	23100	23100	23550	23550	24300	24300	24700	24700
- Coniferous (percent)	%	27*	26	26	21	22	22	23	23	23	23	24	24
NAI/ha	m3o.b./ha	2.0	3.0	3.2	3.5	3.5	3.5	3.5	3.5	3.5	3.5	3.6	3.6
NAI as % of GS	%	2.0	2.1	2.5	2.5	2.4	2.4	2.4	2.4	2.3	2.3	2.2	2.2
- Coniferous	%	1.9*	2.0	2.1	2.0	1.7	1.7	2.0	2.0	2.0	2.0	2.0	2.0
- Non-coniferous	%	2.1*	2.2	2.6	2.8	2.6	2.6	2.5	2.5	2.5	2.5	2.4	2.4
Fellings*	1000 m3 o.b.	35150*	22960*	23790*	19940	23500	23500	26000	26000	28000	28000	30000	30000
- Coniferous*	"	8930	5355	6315	5790	6985	6985	7575	7575	8000	8000	8730	8730
- Non-coniferous*	"	26220	17605	17475	14150	16515	16515	18425	18425	20000	20000	22170	22170
Felling/NAI ratio*	%	239	111	106	69	79	79	85	85	89	89	92	92
- Coniferous*	%	223	101	110	94	107	107	108	108	109	109	112	112
- Non-coniferous*	%	245	114	105	62	71	71	78	78	83	83	86	86
Removals (overbark)	1000 m3 o.b.	27250	17800	18440	14640	18500	18500	20600	20600	22400	22400	24400	24400
- Coniferous	"	6925	4150	4895	4250	5500	5500	6000	6000	6400	6400	7100	7100
- Non-coniferous	"	20325	13650	13545	10390	13000	13000	14600	14600	16000	16000	17300	17300
Removals (underbark)	1000 m3 u.b.	25070	16376	16963	13470	17000	17000	19000	19000	20600	20600	22500	22500
- Coniferous (volume)	"	6370	3820	4505	3910	5000	5000	5600	5600	5900	5900	6600	6600
- Non-coniferous (volume)	"	18700	12556	12458	9560	12000	12000	13400	13400	14700	14700	15900	15900
- Coniferous (percent)	%	25	23	27	29	29	29	29	29	29	29	29	29
Sawlogs as % of total	%	20	23	30	56	56	56	57	57	57	57	59	59
- Coniferous	"	57	53	59	80	93	93	93	93	93	93	93	93
- Non-coniferous	"	8	13	19	44	40	40	42	42	43	43	45	45

Annex table 5.45
Removals forecasts
(1000 m3 underbark)

Country: YUGOSLAVIA

	1949-51	1959-61	1969-71	1980	1990	2000	2010	2020
1. Reported data (3-yr av.) + ETTS III forecasts					ETTS III FORECASTS			
1.1 Total removals	25070	16376	16963	14800	17100	18000		
1.1.1 – Coniferous	6370	3820	4505			5900		
1.1.2 – Non-coniferous	18700	12556	12458			12100		
1.2 Of which: Sawlogs + veneer logs, total	5047	3706	5056					
1.2.1 – Coniferous	3619	2037	2639					
1.2.2 – Non-coniferous	1428	1669	2417					
2. Forecasts for ETTS IV (Lower)				B.p. a/	LOWER FORECASTS			
2.1 From exploitable closed forest, total				13470	17000	19000	20600	22500
2.1.1 – Coniferous				3910	5000	5600	5900	6600
2.1.2 – Non-coniferous				9560	12000	13400	14700	15900
2.2 Other removals b/				316	300*	------→		
2.2.1 – Coniferous				10	–*	------→		
2.2.2 – Non-coniferous				306	300*	------→		
2.3 Total removals	25070	16376	16963	13786	17300	19300	20900	22800
2.3.1 – Coniferous	6370	3820	4505	3920	5000	5600	5900	6600
2.3.2 – Non-coniferous	18700	12556	12458	9866	12300	13700	15000	16200
2.4 Of which: sawlogs + veneer logs, total	5047	3706	5056	7627	9450	10840	11805	13295
2.4.1 – Coniferous	3619	2037	2639	3580	4650	5210	5485	6140
2.4.2 – Non-coniferous	1428	1669	2417	4047	4800	5630	6320	7155
3. Forecasts for ETTS IV (Higher)					HIGHER FORECASTS			
3.1 From exploitable closed forest, total				13470	17000	19000	20600	22500
3.1.1 – Coniferous				3910	5000	5600	5900	6600
3.1.2 – Non-coniferous				9560	12000	13400	14700	15900
3.2 Other removals b/				316	300*	------→		
3.2.1 – Coniferous				10	–*	------→		
3.2.2 – Non-coniferous				306	300*	------→		
3.3 Total removals	25070	16376	16963	13786	17300	19300	20900	22800
3.3.1 – Coniferous	6370	3820	4505	3920	5000	5600	5900	6600
3.3.2 – Non-coniferous	18700	12556	12458	9866	12300	13700	15000	16200
3.4 Of which: sawlogs + veneer logs, total	5047	3706	5056	7627	9450	10840	11805	13295
3.4.1 – Coniferous	3619	2037	2639	3580	4650	5210	5485	6140
3.4.2 – Non-coniferous	1428	1669	2417	4047	4800	5630	6320	7155

a/ Base period (see explanatory notes).
b/ Other removals: e.g. from other wooded land and trees outside the forest. For 1980, data for 1981 taken from table 12 of Forest Resource Enquiry (Part I). For the forecasts for 1990 to 2020, it has been assumed that other removals will remain at around the 1980 level.

Tableau annexe 5.46

Prévisions forestières concernant la forêt dense exploitable

Pays : BULGARIE**

	Unité	1950	1960	1970	Période de base	1990 Faible	1990 Fort	2000 Faible	2000 Fort	2010 Faible	2010 Fort	2020 Faible	2020 Fort
Forêt dense exploitable	1000 ha	2964*	3169*	3184	3300	3300	3350*	3350*	3400*	3400*	3450*	3450*	3500*
Matériel sur pied (GS)	Mill.m3 o.b.	210*	233*	264*	298	303	303	309	308	315	312	321	316
- Résineux (volume)	"	50*	66*	90	101	108	108	115	115	121	122	128	129
- Feuillus (volume)	"	160*	167*	174	197	195	195	194	193	194	190	193	187
- Résineux (pourcentage)	%	24*	28*	34	34	36	36	37	37	38	39	40	41
Matériel sur pied/ha	m3 o.b./ha	71*	74*	83	90	92	90	92	91	93	90	93	90
Accroissement annuel net (NAI)	1000 m3 o.b.	6100*	6300*	6500	6000	6600	7035	6700	7480	6800	7935	6900	8400
- Résineux (volume)	"	1200*	1600*	2000	2600	2180	2390	2210	2620	2245	2855	2275	3110
- Feuillus (volume)	"	4900*	4700*	4500	3400	4420	4645	4490	4860	4555	5080	4625	5290
- Résineux (pourcentage)	%	20*	25*	31	43	33	34	33	35	33	36	33	37
NAI/ha	m3 o.b./ha	2.1*	2.0*	2.0	1.8	2.0	2.1	2.0	2.2	2.0	2.3	2.0	2.4
NAI en pourcentage de GS	%	2.9*	2.7*	2.5	2.0	2.2	2.3	2.2	2.4	2.2	2.5	2.1	2.7
- Résineux	%	2.4*	2.4*	2.2	2.6	2.2	2.2	1.9	2.3	1.9	2.3	1.8	2.4
- Feuillus	%	3.1*	2.8*	2.6	1.7	2.3	2.4	2.3	2.5	2.3	2.7	2.4	2.8
Abattages*	1000 m3 o.b.	6975*	7425*	7020	6030	6070	6540	6165	7030	6255	7540	6350	8065
- Résineux*	"	2045	1855	2135	1475	1525	1695	1570	1915	1615	2140	1660	2395
- Feuillus*	"	4930	5570	4885	4555	4545	4845	4595	5115	4640	5400	4690	5670
Rapport abattage/NAI*	%	114*	118*	108	101	92	93	94	94	92	95	92	96
- Résineux*	%	170*	116*	107	57	70	71	71	73	72	75	73	77
- Feuillus*	%	101*	119*	109	134	103	104	102	105	102	106	101	107
Quantités enlevées (sur écorce)	1000 m3 o.b.	5715	6085	5755	4940	5015	5405	5140	5860	5255	6335	5380	6835
- Résineux	"	1675	1520	1750	1210	1260	1400	1310	1595	1355	1800	1405	2030
- Feuillus	"	4040	4566	4005	3730	3755	4005	3830	4265	3900	4535	3975	4805
Quantités enlevées (sous écorce)	1000 m3 u.b.	5062	5398	5050	4335	4400	4740	4510	5140	4610	5555	4720	5995
- Résineux (volume)	"	1475	1323	1515	1050	1095	1220	1140	1385	1180	1565	1220	1765
- Feuillus (volume)	"	3605	4075	3535	3285	3305	3520	3370	3755	3430	3990	3500	4230
- Résineux (pourcentage)	%	29	25	30	24	25	26	25	27	26	28	26	29
Quantités enlevées de grumes de sciage en % du total	%	30	38	42	38	38	40	38	42	38	44	38	46
- Résineux	"	56	58	64	58	58	60	58	62	58	64	58	66
- Feuillus	"	19	31	33	32	31	33	31	35	31	36	31	38

N.B. o.b. = volume sur écorce ; u.b. = volume sous écorce.　　* Prévisions préparées par le secrétariat.　　** Prévisions préparées par le secrétariat.

Tableau annexe 5.47
Prévisions des quantités enlevées
(1000 m3 sous écorce)

Pays: BULGARIE**

	1949-51	1959-61	1969-71	1980	1990	2000	2010	2020
1. Données fournies (moyenne de 3 ans) + prévisions de la 3ème étude (ETTS III)				Prévisions du ETTS III				
1.1 Quantités enlevées, total	5062	5398	5050	4900	5600	5900	...	...
1.1.1 – Résineux	1457	1323	1515	...	...	2900*		
1.1.2 – Feuillus	3605	4075	3431	...	...	3000**		
1.2 Dont: grumes de sciages et de placage, total	1498	2040	2142					
1.2.1 – Résineux	823	771	965					
1.2.2 – Feuillus	675	1269	1177					
2. Prévisions pour la 4ème étude (ETTS IV) (Faibles)				Période de base	PREVISIONS FAIBLES			
2.1 Des forêts denses exploitables, total				4335	4400	4510	4610	4720
2.1.1 – Résineux				1050	1095	1140	1180	1220
2.1.2 – Feuillus				3285	3305	3370	3430	3500
2.2 Autres quantités enlevées a/				...	—*—	——→	——→	——→
2.2.1 – Résineux				...	—*—	——→	——→	——→
2.2.2 – Feuillus				...	—*—	——→	——→	——→
2.3 Quantités enlevées, total	5062	5398	5050	4438	4400	4510	4610	4720
2.3.1 – Résineux	1457	1323	1515	1197	1095	1140	1180	1220
2.3.2 – Feuillus	3605	4075	3431	3241	3305	3370	3430	3500
2.4 Dont: grumes de sciages et de placage, total	1498	2040	2142	1478	1660	1705	1750	1795
2.4.1 – Résineux	823	771	965	652	635	660	685	710
2.4.2 – Feuillus	675	1269	1177	826	1025	1045	1065	1085
3. Prévisions pour la 4ème étude (ETTS IV) (Fortes)					PREVISIONS FORTES			
3.1 Des forêts denses exploitables, total				4335	4740	5140	5555	5995
3.1.1 – Résineux				1050	1220	1385	1565	1765
3.1.2 – Feuillus				3285	3520	3755	3990	4230
3.2 Autres quantités enlevées, total a/				...	—*—	——→	——→	——→
3.2.1 – Résineux				...	—*—	——→	——→	——→
3.2.2 – Feuillus				...	—*—	——→	——→	——→
3.3 Quantités enlevées, total	5062	5398	5050	4438	4740	5140	5555	5995
3.3.1 – Résineux	1457	1323	1515	1197	1220	1385	1565	1765
3.3.2 – Feuillus	3605	4075	3431	3241	3520	3755	3990	4230
3.4 Dont: grumes de sciages et de placage, total	1498	2040	2142	1478	1890	2175	2435	2770
3.4.1 – Résineux	823	771	965	652	730	860	1000	1165
3.4.2 – Feuillus	675	1269	1177	826	1160	1315	1435	1605

a/ Autres quantités enlevées : p. ex. d'autres terres boisées et d'arbres hors des forêts.
** Prévisions préparées par le secrétariat.

Annex table 5.48

Forestry forecasts relating to exploitable closed forest

Country: CZECHOSLOVAKIA(*)

	Unit	1950	1960	1970	Base period	1990 Lower	1990 Higher	2000 Lower	2000 Higher	2010 Lower	2010 Higher	2020 Lower	2020 Higher
Exploitable closed forest	1000 ha	3983	4119	3890	4185	4185	4190	4185	4200	4185	4200	4185	4200
Growing stock (GS)	Mill.m3 o.b.	388	608	801	923	931	931	934	933	933	932	929	932
- Coniferous (volume)	"	257	449	598	686	690	689	690	687	685	683	679	680
- Non-coniferous (volume)	"	131	159	203	237	241	242	244	246	248	249	250	252
- Coniferous (percent)	%	66	74	75	74	74	74	74	74	73	73	73	73
Growing stock/ha	m3o.b./ha	97	148	206	221	222	222	223	222	223	222	222	222
Net annual increment (NAI)	1000 m3 o.b.	15228	16874	20417	22503	20000	21000	19500	21000	19200	21200	19000	21500
- Coniferous (volume)	"	11752	13022	15473	17055	14500	15300	14000	15100	13600	15100	13300	15200
- Non-coniferous (volume)	"	3476	3852	4944	5448	5500	5700	5500	5900	5000	6100	5700	6300
- Coniferous (percent)	%	77	77	76	76	73	73	72	72	72	72	72	73
NAI/ha	m3o.b./ha	3.8	4.1	5.2	5.4	4.8	5.0	4.7	5.0	4.6	5.0	4.5	5.1
NAI as % of GS	%	3.9	2.8	2.5	2.4	2.1	2.3	2.1	2.3	2.1	2.3	2.0	2.3
- Coniferous	%	4.6	2.9	2.6	2.5	2.1	2.2	2.0	2.2	2.0	2.2	2.0	2.2
- Non-coniferous	%	2.7	2.4	2.4	2.3	2.3	2.4	2.3	2.4	2.3	2.4	2.3	2.5
Fellings*	1000 m3 o.b.	12620	14490	15000	21460	19500	20500	19500	21000	19500	21200	19500	21500
- Coniferous*	"	9710	11410	11190	16470	14300	15300	14300	15500	14200	15500	14000	15500
- Non-coniferous*	"	2910	3080	3810	4990	5200	5200	5200	5500	5300	5700	5500	6000
Felling/NAI ratio*	%	83	86	73	95	98	98	100	100	102	100	103	100
- Coniferous*	%	83	88	72	97	99	100	102	103	104	103	105	102
- Non-coniferous*	%	84	80	77	92	95	91	95	93	95	93	96	95
Removals (overbark)*	1000 m3 o.b.	12620	14490	15000	21460	19500	20500	19500	21000	19500	21200	19500	21500
- Coniferous*	"	9710	11410	11190	16470	14300	15300	14300	15500	14200	15500	14000	15500
- Non-coniferous*	"	2910	3080	3810	4990	5200	5200	5200	5500	5300	5700	5500	6000
Removals (underbark)	1000 m3 u.b.	11362	13046	13500	19317	17550	18450	17550	18900	17550	19080	17550	19350
- Coniferous (volume)	"	8737	10273	10070	14823	12870	13770	12870	13950	12780	13950	12600	13950
- Non-coniferous (volume)	"	2625	2773	3430	4494	4680	4680	4680	4950	4770	5130	4950	5400
- Coniferous (percent)	%	77	79	75	77	73	75	73	74	73	73	72	72
Sawlogs as % of total	%	52	54	51	55	53	54	52	53	51	52	51	52
- Coniferous	"	59	57	54	58	56	57	55	56	54	55	54	55
- Non-coniferous	"	27	43	45	44	44	44	44	44	44	44	43	43

(*) Forecasts prepared by the secretariat on the basis of partial forecasts by country.

Annex table 5.49
Removals forecasts
(million m3 underbark)

Country: CZECHOSLOVAKIA(*)

	1949-51	1959-61	1969-71	1980	1990	2000	2010	2020
1. Reported data (3-yr av.) + ETTS III forecasts				ETTS III forecasts				
1.1 Total removals	11362	13046	13500	14000	15000	16000		
1.1.1 – Coniferous	8737	10273	10070			12000		
1.1.2 – Non-coniferous	2625	2773	3430			4000		
1.2 Of which: Sawlogs + veneer logs, total	5890	7069	6942					
1.2.1 – Coniferous	5180	5876	5390					
1.2.2 – Non-coniferous	710	1193	1582					
2. Forecasts for ETTS IV (Lower)				B. p. b/	LOWER FORECASTS			
2.1 From exploitable closed forest, total				19317	17550	17550	17550	17550
2.1.1 – Coniferous				14823	12870	12870	12780	12600
2.1.2 – Non-coniferous				4494	4680	4680	4770	4950
2.2 Other removals a/								
2.2.1 – Coniferous					–	–	–	–
2.2.2 – Non-coniferous					–	–	–	–
2.3 Total removals	11362	13046	13500	19317	17550	17550	17550	17550
2.3.1 – Coniferous	8737	10273	10070	14823	12870	12870	12780	12600
2.3.2 – Non-coniferous	2625	2773	3430	4494	4680	4680	4770	4950
2.4 Of which: sawlogs + veneer logs, total	5890	7069	6942	10574	9270	9140	9000	8930
2.4.1 – Coniferous	5180	5876	5390	8597	7210	7080	6900	6800
2.4.2 – Non-coniferous	710	1193	1582	1977	2060	2060	2100	2130
3. Forecasts for ETTS IV (Higher)					HIGHER FORECASTS			
3.1 From exploitable closed forest, total				19317	18450	18900	19080	19350
3.1.1 – Coniferous				14823	13770	13950	13950	13950
3.1.2 – Non-coniferous				4494	4680	4950	5130	5400
3.2 Other removals a/								
3.2.1 – Coniferous					–	–	–	–
3.2.2 – Non-coniferous					–	–	–	–
3.3 Total removals	11362	13046	13501	19317	18450	18900	19080	19350
3.3.1 – Coniferous	8737	10273	10070	14823	13770	13950	13950	13950
3.3.2 – Non-coniferous	2625	2773	3431	4494	4680	4950	5130	5400
3.4 Of which: sawlogs + veneer logs, total	5890	7069	6942	10574	9910	9990	9930	9990
3.4.1 – Coniferous	5180	5876	5390	8597	7850	7810	7670	7670
3.4.2 – Non-coniferous	710	1193	1582	1977	2060	2180	2260	2320

a/ Other removals: e.g. from other wooded land and trees outside the forest.
b/ Base period (see explanatory notes).
(*) Forecasts prepared by secretariat on the basis of partial forecasts by country.

Annex table 5.50

Forestry forecasts relating to exploitable closed forest

Country: GERMAN DEMOCRATIC REPUBLIC **

	Unit	1950	1960	1970	Base period	1990		2000		2010		2020	
						Lower	Higher	Lower	Higher	Lower	Higher	Lower	Higher
Exploitable closed forest	1000 ha	2749	2680*	2680	2590	2590	2620	2590	2650	2590	2680	2590	2710
Growing stock (GS)	Mill. m3 o.b.	253*	350	350	440	462	467	475	489	486	506	493	517
- Coniferous (volume)	"	196*	257	257	335	347	352	353	365	359	376	364	383
- Non-coniferous (volume)	"	57*	93	93	105	115	115	122	124	127	130	129	134
- Coniferous (percent)	%	77	73	73	76	75	75	74	75	74	74	74	74
Growing stock/ha	m3o.b./ha	92	131	131	170	178	178	183	184	188	189	190	191
Net annual increment (NAI)	1000 m3 o.b.	5410	13200	13200	15000	14765	15195	14505	15370	14245	15545	13985	15720
- Coniferous (volume)	"	4545	11100*	11100*	11500	11370	11700	11170	11990	10970	12125	10770	12420
- Non-coniferous (volume)	"	865	2100*	2100*	3500	3395	3495	3335	3380	3275	3420	3215	3300
- Coniferous (percent)	%	84	84	84	77	77	77	77	78	77	78	77	79
NAI/ha	m3 o.b./ha	2.0	4.9	4.9	5.8	5.7	5.8	5.6	5.8	5.5	5.8	5.4	5.8
NAI as % of GS	%	2.1	3.2	3.2	3.4	3.2	3.3	3.1	3.1	2.9	3.1	2.8	3.0
- Coniferous	%	2.3	4.3	4.3	3.4	3.3	3.3	3.2	3.3	3.1	3.2	3.0	3.2
- Non-coniferous	%	1.5	2.3	2.3	3.3	3.0	3.0	2.7	2.7	2.6	2.6	2.5	2.5
Fellings*	1000 m3 o.b.	15775	9070	9360	12040	13345	12800	13280	13495	13390	14180	13420	14895
- Coniferous*	"	13780	7665	7735	9660	10800	10180	10610	10790	10440	11275	10440	11925
- Non-coniferous*	"	1995	1405	1625	2380	2545	2620	2670	2705	2950	2905	2980	2970
Felling/NAI ratio*	%	292	69	71	80	90	85	92	88	94	91	96	95
- Coniferous*	%	303	69	70	84	95	87	95	90	95	93	97	96
- Non-coniferous*	%	231	67	77	68	75	75	80	80	90	85	93	90
Removals (overbark)	1000 m3 o.b.	15025	8640	8915	11465	12680	12160	12615	12820	12720	13470	12750	14150
- Coniferous	"	13135	7300	7365	9200	10260	9670	10080	10250	9920	10710	9920	11330
- Non-coniferous (volume)	"	1890	1340	1550	2265	2420	2490	2535	2570	2800	2760	2830	2830
Removals (underbark)	1000 m3 u.b.	13043	7509	7758	9990	11045	10600	10995	11175	11100	11740	11120	12330
- Coniferous (volume)	"	11322	6291	6351	7933	8845	8335	8690	8835	8550	9230	8550	9765
- Non-coniferous (volume)	"	1720	1218	1407	2057	2200	2265	2305	2340	2550	2510	2570	2565
- Coniferous (percent)	%	87	84	82	79	80	79	79	79	77	79	77	79
Sawlogs as % of total	%	48	56	42	41	40	40	40	40	40	40	40	40
- Coniferous	"	49	54	39	40	40	40	40	40	40	40	40	40
- Non-coniferous	"	46	64	55	39	40	40	40	40	40	40	40	40

** Forecasts prepared by the secretariat.

Annex table 5.51
Removals forecasts
(million m3 underbark)

Country: GERMAN DEMOCRATIC REPUBLIC**

	1949-51	1959-61	1969-71	1980	1990	2000	2010	2020
1. Reported data (3-yr av.) + ETTS III forecasts				*ETTS III forecasts*				
1.1 Total removals	13043	7509	7758	8000	9000	10000		
1.1.1 – Coniferous	11322	6291	6351			8500		
1.1.2 – Non-coniferous	1720	1218	1407			1500		
1.2 Of which: Sawlogs + veneer logs, total	6290	4175	3264					
1.2.1 – Coniferous	5505	3398	2485					
1.2.2 – Non-coniferous	785	777	779					
2. Forecasts for ETTS IV (Lower)				B. p. b/		*LOWER FORECASTS*		
2.1 From exploitable closed forest, total				9990	11045	10995	11100	11120
2.1.1 – Coniferous				7933	8845	8690	8550	8550
2.1.2 – Non-coniferous				2057	2200	2305	2550	2570
2.2 Other removals a/					–		↑	↑
2.2.1 – Coniferous					–		↑	↑
2.2.2 – Non-coniferous					–		↑	↑
2.3 Total removals	13043	7509	7758	9990	11045	10995	11100	11120
2.3.1 – Coniferous	11322	6291	6351	7933	8845	8690	8550	8550
2.3.2 – Non-coniferous	1720	1218	1407	2057	2200	2305	2550	2570
2.4 Of which: sawlogs + veneer logs, total	6290	4175	3264	4046	4420	4400	4440	4450
2.4.1 – Coniferous	5505	3398	2485	3240	3540	3480	3420	3420
2.4.2 – Non-coniferous	785	777	779	806	880	920	1020	1030
3. Forecasts for ETTS IV (Higher)						*HIGHER FORECASTS*		
3.1 From exploitable closed forest, total				9990	10600	11175	11740	12330
3.1.1 – Coniferous				7933	8335	8835	9230	9765
3.1.2 – Non-coniferous				2057	2265	2340	2510	2565
3.2 Other removals a/					–		↑	↑
3.2.1 – Coniferous					–		↑	↑
3.2.2 – Non-coniferous					–		↑	↑
3.3 Total removals	13043	7509	7758	9990	10600	11175	11740	12330
3.3.1 – Coniferous	11322	6291	6351	7933	8335	8835	9230	9765
3.3.2 – Non-coniferous	1720	1218	1407	2057	2265	2340	2510	2565
3.4 Of which: sawlogs + veneer logs, total	6290	4175	3264	4046	4240	4470	4700	4930
3.4.1 – Coniferous	5505	3398	2485	3240	3330	3530	3690	3910
3.4.2 – Non-coniferous	785	777	779	806	910	940	1010	1020

a/ Other removals: e.g. from other wooded land and trees outside the forest.
b/ Base period (see explanatory notes).
** Forecasts prepared by the secretariat.

Annex table 5.52

Forestry forecasts relating to exploitable closed forest

Country: HUNGARY

	Unit	1950	1960	1970	Base period	1990 Lower	1990 Higher	2000 Lower	2000 Higher	2010 Lower	2010 Higher	2020 Lower	2020 Higher
Exploitable closed forest	1000 ha	1173*	1214	1466	1596	1656	1656	1726	1736	1796	1816	1866	1896
Growing stock (GS)	Mill.m3 o.b.	85*	157	215	253	276	270	303	283	332	299	362	318
- Coniferous (volume)	"	6*	10	16	29	37	36	44	43	49	47	53	49
- Non-coniferous (volume)	"	79*	147	199	224	239	234	259	240	283	252	309	269
- Coniferous (percent)	%	7*	6	7	11	13	13	15	15	15	16	15	15
Growing stock/ha	m3o.b./ha	72*	129	147	159	167	163	176	163	185	165	194	168
Net annual increment (NAI)	1000 m3 o.b.	2900*	4692	9000	9710	9880	9940	10040	10280	10130	10540	10390	10790
- Coniferous (volume)	"	300*	333	910	1240	1400	1430	1450	1510	1490	1590	1570	1630
- Non-coniferous (volume)	"	2600*	4359	8090	8470	8480	8510	8590	8770	8640	8950	8820	9160
- Coniferous (percent)	%	10*	7	10	13	14	14	14	15	15	15	15	15
- Non-coniferous (percent)													
NAI/ha	m3o.b./ha	2.5*	3.9	6.1	6.1	6.0	6.0	5.8	5.9	5.6	5.8	5.6	5.7
NAI as % of GS	%	3.4*	3.0	4.2	3.8	3.6	3.7	3.3	3.6	3.1	3.5	2.9	3.4
- Coniferous	%	5.0*	3.3	5.7	4.3	3.8	4.0	3.3	3.5	3.0	3.4	3.0	3.3
- Non-coniferous	%	3.3*	3.0	4.1	3.8	3.5	3.6	3.3	3.7	3.1	3.6	2.9	3.4
Fellings*	1000 m3 o.b.	3000	3810	5950	7540	7350	8750	7180	8760	7215	8800	7230	8820
- Coniferous*	"	110	250	300	530	590	785	790	965	1080	1320	1230	1500
- Non-coniferous*	"	2890	3560	5650	7010	6760	7965	6390	7795	6135	7480	6000	7320
Felling/NAI ratio*	%	103	81	66	78	74	88	72	85	71	83	70	82
- Coniferous*	"	37	75	60	43	42	55	54	64	72	83	78	92
- Non-coniferous*	"	111	82	90	83	80	93	74	89	71	84	68	80
Removals (overbark)	1000 m3 o.b.	2800	3660	5310	6550	6530	7900	6490	7960	6570	8020	6600	8070
- Coniferous	"	100	230	270	400	440	600	600	750	860	1050	1010	1240
- Non-coniferous	"	2700	3430	5040	6150	6090	7300	5890	7210	5710	6970	5590	6830
Removals (underbark)	1000 m3 u.b.	2630	3440	4990	6160	6140	7350	6100	7480	6180	7540	6210	7585
- Coniferous (volume)	"	90	210	240	380	415	560	565	700	810	990	950	1170
- Non-coniferous (volume)	"	2540	3230	4750	5780	5725	6790	5535	6780	5370	6550	5260	6415
- Coniferous (percent)	%	3	6	5	6	7	8	9	9	13	13	15	15
Sawlogs as % of total	%	16	22	25	30	29	28	28	27	27	26	26	25
- Coniferous	"	48	23	35	45	45	45	45	46	46	47	46	47
- Non-coniferous	"	15	22	25	29	28	27	26	25	24	23	22	22

Country: HUNGARY

Annex table 5.53
Removals forecasts
(million m3 underbark)

	1949-51	1959-61	1969-71	1980	1990	2000	2010	2020
1. Reported data (3-yr av.) + ETTS III forecasts				ETTS III forecasts				
1.1 Total removals	2630	3440	4990	6300	6700	7200		
1.1.1 – Coniferous	90	210	240			900		
1.1.2 – Non-coniferous	2540	3230	4750			6300		
1.2 Of which: Sawlogs + veneer logs, total	417	760	1261					
1.2.1 – Coniferous	43	48	83					
1.2.2 – Non-coniferous	374	712	1178					
2. Forecasts for ETTS IV (Lower)				B. p. b/	LOWER FORECASTS			
2.1 From exploitable closed forest, total				6160	6140	6100	6180	6210
2.1.1 – Coniferous				380	415	565	810	950
2.1.2 – Non-coniferous				5780	5725	5535	5370	5260
2.2 Other removals a/					–*			→
2.2.1 – Coniferous					–*			→
2.2.2 – Non-coniferous					–*			→
2.3 Total removals	2630	3440	4990	6160	6140	6100	6180	6210
2.3.1 – Coniferous	90	210	240	380	415	565	810	950
2.3.2 – Non-coniferous	2540	3230	4750	5780	5725	5535	5370	5260
2.4 Of which: sawlogs + veneer logs, total	417	760	1261	1839	1790	1695	1660	1595
2.4.1 – Coniferous	43	48	83	166	185	255	370	440
2.4.2 – Non-coniferous	374	712	1178	1673	1605	1440	1290	1155
3. Forecasts for ETTS IV (Higher)					HIGHER FORECASTS			
3.1 From exploitable closed forest, total				6160	7350	7480	7540	7585
3.1.1 – Coniferous				380	560	700	990	1170
3.1.2 – Non-coniferous				5780	6790	6780	6550	6415
3.2 Other removals a/					–*			→
3.2.1 – Coniferous					–*			→
3.2.2 – Non-coniferous					–*			→
3.3 Total removals	2630	3440	4990	6160	7350	7480	7540	7585
3.3.1 – Coniferous	90	210	240	380	560	700	990	1170
3.3.2 – Non-coniferous	2540	3230	4750	5780	6790	6780	6550	6415
3.4 Of which: sawlogs + veneer logs, total	417	760	1261	1823	2085	2015	1970	1960
3.4.1 – Coniferous	43	48	83	170	250	320	465	550
3.4.2 – Non-coniferous	374	712	1178	1653	1835	1695	1505	1410

a/ Other removals: e.g. from other wooded land and trees outside the forest.
b/ Base period (see explanatory notes).

Annex table 5.54

Forestry forecasts relating to exploitable closed forest

Country: POLAND

	Unit	1950	1960	1970	Base period	1990 Lower	1990 Higher	2000 Lower	2000 Higher	2010 Lower	2010 Higher	2020 Lower	2020 Higher
Exploitable closed forest	1000 ha	7103*	7524	8371	8410	8665	8785	8755	8880	8820	9005	8880	9100
Growing stock (GS)	Mill.m3 o.b.		723	1049	1162	1218	1270	1266	1373	1282	1459	1278	1524
- Coniferous (volume)	"		605	857	897	950	999	989	1096	1002	1180	1002	1248
- Non-coniferous (volume)	"		118	192	265	268	271	277	277	280	279	276	276
- Coniferous (percent)	%		84	82	77	78	79	78	80	78	81	78	82
Growing stock/ha	m3o.b./ha		96	125	138	141	145	145	155	145	162	144	167
Net annual increment (NAI)	1000 m3 o.b.	14300*	20412	34380	28450	32400	36900	30000	37160	30200	37460	29800	37000
- Coniferous (volume)	"	12700*	17900	28880	23430	25900	30220	23500	30600	23700	30860	23800	30470
- Non-coniferous (volume)	"	1600*	2512	5500	5020	6500	6670	6560	6560	6500	6600	6000	6230
- Coniferous (percent)	%	89*	88	84	82	80	82	78	82	78	82	80	82
NAI/ha	m3o.b./ha	2.0*	2.7	4.1	3.4	3.7	4.2	3.4	4.2	3.4	4.2	3.4	4.1
NAI as % of GS	%		2.8	3.3	2.4	2.7	2.9	2.4	2.7	2.4	2.6	2.3	2.4
- Coniferous	%		3.0	3.7	2.6	2.7	3.0	2.4	2.8	2.4	2.6	2.4	2.4
- Non-coniferous	%		2.1	2.9	1.9	2.4	2.5	2.3	2.4	2.3	2.4	2.2	2.3
Fellings*	1000 m3 o.b.	16030	19710	22140	25300	25740	25740	27720	27720	29290	29700	31700	31700
- Coniferous*	"	14620	16980	18150	20400	20040	20040	21620	21630	22910	23320	24920	24910
- Non-coniferous*	"	1410	2730	3990	4900	5700	5700	6100	6090	6380	6380	6780	6790
Felling/NAI ratio*	%	112	97	64	89	79	70	92	75	97	79	106	86
- Coniferous*	%	115	95	63	87	77	66	92	71	97	76	105	82
- Non-coniferous*	%	88	109	73	98	88	85	94	93	98	97	113	109
Removals (overbark)	1000 m3 o.b.	15380	18910	21240	24260	24600	24690	26600	26590	28000	28480	30400	30410
- Coniferous	"	14020	16290	17410	19560	19150	19220	20750	20750	21900	22380	23900	23900
- Non-coniferous (volume)	"	1360	2620	3830	4700	5450	5470	5850	5840	6100	6100	6500	6510
Removals (underbark)	1000 m3 u.b.	13146	16167	18150	20744	21000	21100	22750	22730	23950	24340	26000	26000
- Coniferous (volume)	"	11987	13924	14877	16716	16360	16420	17750	17730	18700	19120	20400	20430
- Non-coniferous (volume)	"	1159	2243	3273	4028	4650	4680	5000	5000	5250	5220	5600	5570
- Coniferous (percent)	%	91	86	82	81	78	78	78	78	78	79	78	79
Sawlogs as % of total	%	49	57	54	53	49	51	49	50	47	49	45	48
- Coniferous	"	49	60	57	55	53	55	52	54	50	53	48	52
- Non-coniferous	"	46	39	39	41	36	36	36	36	35	35	34	34

Annex table 5.55
Removals forecasts
(1000 m3 underbark)

Country: POLAND

	1949-51	1959-61	1969-71	1980	1990	2000	2010	2020
1. Reported data (3-yr av.) + ETTS III forecasts				ETTS III FORECASTS				
1.1 Total removals	13146	16167	18150	22400	23600	26200		26000
1.1.1 – Coniferous	11987	13924	14877			21200		20400
1.1.2 – Non-coniferous	1159	2243	3273			5000		5600
1.2 Of which: Sawlogs + veneer logs, total	6450	9234	9769					
1.2.1 – Coniferous	5920	8351	8483					
1.2.2 – Non-coniferous	530	883	1286					
2. Forecasts for ETTS IV (Lower)				B.p. a/	LOWER FORECASTS			
2.1 From exploitable closed forest, total				20744	21000	22750	23950	26000
2.1.1 – Coniferous				16716	16360	17750	18700	20400
2.1.2 – Non-coniferous				4028	4650	5000	5250	5600
2.2 Other removals b/				456	450*	→	→	→
2.2.1 – Coniferous				152	150*	→	→	→
2.2.2 – Non-coniferous				304	300*	→	→	→
2.3 Total removals	13146	16167	18150	21200	21450	23200	24400	26450
2.3.1 – Coniferous	11987	13924	14877	16868	16510	17900	18850	20550
2.3.2 – Non-coniferous	1159	2243	3273	4332	4940	5300	5550	5900
2.4 Of which: sawlogs + veneer logs, total	6450	9234	9769	10787	10340	11030	11190	11690
2.4.1 – Coniferous	5920	8351	8483	9223	8670	9230	9350	9790
2.4.2 – Non-coniferous	530	883	1286	1564	1670	1800	1840	1900
3. Forecasts for ETTS IV (Higher)					HIGHER FORECASTS			
3.1 From exploitable closed forest, total				20744	21200	22730	24340	26000
3.1.1 – Coniferous				16716	16420	17730	19120	20430
3.1.2 – Non-coniferous				4028	4680	5000	5220	5570
3.2 Other removals b/				456	1350*	2230*	3120*	4000*
3.2.1 – Coniferous				152	450*	740*	1030*	1320*
3.2.2 – Non-coniferous				304	900*	1490*	2090*	2680*
3.3 Total removals	13146	16167	18150	21200	22450	24960	27460	30000
3.3.1 – Coniferous	11987	13924	14877	16868	16870	18470	20150	21750
3.3.2 – Non-coniferous	1159	2243	3273	4332	5580	6490	7310	8250
3.4 Of which: sawlogs + veneer logs, total	6450	9234	9769	10840	10710	11370	11960	12510
3.4.1 – Coniferous	5920	8351	8483	9190	9030	9570	10130	10620
3.4.2 – Non-coniferous	530	883	1286	1650	1680	1800	1830	1890

a/ Base period (see explanatory notes).
b/ Other removals: e.g. from other wooded land and trees outside the forest. Reply to outlook enquiry suggests that supplies from these sources may not change (taken for lower forecast) or may rise to 4 million m3 by 2020 (taken for higher forecast).

Tableau annexe 5.56

Prévisions forestières concernant la forêt dense exploitable

Pays : ROUMANIE**

	Unité	1950	1960	1970	Période de base	1990 Faible	1990 Fort	2000 Faible	2000 Fort	2010 Faible	2010 Fort	2020 Faible	2020 Fort
Forêt dense exploitable	1000 ha	6326*	5008	5861	5723	5720	5770	5720	5820	5720	5870	5720	5920
Matériel sur pied (GS)	Mill.m3 o.b.		1130	1268	1202	1304	1306	1395	1399	1474	1483	1542	1556
- Résineux (volume)	"		424	487	482	512	513	539	543	562	570	582	594
- Feuillus (volume)	"		706	781	720	792	793	856	856	912	913	960	962
- Résineux (pourcentage)	%		38	38	40	39	39	39	39	38	38	38	38
Matériel sur pied./ha	m3 o.b./ha		226	216	210	228	226	244	240	258	253	270	263
Accroissement annuel net (NAI)	1000 m3 o.b.			26900	31594	31460	32310	31460	33170	31460	34050	31460	34930
- Résineux (volume)	"			9700	10992	11010	11630	11010	12270	11010	12940	11010	13620
- Feuillus (volume)	"			17200	20602	20450	20680	20450	20900	20450	21110	20450	21310
- Résineux (pourcentage)	%			36	35	35	36	35	37	35	38	35	39
NAI/ha	m3 o.b./ha			4.6	5.5	5.5	5.6	5.5	5.7	5.5	5.8	5.5	5.9
NAI en pourcentage de GS	%			2.1	2.6	2.4	2.5	2.3	2.4	2.1	2.3	2.0	2.2
- Résineux	%			2.0	2.3	2.2	2.3	2.0	2.3	2.0	2.3	1.9	2.3
- Feuillus	%			2.2	2.9	2.6	2.6	2.4	2.4	2.2	2.3	2.1	2.2
Abattages*	1000 m3 o.b.	18700	21600	26700	20700	21800	22500	23000	24200	24100	26200	25200	28100
- Résineux*	"	6100	6800	8620	7800	8100	8600	8500	9400	8800	10400	9100	11300
- Feuillus*	"	12600	14800	18080	12900	13700	13900	14500	14800	15300	15800	16100	16800
Rapport abattage/NAI*	%			99	66	69	70	73	73	77	77	80	80
- Résineux*	%			89	71	74	74	77	77	80	80	83	83
- Feuillus*	%			105	63	67	67	71	71	75	75	79	79
Quantités enlevées (sur écorce)	1000 m3 o.b.	18150	21010	25920	20070	21170	21850	22330	23500	23400	25440	24470	27290
- Résineux	"	5920	6570	8370	7570	7870	8350	8250	9130	8540	10100	8840	10970
- Feuillus	"	12230	14440	17550	12500	13300	13500	14080	14370	14860	15340	15630	16320
Quantités enlevées (sous écorce)	1000 m3 u.b.	16654*	19276	23780	18417	19420	20050	20490	21560	21470	23340	22450	25040
- Résineux (volume)	"	5433	6027	7680	6942	7220	7760	7570	8380	7830	9270	8110	10060
- Feuillus (volume)	"	11221*	13249	16100	11475	12220	12290	12920	13180	13640	14070	14340	14980
- Résineux (pourcentage)	%	38	31	32	38	37	39	37	39	36	40	36	40
Quantités enlevées de grumes de sciage en % du total	%	36*	37	41	47	47	47	47	47	47	47	47	47
- Résineux	%	78*	65	54	48	48	48	48	48	48	48	48	48
- Feuillus	%	11*	24	34	46	46	46	46	46	46	46	46	46

** Prévisions préparées par le secrétariat.

Tableau annexe 5.57
Prévisions des quantités enlevées
(1000 m3 sous écorce)

Pays: ROUMANIE**

	1949-51	1959-61	1969-71	1980	1990	2000	2010	2020
1. Données fournies (moyenne de 3 ans) + prévisions de la 3ème étude (ETTS III)				Prévision du ETTS III				
1.1 Quantités enlevées, total	16654	19276	23776	21900	23100	23500	...	...
1.1.1 – Résineux	5433	6027	7680	...	...	7300	...	...
1.1.2 – Feuillus	11221	13249	16096	...	...	16200	...	...
1.2 Dont: grumes de sciages et de placage, total	5177	7056	9664	...	...	...	...	...
1.2.1 – Résineux	4200	3918	4174	...	...	...	...	...
1.2.2 – Feuillus	977	3138	5490	...	...	...	...	...
2. Prévisions pour la 4ème étude (ETTS IV) (Faibles)				Période de base	PREVISIONS FAIBLES			
2.1 Des forêts denses exploitables, total	...	...	...	18417	19420	20490	21470	22450
2.1.1 – Résineux	...	...	...	6942	7220	7570	7830	8110
2.1.2 – Feuillus	...	...	...	11475	12220	12920	13640	14340
2.2 Autres quantités enlevées a/	...	...	...	...	–	–	↑	↑
2.2.1 – Résineux	...	...	...	...	–	–	↑	↑
2.2.2 – Feuillus	...	...	...	...	–	–	↑	↑
2.3 Quantités enlevées, total	16654	19276	23776	18417	19420	20490	21470	22450
2.3.1 – Résineux	5433	6027	7680	6942	7220	7570	7830	8110
2.3.2 – Feuillus	11221	13249	16096	11475	12220	12920	13640	14340
2.4 Dont: grumes de sciages et de placage, total	5177	7056	9664	8577	9090	9570	10030	10490
2.4.1 – Résineux	4200	3918	4174	3340	3470	3630	3760	3890
2.4.2 – Feuillus	977	3138	5490	5237	5620	5940	6270	6600
3. Prévisions pour la 4ème étude (ETTS IV) (Fortes)					PREVISIONS FORTES			
3.1 Des forêts denses exploitables, total	...	...	...	18417	20050	21560	23340	25040
3.1.1 – Résineux	...	...	...	6942	7760	8380	9270	10060
3.1.2 – Feuillus	...	...	...	11475	12290	13180	14070	14980
3.2 Autres quantités enlevées, total a/	...	...	...	...	–	–	↑	↑
3.2.1 – Résineux	...	...	...	...	–	–	↑	↑
3.2.2 – Feuillus	...	...	...	...	–	–	↑	↑
3.3 Quantités enlevées, total	16654	19276	23776	18417	20050	21560	23340	25040
3.3.1 – Résineux	5433	6027	7680	6942	7760	8380	9270	10060
3.3.2 – Feuillus	11221	13249	16096	11475	12290	13180	14070	14980
3.4 Dont: grumes de sciages et de placage, total	5177	7056	9664	8577	9370	10080	10920	11720
3.4.1 – Résineux	4200	3918	4174	3340	3720	4020	4450	4830
3.4.2 – Feuillus	977	3138	5490	5237	5650	6060	6470	6890

a/ Autres quantités enlevées : p. ex. d'autres terres boisées et d'arbres hors des forêts.
** Prévisions préparées par le secrétariat.

Annex table 7.1/Tableau annexe 7.1

Europe: apparent consumption of sawnwood and wood-based panels, 1913 to 1980
Europe: consommation apparente de sciages et de panneaux dérivés du bois, 1913 à 1980

(million m3)/(millions de m3)

	Total sawnwood a/ Sciages total a/	Coniferous sawnwood Sciages résineux	Non-coniferous sawnwood Sciages feuillus	Wood-based panels b/ Panneaux dérivés du bois b/	Plywood Contre-plaqués	Particle board Panneaux de particules	Fibreboard Panneaux de fibres
1913	54.5	47.0	7.5	0.1	0.1	-	-
1920–24	48.6	40.9	7.7	0.2	0.2	-	-
1925–29	56.5	48.1	8.4	0.5	0.5	-	-
1930–34	53.8	45.6	8.2	0.9	0.8	-	0.1
1935–38	61.2	50.9	10.3	1.7	1.3	-	0.4
1947–49	55.3	43.7	11.6	2.0	1.0	-	1.0
1949–51	62.1	48.5	10.3	3.0	1.4	-	1.1
1954–56	70.7	55.3	11.8	5.1	2.0	0.4	1.9
1959–61	76.4	59.1	14.3	8.7	2.8	2.2	2.6
1964–66	86.2	66.4	17.3	14.8	3.8	6.1	3.6
1969–71	93.2	71.8	19.3	23.2	4.9	12.5	4.1
1974–76	94.3	72.2	20.0	31.9	5.1	20.2	4.6
1979–81	102.3	78.2	22.4	35.6	5.4	23.8	4.4

Sources: For 1913 to 1947–49, European Timber Statistics, 1913–1950. For subsequent years, ECE/FAO data base. The two series may not be strictly comparable.

De 1913 à 1947–49, Statistiques européennes du bois, 1913–1950. Pour les années suivantes, base de données CEE/FAO. Les deux séries ne sont peut-être pas exactement comparables.

a/ Includes sleepers as well as coniferous and non-coniferous sawnwood.

Outre des sciages résineux et feuillus, comprend des traverses .

b/ Includes veneer sheets as well as plywood, particle board and fibreboard.

Outre des contreplaqués et des panneaux de particules et de fibres, comprend des feuilles de placage.

Annex table 7.2/Tableau annexe 7.2

Europe: apparent consumption of paper and paperboard, 1913 to 1980
Europe: consommation apparente des papiers et cartons, 1913 à 1980

(million m.t.)/(millions de t.m.)

	Paper and paperboard	Newsprint	Other paper	Paperboard	Printing and writing	Other paper and paperboard
	Papiers et cartons	Papier journal	Autres papiers	Cartons	Impression et écriture	Autres papiers et cartons
1913	5.2	1.1	3.3	0.8	..	..
1920–24	4.6	1.1	2.8	0.6	..	..
1925–29	6.8	1.7	4.0	1.0	..	..
1930–34	8.2	2.2	4.5	1.4	..	..
1935–38	10.1	2.6	5.7	1.9	..	..
1947–49	7.8	1.5	4.6	1.7	..	..
1949–51	10.5	1.9	..	..	2.3	6.3
1954–56	15.1	2.9	..	..	3.2	9.0
1959–61	21.4	3.9	..	..	4.5	13.0
1964–66	29.0	4.8	..	..	6.4	18.0
1969–71	38.4	5.7	..	..	9.3	23.3
1974–76	42.9	5.9	..	..	10.7	26.3
1979–81	49.2	6.5	..	..	13.7	29.0

Sources: As for annex table 7.1. The classification used in the earlier publication is different from that used at present.

Voir tableau annexe 7.1. La classification utilisée précédemment diffère de celle utilisée actuellement.

Annex table 7.3/Tableau annexe 7.3

Europe: apparent consumption of fuelwood, pitprops and other industrial wood
Europe: consommation apparente de bois de chauffage, de bois de mine
et d'autres bois d'oeuvre et d'industrie

(million m3)/(millions de m3)

	Fuelwood Bois de chauffage	Pitprops Bois de mine	Other industrial wood Autres bois d'oeuvre et d'industrie	Pitprops and other industrial wood Bois de mine et autres bois d'oeuvre et d'industrie
1913	136.6	14.0	11.0	25.0
1920–24	146.4	12.5	14.3	26.8
1925–29	143.8	15.4	15.8	31.2
1930–34	130.5	13.0	15.5	28.4
1935–38	128.5	16.4	17.7	34.1
1947–49	138.4	15.5	18.5	34.0
1949–51	121.8	15.7	22.5	38.3
1954–56	107.5	16.4	21.4	37.8
1959–61	93.6	14.1	21.3	35.4
1964–66	80.5	12.4	19.4	31.8
1969–71	69.3	9.7	20.1	29.8
1974–76	54.9	7.5	19.6	27.1
1979–81	53.8	6.5	16 8	23.3

Sources: As for annex table 7.1.

Voir tableau annexe 7.1.

Annex table 7.4/Tableau annexe 7.4

Europe: Per caput consumption of forest products 1920-24 to 1979-81
Europe: Consommation de produits forestiers par tête, 1920-24 à 1979-81

	Population	Per caput consumption Consommation par tête			
		Fuelwood Bois de chauffage	Sawnwood Sciages	Wood-based panels Panneaux dérivés du bois	Paper and paperboard Papiers et cartons
	(million)	(m3/1000 capita) (m3/1000 h.)			(m.t./1000 cap.) (t.m./1000 h.)
1920-24	344.9	425	141	1	13
1925-29	360.6	399	157	1	19
1930-34	375.7	347	143	2	22
1935-38	388.4	331	158	4	26
1947-49	410.8	337	135	5	19
1949-51	412.8	295	150	7	25
1954-56	432.0	249	164	12	35
1959-61	452.7	207	169	19	47
1964-66	478.7	169	181	31	61
1969-71	498.2	139	188	46	77
1974-76	517.5	106	182	62	83
1979-81	533.8	100	192	67	91

Sources: As for annex table 7.1.

Voir tableau annexe 7.1.

Annex table 8.1/Tableau annexe 8.1

Sawn softwood: Broad end-use patterns
Sciages résineux : utilisations finales

(Percentage of total apparent consumption)
(Pourcentage de la consommation apparente totale)

	Year Année	Construction B.T.P.	Furniture Ameublement	Packaging Emballage	Other Autres
As reported in country replies/Selon réponses nationales					
Finland/Finlande a/	1970	57	7	9	27b/
	1975	69	5	5	21b/
	1980	65	10	14	21b/
Norway/Norvège	1980	73	4	9	14
Poland/Pologne	1970	56	7.4	8.2	28.4b/
	1980	61.6	8.0	8.3	22.1b/
Portugal/Portugal	1972	89	6	1	4
	1980	92	3	2	3
Sweden/Suède	(1980)	75	5	10	10
Switzerland/Suisse	1979	65	..	..	..
Yugoslavia/Yougoslavie	1978	..	9	..	..
From other sources/Selon d'autres sources					
France c/	1979	81.5	4.3	12.3	1.9
Italy/Italie c/	1979	82	5	10	3
United Kingdom/ Royaume-Uni	1977/80	70	3	13	14
United States/ Etats-Unis d/	1981	66	5	5	24

a/ Including sawn hardwood which accounts for about 3% of sawnwood consumption and manufacturing for export (wooden building elements, houses, etc.) but excluding sawnwood used for production of blockboard (core plywood) / Comprend les sciages feuillus (3% de la consommation) et la production d'éléments et maisons préfabriquées pour l'exportation, mais ne comprend pas les sciages utilisés pour la fabrication de contreplaqués à âme.

b/ Including sleepers / Comprend les traverses.

c/ Secretariat estimates based on various sources / Estimations du secretariat, qui s'est fondé sur des sources d'information diverses.

d/ Spelter H. and R.B. Phelps: Changes in post war US lumber consumption patterns, Forest Products Journal, Vol. 34, No. 2, February 1984.

Annex table 8.2/Tableau annexe 8.2

Sawn hardwood: end-use patterns
Sciages feuillus : utilisations finales

(Percentage of total apparent consumption)
(Pourcentage de la consommation apparente totale)

	Year Année	Construction B.T.P.	Furniture Ameublement	Packaging Emballage	Other Autres
As reported in country replies/Selon réponses nationales					
Poland/Pologne	(1980)	60a/	15*	..	..
Portugal	1972	70	23b/	3	4
	1980	73	22b/	2	3
From other sources/Selon d'autres sources					
France c/	1979	37.2	31.3	25.8	5.7
Germany, Fed.Rep. of/ Allemagne, Rép.féd. d' c/	late 1970s	25	30	25	20
Italy/Italie c/	1979	20	40	30	10
Switzerland/Suisse c/	1979	35	..	..	..
United Kingdom/ Royaume-Uni c/	1981	32d/	40	6	22
United States/ Etats-Unis e/	1981	7	30	51	12

a/ The single most important use of sawn hardwood is said to be flooring, accounting for more than half of total apparent consumption of sawn hardwood; followed by joinery and furniture / L'utilisation finale la plus importante serait le parquet (environ la moitié de la consommation des sciages feuillus), suivi de la menuiserie et de l'ameublement.

b/ Comment in reply that this percentage is too low / Commentaire, dans la réponse nationale, que ce pourcentage est trop bas.

c/ Secretariat estimates / Estimations du secrétariat.

d/ Including shopfitting, fencing and related uses but excluding mining / Y compris l'agencement des magasins, les clôtures et utilisations similaires, mais à l'exclusion de l'utilisation dans les mines.

e/ Source: as in annex table 8.1 / Source : voir tableau annexe 8.1.

Annexe table 8.3/Tableau annexe 8.3

Plywood: end-use patterns
Contreplaqués : utilisations finales

(Percentage of total apparent consumption)
(Pourcentage de la consommation apparente totale)

	Year Année	Construction B.T.P.	Furniture Ameublement	Packaging Emballage	Other Autres
		As reported in country replies/Selon réponses nationales			
Finland/Finlande a/	1976	60	13	5	22b/
Norway/Norvège	1980	57	10	.....33.....	
Poland/Pologne					
– veneer plywood	1970	3	39	24c/	34
– contreplaqué à plis	1980	14	25	29c/	32
– core plywood	1970	3	45	39c/	13
– contreplaqué à âme	1980	10	52	29c/	9
Portugal d/	1980	65	20	.....15.....	
Sweden/Suède a/	..	58	15	20	7
		From other sources/Selon d'autres sources			
Italy/Italie f/	1979	45	13	25	17

a/ Veneer plywood and core plywood, the latter accounting for less than 10% of total plywood consumption / Contreplaqués à plis et à âme. Ce dernier prend moins de 10% de la consommation totale de contreplaqué.

b/ Of which transport 7%, advertising signs 4%, D-I-Y 4%, other 7% / Dont transport 7%, pancartes publicitaires 4%, bricolage 4%, autres utilisations 7%.

c/ Including transport / Y compris les transports.

d/ Veneer plywood / Contreplaqué à plis.

e/ Industrial uses / Utilisations industrielles.

f/ Secretariat estimate / Estimations du secrétariat.

Annex table 8.4/Tableau annexe 8.4

Particle board: end-use patterns
Panneaux de particules : utilisations finales

(Percentage of total apparent consumption)
(Pourcentage de la consommation apparente totale)

	Year Année	Construction B.T.P.	Furniture Ameublement	Packaging Emballage	Other Autres
As reported in country replies/Selon réponses nationales					
Finland/Finlande a/	1975	75	15	.....10.....	
	1982	70	20	.....10.....	
Norway/Norvège	1980	58	20	.....20.....	
Poland/Pologne	1970	9	81	–	10
– of wood/de bois	1980	8	79	1	12
– of flax/de lin	1970	25	50	2	23
	1980	36	31	2	31
Portugal	1980	10	87	......3.....	
From other sources/Selon d'autres sources					
Italy/Italie a/	1979	8	90	1	1

a/ Secretariat estimate / Estimation du secrétariat.

Annex table 8.5/Tableau annexe 8.5

Fibreboard: end-use patterns
Panneaux de fibres : utilisations finales

(Percentage of total apparent consumption)
(Pourcentage de la consommation apparente totale)

	Year Année	Construction B.T.P.	Furniture Ameublement	Packaging Emballage	Other Autres
Hardboard/Panneaux durs					
As reported in country replies/Selon réponses nationales					
Finland/Finlande	1980	84	10	5	1
Poland/Pologne a/	1970	25	25	14b/	36
	1980	20	41	11b/	28
Portugal	1980	45	35	7	13
Sweden/Suède	1982	75	6	8	11
From other sources/Selon d'autres sources					
United Kingdom/ Royaume-Uni	1981	38	32	2	28
Insulating board/Panneaux isolants					
As reported in country replies/Selon réponses nationales					
Finland/Finlande	1980	100	–	–	–
Poland/Pologne	1970	75	–	–	25
	1980	71	–	–	29
Sweden/Suède	1982	100	–	–	–
From other sources/Selon d'autres sources					
United Kingdom/ Royaume-Uni	1981	98	–	–	2

a/   A similar pattern seems to apply to Czechoslovakia for fibreboard utilization / Les mêmes pourcentages seraient applicables à l'utilisation des panneaux de fibres en Tchécoslovaquie.

b/   Including transport / Y compris les transports.

Annex table 8.6/Tableau annexe 8.6

Summary of replies to enquiry
concerning use of forest products in construction

Résumé des réponses nationales à l'enquête
sur les utilisations des produits forestiers dans la construction

FINLAND/FINLANDE

Share of different construction sectors in consumption of forest products
Part des différents secteurs de la construction
dans la consommation des produits forestiers

	Sawn softwood Sciages résineux a/		Plywood Contre-plaqués	Particle board Panneaux de particules	Hardboard Panneaux durs	Insulating board Panneaux isolants
	1970	1980	1976	1982	1980	1980
New building Bâtiment nouveau	37.2	40.9	60 b/	26	20	60
Joinery Menuiserie	6.6	10.4		12	40	
Repairs and maintenance Entretien et réparation	4.4	8.7		25	20	40
Other construction and works Autre bâtiment et travaux publics	9.1	5.3		7	4	
Construction, total	57.3	65.3	60	70	84	100

a/ Including sawn hardwood / Y compris les sciages feuillus.

b/ Of which for agricultural buildings: 8% / Dont pour les bâtiments agricoles : 8%.

NORWAY/NORVEGE

Share of different construction sectors
in consumption of sawn softwood and particle board
Part des différents secteurs de la construction
dans la consommation des sciages résineux et des panneaux de particules

	Sawn softwood Sciages résineux		Particle board Panneaux de particules	
	1972	1980	1974	1980
	(%)			
New building Bâtiment neuf	56	38	48	52
of which/dont :				
residential/résidentiel	42	30	45	50
of which/dont :				
low-rise/1-2 étages	34	25	42	48
high-rise/ plus de 2 étages	4	1	1	1
secondary residence/résidences secondaires	4	4	2	1
non-residential/non-résidentiel	14	8	3	2
of which/dont :				
commercial	3	1		
industrial/industriel	3	1		
agricultural/agricole	5	3		
other/autre	3	3		
Repairs and maintenance/Entretien et réparation	14	28	7	6
Other construction and works/ Autre B.T.P.	6	7		
Construction, total	76	73	55	58

SWEDEN/SUEDE

Sawn softwood (around 1980)/Sciages résineux (vers 1980)
(% of total consumption)/(% de la consommation totale)

Building industry/Industrie du bâtiment	50
Prefabricated houses/Maisons préfabriquées	10
Building components/Elements de construction	4
Other/autres	11
Construction, total	75

	Hardboard Panneaux durs	Insulating board Panneaux isolants
	(% of total consumption – 1982) (% de la consommation totale – 1982)	
New building/Bâtiment neuf	22	45
Joinery/Menuiserie	24	
Repairs and maintenance/Entretien et réparation	27	55
Construction, total	73	100

SWITZERLAND/SUISSE

Share of different building types in consumption of forest products and in building volume (new building only) / Part des différents types de bâtiment dans la consommation de produits forestiers et dans le volume bâti (bâtiment neuf seulement).

	Net use of forest products Utilisation nette des produits forestiers a/	Building volume Volume bâti
	(%)	
Residential building/Bâtiments résidentiels of which/dont :	70	53
one-family/pour une famille	49	28
multi-family/pour plusieurs familles	21	25
Buildings for commercial purposes / Bâtiments à usage commercial	17	26
Buildings for social, cultural and related purposes/Bâtiments à des fins sociales, culturelles, etc.	13	21
TOTAL	100	100

a/ 630 000 m3 (sawnwood 550 000 m3 + panels 80 000 m3)/
    630 000 m3 (sciages 550 000 m3 + panneaux 80 000 m3).

UNITED KINGDOM/ROYAUME-UNI

	Estimated share of sawn softwood consumption Part estimée de la consommation de sciages résineux
New residential construction/ Habitations neuves	20 a/
Other construction/ Autres B.T.P.	50
Non construction/ Autres que les B.T.P.	30

a/ This percentage may be an underestimate / Il pourrait s'agir d'une sous-estimation

Note: Views differ on the important of repair and maintenance work and D-I-Y
Les avis divergent en ce qui concerne l'importance de l'entretien de la réparation et du bricolage.

UNITED STATES/ETATS-UNIS

Share in consumption of different types of construction
Part de la consommation prise par différents types de construction

	Sawn softwood Sciages résineux		Sawn hardwood Sciages feuillus	
	1969	1981	1969	1981
	(%)			
Residential building/Bâtiment résidentiel	54	57	10	5
of which/dont :				
single-family/une famille	30	27	9	5
multi-family/plusieurs familles	8	6	1	
mobile-homes	2	2	–	
additions, repairs/élargissements, réparations	14	22		
Non-residential/Non résidentiel	11	9	2	2
of which/dont :				
buildings/bâtiments	5	4	2	2
utilities/électricité, eau, gaz, etc.	2	2		
farms/agricoles	2	2		
miscellaneous/autres	2	1		
Construction, total	65	66	12	7

Source: See annex table 8.1, footnote d/ / Voir tableau annexe 8,1, note d/.

Annex 8.7

Use of sawnwood and wood-based panels
in new building construction in Finland.

It was considered of interest to present briefly the main findings of a very detailed study 1/ carried out in Finland. It should however be borne in mind that these may not be automatically extrapolated to other countries with different circumstances.

The study was part of a larger research project undertaken by the Finnish Forest Research Institute, and it had as its purpose to describe the use of sawnwood and wood-based panels in new building construction (including enlargement of old buildings). It was based on a systematic sample survey of buildings completed in 1975.

The pattern of sawnwood use (including sawn hardwood, which accounted for a few percent, at most, of total sawnwood consumption) in construction in 1975 was as follows:

	Volume a/ (1000 m3)	Percentage share in total consumption (%)
Construction, total of which:	1510	69.0
new building	1080	49.3
building repair	50	2.3
joinery	200	9.1
harbours, bridges, etc.	80	3.7
other (including wooden building elements glue-laminated structures, etc.)	100	4.6

a/ Including volume lost during construction.

The study covered new buildings and enlargements of old buildings (statistically recorded, i.e. for which building permission had been granted). Annex table 8.7.1 shows completions in 1975 by type of building and the corresponding use of sawnwood and wood-based panels. The volumes indicated in this, as well as the following tables represent the net use i.e. the volumes actually put in place (structural use: sawnwood 830.020 m3; wood-based panels 13.900 m3) and temporarily used (proportionally to the assumed lifespan: sawnwood 146.480 m3; wood-based panels 13.900 m3).

1/ Valtonen, K. : Use of sawnwood and wood-based panels in new building construction in the 1970s, Folia Forestalia 529: 1-42 Helsinki, Finland, 1982 (in Finnish, with English summary).

Volumes lost during the building work, such as cut-offs, rejects, etc. were excluded. These represented an estimated 10% of net volumes and represent the difference between the above figure of 1080 thousand m3 and the net volume in Annex table 8.7.1 of 976.500. The data in the table also exclude joinery installed, but not manufactured on site, mostly doors, windows and prefabricated built-in fittings (only 33.200 m3 of sawnwood, and 7.400 m3 of panels were used for built-in furniture etc. made on site, as annex table 8.7.3 shows). The whole of the volume of 200 000 m3 included under "joinery" was destined however to the building sector, mostly new buildings (about 70%), a smaller part (about 25%) being used in building repair work. Export of joinery was not more than 5% to 6% of the total volume.

The pattern of new building activity is characterized by the significant share of wood-framed buildings which accounted for about a third of the total volume of new buildings completed in 1975 but for two thirds of the use of both sawnwood and wood-based panels. They showed in all main categories of buildings a use of sawnwood about four times that of corresponding non-wood-framed buildings, in terms of unit of volume built, while the use of wood-based panels is also higher, but with greater variation between different types of building.

The difference in wood use is equally marked between different types of non-wood-framed buildings. While, per unit of volume built, non-wood-framed, low-rise residential buildings use only about a fourth less sawnwood than wood-framed ones, in high-rise residential buildings it is only just over an eighth. The highest relative consumption levels were shown by the category of summer and vacation houses, and saunas.

Regarding the share of wood-based panels in total wood use, it was about 20% in residential buildings (except for summer houses) and about 10% in non-residential buildings, but with marked differences between individual types of buildings.

Annex table 8.7.1

Finland: Pattern of new building activity and corresponding use
(structural and temporary) of sawnwood and wood-based panels, 1975

Type of building	Number of buildings	Building			Use of forest products								
		Volume	Percentage share of grand total	Average volume per building	Sawnwood			Wood-based panels			Sawnwood plus wood-based panels, per 100 m3 built		
					Total volume	Percentage share of grand total	Per 100m3 built	Total volume	Percentage share of grand total	Per 100m3 built	Volume	Percentage share of: Sawnwood	Wood-based panels
		(100 m3)	(%)	(100 m3)	(m3)	(%)	(m3)	(m3)	(%)	(m3)	(m3)	(%)	(%)
Residential of which:	40312	220530	46.1	5.5	687200	70.4	3.12	164150	82.7	0.74	3.86	80.8	19.2
High-rise	1480a/	97260	20.3	65.7	64350	6.6	0.66	14600	7.4	0.15	0.81	81.5	18.5
Low-rise, non-wood-framed	3340b/	26560	5.6	7.9	99100	10.1	3.73	21900	11.0	0.82	4.55	82.0	18.0
Low-rise, wood-framed	19018c/	82060	17.1	4.3	400000	41.0	4.87	119050	60.0	1.45	6.32	77.1	22.9
Summer houses, saunas d/	16474	14650	3.1	0.9	123750	12.7	8.45	8600	4.3	0.59	9.04	93.5	6.5
Non-residential of which:	10184	258020	53.9	25.3	289300	29.6	1.12	34450	17.3	0.13	1.35	89.6	10.4
Industrial, non-wood-framed	1413	114770	24.0	81.2	42000	4.3	0.37	8650	4.4	0.08	0.45	82.2	17.8
Industrial wood-framed	879	20560	4.3	23.4	27100	2.8	1.32	2450	1.2	0.12	1.44	91.7	8.3
Farm buildings	3344	27970	5.8	8.4	72750	7.4	2.60	4050	2.0	0.14	2.74	94.9	5.1
Other buildings, non-wood-framed	1282	74700	25.6	58.3	71400	7.3	0.96	9700	4.9	0.13	1.09	88.1	11.9
Other buildings, wood-framed	3266	20020	4.2	6.1	76050	7.8	3.80	9600	4.8	0.48	4.28	88.8	11.2
Grand total	50496	478550	100.0	9.5	976500	100.0	2.04	198600	100.0	0.41	2.45	83.3	16.7

Note: Low-rise residential buildings have a maximum of three storeys.

a/ With 40228 dwellings.
b/ With 7378 dwellings.
c/ With 21802 dwellings.
d/ Not considered as residential buildings in Finnish housing statistics in the 1970s, except if inhabited all year.

Annex table 8.7.2

Finland: Share of structural and of temporary uses in total use of sawnwood and wood-based panels, 1975

| | Shares of structural and temporary uses in total use of sawnwood and wood-based panels | | | | | |
| | Sawnwood | | | Wood-based panels | | |
	Total use (m3)	Structural use (%)	Temporary use (%)	Total use (m3)	Structural use (%)	Temporary use (%)
Residential	687200	88	12	164150	95	5
of which:						
High-rise	64350	64	36	14600	58	42
Low-rise non-wood-framed	99100	87	13	21900	94	6
Low-rise wood-framed	400000	90	10	119050	99	1
Summerhouses, saunas	123750	95	5	8600	99	1
Non-residential	289300	78	22	34450	84	16
of which:						
Industrial non-wood-framed	42000	62	38	8650	80	20
Industrial wood-framed	27100	87	13	2450	99	1
Farm buildings	72750	85	15	4050	97	3
Other buildings non-wood-framed	71400	63	37	9700	64	36
Other buildings wood-framed	76050	90	10	9600	100	–
Grand total	976500	85	15	198600	93	7

Annex table 8.7.3

Finland: Share of different end-uses in total structural use, 1975 in all types of buildings

| End-use | Sawnwood | | Wood-based panels | |
	Volume (m3)	Share (%)	Volume (m3)	Share (%)
Roof	315410	38	5540	3
Uppermost floor	107900	13	33250	18
Intermediate floor	16600	2	5540	3
Ground floor	58100	7	11080	6
External wall	190900	23	35090	19
Internal wall	107900	13	86810	47
Built-in furniture etc.	33200	4	7390	4
Structural use, total	830020	100	184700	100

The shares of structural (actually put in place) and temporary uses for the different types of buildings are shown in annex table 8.7.2. Here, the important role of formwork for some types of buildings is clearly evident.

The volumes and corresponding shares of sawnwood and panels (excluding temporary uses) in the different main end-uses are shown in annex table 8.7.3.

In all types of buildings, the roof was a major end-use of sawnwood, acounting for between 29% (summerhouses and saunas) and 54% (wood-framed industrial buildings) of total structural sawnwood use, followed by external walls (23%), top floors and internal walls (13%). Internal walls were the most important end-use for panels (47%), followed by external walls (19%) and top floor (18%).

The use of sawnwood and wood-based panels per dwelling in 1975 was as follows:

	Sawnwood	Wood-based panels
	(m3)	
In high-rise residential buildings	1.60	0.36
In low-rise residential buildings, non-wood-framed	13.43	2.97
In low-rise residential buildings, wood-framed	18.35	5.46

How the changes in the pattern of new building activity influences the corresponding use of sawnwood and wood-based panels between 1970 and 1979 is described by the author in the following terms: "The total use of sawnwood in new building construction in 1970 was 770 000 m3 and in 1979 about 1 million m3. The use of wood-based panels in 1970 was 143 000 m3 and 210 000 m3 in 1979. The total use of sawnwood and wood-based panels in new building construction has increased during the whole time of the 1970s, although the total building volume produced has clearly decreased after the middle of the decade. The reason is that in the latter half of the decade the production of those building types has increased in which the use of sawnwood and wood-based panels per unit of volume built is relatively large. The most important factor in increasing the use of wood products has been the high increase in production of wood-framed, low-rise residential buildings. The building volume of the wooden, low-rise, residential buildings produced in 1970 was 5.1 million m3 and in 1979 nearly 10 million m3".

Annex table 10.1
Forest product prices in real terms:
combined series

Tableau annexe 10.1
Prix des produits forestiers en
termes réels : séries combinées

(1965=100)

Year / Année	Sawnwood / Sciages	Wood-based pannels / Panneaux dérivés du bois	Sawn softwood / Sciages résineux	Sawn hardwood / Sciages feuillus	Plywood / Contre-plaqués	Particle board / Panneaux de particules	Fibreboard / Panneaux de fibres	Newsprint / Papier journal	Printing & writing / Impression & écriture	Other paper & paper-board/Autres papiers & cartons	Paper & paperboard / Papiers & cartons
1964	99.53	99.93	99.32	99.68	102.04	101.43	98.99	101.86	100.16	103.51	101.74
1965	100.00	100.00	100.00	100.00	100.00	100.00	100.00	100.00	100.00	100.00	100.00
1966	98.51	96.00	98.33	99.47	100.25	93.29	98.09	98.52	97.50	99.61	98.51
1967	95.67	90.20	94.81	99.25	91.95	85.76	94.87	99.13	97.04	99.49	98.39
1968	99.07	90.54	98.18	102.59	91.12	86.64	93.17	96.53	95.50	92.87	94.90
1969	102.34	92.22	102.25	103.13	93.61	88.00	92.80	92.25	96.14	85.67	91.69
1970	106.95	92.37	106.81	107.48	94.36	87.55	94.28	93.38	102.80	88.63	95.65
1971	105.64	92.32	105.01	108.64	90.83	87.84	94.74	95.10	103.87	94.70	95.39
1972	102.99	86.67	101.45	109.23	93.47	79.71	93.88	90.75	101.42	80.43	92.23
1973	131.51	87.31	131.61	131.79	100.16	77.34	101.40	89.19	101.48	79.60	91.61
1974	148.92	89.08	150.72	142.54	105.01	77.86	106.82	116.37	134.22	99.24	117.02
1975	118.20	78.20	117.74	121.11	86.79	70.00	96.51	126.78	129.01	100.62	117.17
1976	122.36	79.63	122.58	122.44	89.61	71.16	94.31	115.20	117.24	85.14	104.69
1977	124.65	77.77	124.49	126.18	94.02	67.71	94.99	112.91	114.30	71.37	99.87
1978	119.77	73.20	117.01	130.55	84.45	67.59	86.38	123.35	110.73	66.67	100.01
1979	124.06	76.43	120.34	138.65	85.40	67.54	92.39	122.26	114.62	71.67	102.46
1980	130.57	80.69	128.68	139.47	87.88	72.41	94.08	110.64	109.93	77.98	99.92
1981	117.18	77.80	117.44	117.20	84.27	69.30	87.51	109.86	118.42	81.23	104.87

N.B. For explanation of how series were constructed, see text / Pour savoir comment ces séries ont été construites, voir texte.

Note: The original price series are available on request from the secretariat / Les séries originales de prix sont disponibles sur demande auprès du secrétariat.

Annex 11.1

Presentation of the "end-use elasticities" model

Introduction

The background and objectives of the model are set out in chapter 11 and are not repeated here. Chapter 11 also presents the country groups used for the model.

The model

According to the theory of modelling, forecasts for supply and demand should be made simultaneously. By the price mechanism, demand affects supply and _vice versa_. Finally, an equilibrium is achieved.

For practical reasons (lack of resources and time) it has not been possible to form a more complete simultaneous model. There are in fact different opinions about the advantages of using simultaneous equations. Scott Armstrong, for example, claims that simultaneous equations do not give better results - perhaps even worse (Scott Armstrong: Long Range Forecasting, Wiley 1978 pp 179 - 180).

The model was developed in stages. During the development stage only the four major consumer countries were used, i.e. France, the Federal Republic of Germany, Italy and the United Kingdom.

Initially, the following general form was tried for these four countries and each product:

$$IAC_t = a \cdot ENDR_t^b \cdot RSUB_t^c \cdot RSUB_{t-1}^d \cdot e^{fT}$$

IAC = Index of apparent consumption

ENDR = Index of end-use activity (see chapter 11 for explanation of this variable)

RSUB = Ratio of price of the forest product in question to index of prices for substitutes

T = Time trend.

After taking natural logs of both sides of the equation, the ordinary least-square method was used. The parameters b, c and d are thus estimates of the elasticities according to end-use activity and price respectively. The parameter f is an estimate of the trend factor, which is meant to represent shifts in the consumption depending, for example, on technical changes within the end-use sector. Thus this parameter expresses the percentage change each year.

Prices of substitutes in the denominator of the price variable were weighted according to rough estimates.

The most important part of the end-use variable is residential investment. But it should be noted that total residential investment used in the model for the construction sector as a whole is a rather rough description of changes within the residential construction sector.

Total residential investment can be broken down into new residential investment and repair and maintenance. In addition, new residential investment may be broken down into one or two family dwellings and multi-family dwellings. In each of these parts the need for a specific wood product is quite different (see discussion in chapter 9). In order to get a better end-use variable it would be theoretically desirable to split total residential investment and weight the separate parts. It was, however, not possible to do this.

The results in the first stage were quite encouraging, although in quite a few cases none of the price variables was significant . The unlagged price variable was not significant in any case. The time variable was only significant in one case.

Reactions to and discussions of the results produced some valuable ideas and suggestions. It was therefore decided to test some new equations.

(a) $AC_t = a.ENDR_t^b.WP_t^c.e^{dT}$

(b) $\dfrac{AC_t}{\overline{ENDR}_t} = a.WP_t^b.e^{dt}$

(c) Pooling time-series and cross-sectional data.

(d) $AC = a.GDP^b.WP^c.e^{dT}$ (i.e. use GDP as the demand variable)

where

$WP_t = 0.2.\dfrac{P_t}{PP_t} + 0.5\dfrac{P_{t-1}}{PP_{t-1}} + 0.3\dfrac{P_{t-2}}{PP_{t-2}}$

P = price index for the products in question

PP = producer price index (for practical reasons the weighted price index for substitutes was replaced by the producer price index. It is difficult to get appropriate weights and a great part of the substitute price is in fact measured by the producer price index).

The idea behind "pooling" is that certain countries are assumed to have a common end-use and price elasticity respectively. The data for these countries are therefore put together into a common data set. Thus if, for instance, we have 16 observations for each country and five countries, the new data set consists of 5 X 16 = 80 observations (with a correspondingly higher number of degrees of freedom). The estimations are then done as usual. We assume each country to have its own intercept. To allow different intercepts, dummy variables are added to the equation.

Meanwhile, better information was obtained about the end-use pattern and the weights for the end-use variable were changed accordingly. The weights used are given in table 8.2.

The following remarks may be made about each of the "new" equations:

(a) in this equation a new price variable is introduced. In the first estimation tried, only the lagged variable was significant - and only in a few cases. From a theoretical point of view it is not reasonable to think that demand is only influenced by the price one year ago. It is more reasonable to assume that price changes influence demand in a more gradual manner. Therefore, the prices for the present year, the previous year and the year before that were weighted together. It is reasonable to assume that the price for the previous year (t - 1) has the highest weight.

(b) To put the end-use variable on the left side is a restriction, as end-use elasticity is assumed to be 1. This restriction sounds theoretically reasonable, but in practice it is not sure, as the end-use variable is neither measured nor defined adequately.

(c) Pooling time-series and cross-sectional data

Advantages   - the number of observations increases and so the number of degrees of freedom is increased markedly. When estimating the countries separately, the number of observations is just 16, which is a bit on the small side.
                - the problem with multicollinearity decreases.

Disadvantages - the price- and demand elasticities are supposed to be the same for different countries. Theoretically this might be a reasonable assumption but due to variations in measuring and defining data this presumption may well not be valid.

<u>Some comments on the coefficient of determination (multiple correlation) and the standard error of the residuals</u>.

A high value for R^2 (the coefficient of determination) e.g. close to 1, is often interpreted as a proof of a good approach and, <u>vice versa</u> a low R^2 gives cause to reject an equation.

As R^2 is defined, however, even a rather bad equation gives a rather high R^2 if the dispersion for the dependent variable is high, or, to put it in another way, if the dependent variable has a strong trend. A better approach but with a weak trend for the dependent variable can, on the other hand, give a markedly lower R^2.

Sometimes it is, in fact, possible to "arrange" a high R^2 by handling the data adequately. Let us look at an example from the data we have studied.

If we pool the time series for separate countries to one single series, we may do it in two different ways. One way is to pool the <u>index</u> series (1975=100) of the countries for the dependent variable. The other way is to pool <u>real</u> values (1000 m3). If we then use country-dummies, we get exactly the same estimates for the price- and income-elasticities in both cases. However, R^2 will automatically be much higher when using real values instead of indices, purely as a consequence of the difference in level. The interpretation of R^2 will then be quite nonsensical.

The standard error of the residuals (S_e) is in these circumstances a good substitute for R_2. With the exponential approach used in this study, S_e x 100, is the standard deviation for the differences between actual and fitted values expressed in percentage. This measure is totally independent of the composition of the dependent variable.

Results of the tests

The four functions mentioned above were tested for the four major consuming countries. The following results were obtained.

When using prices with the weighted lag instead of with a lag of only one year, the price variable became significant in some more cases. As mentioned above the "new" price variable is also preferable from a theoretical point of view.

By moving the end-use variable to the left side (equation (b)) the results, in several cases, were not as good as those obtained with the "original" form of the equation. This is natural as we force the end-use elasticity to be equal to 1. When the estimated end-use elasticity is 1 in the "original" form, both forms give about the same estimate.

The technique of pooling time-series and cross-sectional data was tested not only in the four major consuming countries but also for all countries together and each of the three country groups (defined on page 1). The pooling technique seemed to be quite promising.

This example (annex table 11.2) gives a good picture of the different pooling assumptions tested for each product. It should be noted that when pooling all countries, it does not make sense to use price elasticity dummies and end-use elasticity dummies at the same time, as we then get the same result as if we estimated each country group separately.

When all countries are pooled together, we can see from the table that the price variable is significant and has a correct sign only without any dummies. In this case S_e is remarkably high, about 20%. Besides the Durbin/Watson coefficient is low, which might indicate a specification error. When dummies were used, the end-use elasticities are strikingly stable.

The further results of pooling are commented on below when the results for each group are shown separately.

To use GDP as the demand variable instead of ENDR gave a much less satisfactory result. Besides, it is theoretically better to use a less general demand variable than GDP if this variable can be measured and forecast adequately.

As a result of these tests, two forms of equations were chosen, namely the "original" form with weighted prices and pooling of time-series and cross-sectional data country group by country group.

Final results, country group by country group

(More general comments are given in the text of chapter 11 and are not repeated here).

In the equations with pooled data, country dummies were used. The results are mixed, as can be seen. In some cases, no significant results were obtained. The best results are throughout obtained for country group I, both by country and for the group as a whole. This is hardly surprising, as this country group is the most homogeneous as regards the pattern of consumption and has got well developed statistics. Furthermore, for this group apparent consumption is not affected by stock changes to the same extent as country group III. This remark is particularly valid for Sweden and Finland, as a large share of the production of these countries is exported.

Annex figure 11.1.1

Apparent consumption of sawn softwood

(1975 = 100)

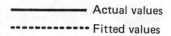

—————— Actual values
------------ Fitted values

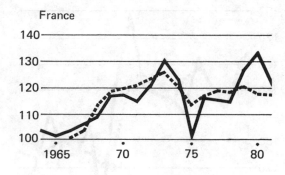

France

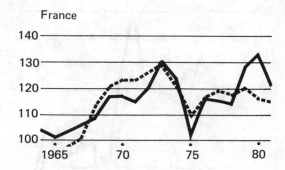

France

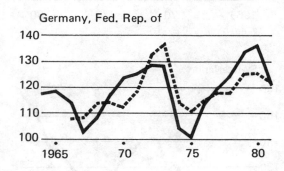

Germany, Fed. Rep. of

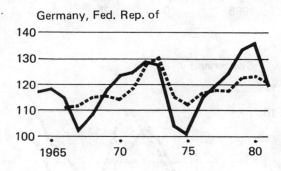

Germany, Fed. Rep. of

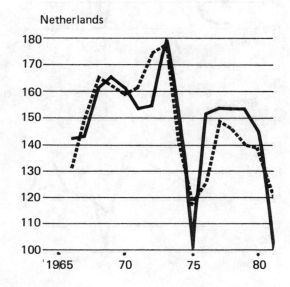

Netherlands

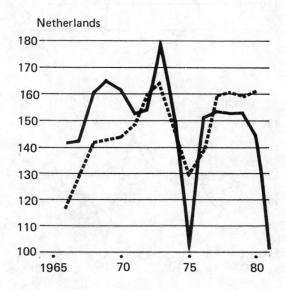

Netherlands

"Country by country" version   "Pooled" version

Annex figure 11.1.1 (continued)

Apparent consumption of sawn softwood

(1975 = 100)

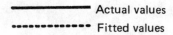

Actual values

Fitted values

United Kingdom

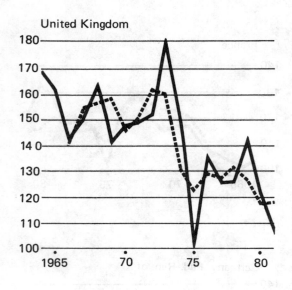

United Kingdom

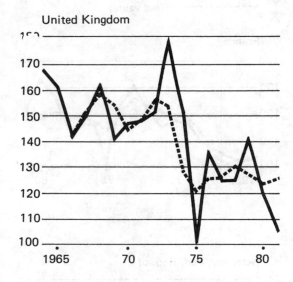

Switzerland

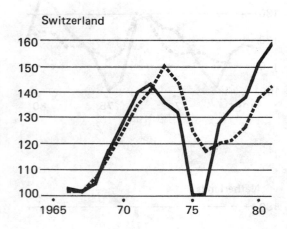

Switzerland

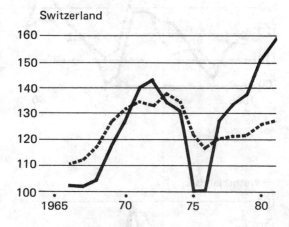

Group 1

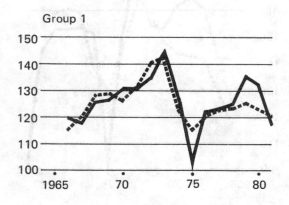

Group 1

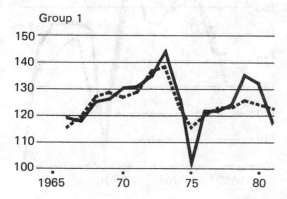

"Country by country" version

"Pooled" version

That results are sometimes not so good and are even missing in some cases might be an indication that the information available is not sufficient to explain the variations in consumption. In some cases the explanation could be that econometric methods are not suitable. Further comments are given below.

Some comments on the Durbin/Watson (D/W) coefficients are necessary. The D/W is calculated to test whether autocorrelation is present or not. If there is autocorrelation this might be an indication of a specification error, although even a well-specified equation can be auto-regressive. One way to avoid the negative impact on the results of auto-regressive disturbances, when using the ordinary least square technique, is to use the Cochran-Orcutt iterative procedure. In this model, when the D/W value indicates auto-regression, we have tried the Cochran-Orcutt procedure. If the fit has clearly improved, the estimated elasticities have changed markedly and the new "result" is sensible from a theoretical and pragmatic point of view, the "new" estimation has been accepted. Otherwise, we have continued to use the first estimation. When the Cochran-Orcutt procedure is used, this is noted in the tables.

- Sawn softwood

When estimating separate countries in Group I, the end-use elasticity was strikingly even (about 1) for four of the countries. Three of these countries have a correct sign for the price elasticity and its value did not differ very much between countries. Both price and end-use elasticity for France differ a lot from the values for the other countries.

Reestimation of those countries whose D/W values indicated auto-correlation had little effect on the value of the parameters and the quality of fit. So the first estimations were accepted.

When estimating group I as a whole we got an end-use elasticity markedly below those for the individual countries, while the price elasticity is about the average for the individual countries. As expected, the estimates of the parameters have lower standard errors for the country group as a whole than for separate countries. The fit for each country separately is obviously less good in the pooled version. For the country group as a whole the two approaches give about equivalent fits.

In group II (excluding Portugal, for which no results were significant), the price variable was not significant for either Italy or Spain. The countries of group III (exporters) have a high degree of self-sufficiency for sawn softwood. Its share of total wood consumption is 70% for Austria and about 80% for the Nordic countries. The price variable was not significant in any case. This result may perhaps be explained by two circumstances. In the first place, wood products and non-wood products do not very often compete in these countries. This might to some extent be explained by a firm tradition within the building sector, which often reflects marked cost differences between wood and non-wood products in many applications. Secondly, among wood products sawn softwood is very dominant, so even a very strong increase for another wood product (for instance particle board) only affects sawn softwood marginally.

For Finland and Sweden no significant results were achieved. This is probably explained by the fact that apparent consumption reflects real consumption very badly as it is a consequence of the extremely strong stock changes at the sawmills. Semi processed goods like sawn softwood often have very large stock changes.

For forecasts, the equation for the group as a whole is recommended.

- ## Sawn hardwood

For most countries in Group I, the results are quite satisfactory, even though the price variable is not significant for the Federal Republic of Germany. For the Netherlands, the price elasticity was not significant. The results for Switzerland, however, were not satisfactory. Furthermore the low D/W value indicated auto-correlation and therefore the equation was reestimated. The fit obviously improved, but was still not very good: the end-use elasticity was significant, but not the price elasticity. The standard errors for the elasticities are relatively large. All these factors put together show a picture of uncertainty. This might well be explained by the fact that sawn hardwood is a marginal product in Switzerland - its share is just 7% of total wood consumption. The pattern of consumption for a marginal product is often unstable - for example, changes in fashion can have remarkable effects on the level of consumption. Because of the mixed results for separate countries if is reasonable to use the result for the group as a whole when forecasting.

In Group II, the price elasticity is unreasonably high for Italy. It is therefore advisable to use the group result.

Sawn hardwood is a marginal product in the Nordic countries (Group III). Therefore no useful results were achieved for these countries or for the group as a whole. Forecasts for the Nordic countries must, therefore, be done in another way. The result for Austria may be usable.

- ## Plywood

In the United Kingdom the consumption of plywood is about 10% of total wood consumption and in the Netherlands 12 - 13%. In the other countries of Group I, the share is just a few per cent. Therefore, in these countries it is doubtful, as mentioned before, whether such rough explanatory variables can be used. The result for the United Kingdom is hardly reliable with a demand elasticity of 2.8 and a price elasticity of - 1.4. These elasticities must be regarded as unreasonably high. The country estimates vary widely between countries. In that light it seems advisable to use the group result for forecasts.

A similar picture applies to Groups II and III. It must be considered questionable whether this approach is appropriate for marginal products. In any case results by country group should be used for forecasting.

- ## Particle board

In view of the importance of the technical problems associated with modelling demand for particle board, they are discussed fully in the text of chapter 11.

- ## Fibreboard

Consumption of fibreboard is quite low in all countries within Group I and its share of total wood consumption is just 2 to 3%. In all countries but Switzerland the consumption has tended to fall. That can to some extent certainly be explained by the progress of particle board. As said before, it is doubtful if such rough explanatory variables can be used for a marginal product.

The results of the regressions differ a lot between countries. The price variable is significant for three of the countries in Group I, - although unreasonably high for France. The trend variable is significant for three countries. The parameter is positive for Switzerland - for the other countries it is negative. The end-use elasticity varies from 0,8 to 1.8. For Group I as a whole, the fit is quite good after Cochran_Orcutt correction for low Durbin/Watson values.

This product is also marginal in Group II. The growth rate was very strong during the estimation period (in the beginning of the period consumption was close to zero in Portugal). In Spain the consumption grew very strongly during the latter part of the 1970s. The product seems to be in an introductory phase in these countries. The methodology used is not therefore very advisable.

In Group III the consumption of fibreboard decreased steadily during the estimation period. The cost advantages of changing to particle board have for many applications negatively affected the use of fibreboard. So it s reasonable to use the same type of equation for fibreboard as for particle board, although the sign for the trend variable is expected to be negative.

It should be pointed out that as consumption of MDF was negligible in the estimation period, no conclusions can be drawn from these data as to future developments for MDF.

Annex table 11.2   **Tableau annexe 11.2**
Regression results   **Résultats des régressions**

Set out below are detailed results of the regressions described in the text and annex 11.1 by country and by country group.

Ce tableau présente les résultats détaillés des régressions décrites dans le texte et l'annexe 11.1 par pays et par groupe de pays.

	Abbreviations	Abréviations
ENDR	Elasticity with respect to activity in user sector	Elasticité par rapport à l'activité dans les secteurs utilisateurs
WP	Elasticity with respect to lagged price variable	Elasticité par rapport à la variable de prix (décalée)
T	Elasticity with respect to time trend	Elasticité par rapport à la tendance dans le temps
R^2	Coefficient of determination	Coefficient de détermination
S_e	Standard error of residuals	Ecart type des résidus
D/W	Durbin/Watson coefficient	Coefficient Durbin/Watson
n/n	Number of observations/number of independent variables	Nombre d'observations/Nombre de variables indépendantes
C-O	With Cochran-Orcutt correction	Avec correction Cochran-Orcutt
( )	Figure in brackets is the standard deviation of the coefficient	Le chiffre entre parenthèses est l'écart type du coefficient

Sawn softwood/Sciages résineux

	ENDR	WP	T	R^2	S_e	D/W	n/n
Group I / Groupe I	0.708 (0.113)	-0.578 (0.126)		0.98	0.108	1.28	80/6
France	0.513 (0.148)	-0.363 (0.163)		0.46	0.059	1.49	16/2
Germany, Fed. Rep. of / Allemagne, Rép. féd. d'	1.036 (0.270)	-0.795 (0.443)		0.49	0.067	0.96	16/2
Netherlands / Pays-Bas	1.056 (0.393)	-0.836 (0.460)	-0.031 (0.012)	0.55	0.107	2.11	16/3
Switzerland / Suisse	1.118 (0.232)			0.60	0.100	0.64	16/1
United Kingdom / Royaume-Uni	1.083 (0.338)	-0.613 (0.179)		0.54	0.105	2.45	16/2
Group II / Groupe II	0.744 (0.153)			0.91	0.107	1.21	32/2
Italy / Italie	0.870 (0.266)			0.39	0.114	0.72	16/1
Spain / Espagne	0.670 (0.186)			0.44	0.103	1.77	16/1
Group III / Groupe III	0.352 (0.097)			0.88	0.131	1.48	64/4
Austria / Autriche	0.610 (0.185)			0.40	0.138	1.23	16/1
Finland/Finlande	Results not significant/Résultats non significatifs						
Norway / Norvège	0.321 (0.100)			0.39	0.076	0.96	16/1
Sweden/Suède	Results not significant/Résultats non significatifs						

Sawn hardwood/Sciages feuillus

	ENDR	WP	T	R²	Se	D/W	n/n
Group I / Groupe I	1.151 (0.144)	-0.463 (0.145)		0.99	0.127	0.69	80/6
France	0.753 (0.163)	-0.279 (0.153)		0.71	0.058	1.75	16/2
Germany, Fed. Rep. of / Allemagne, Rép. féd. d'	0.974 (0.229)	-0.342 (0.316)		0.65	0.072	1.21	16/2
Netherlands / Pays-Bas	1.508 (0.254)			0.70	0.156	0.78	16/1
Switzerland (C-O) / Suisse	1.481 (0.633)	-1.070 (0.798)		0.37	0.117	1.71	16.2
United Kingdom / Royaume-Uni	1.211 (0.293)	-0.486 (0.133)		0.57	0.072	2.43	16/2
Group II / Groupe II	0.797 (0.160)	-0.889 (0.352)		0.90	0.139	1.31	32/3
Italy / Italie	0.949 (0.225)	-1.876 (0.507)		0.51	0.121	2.00	16/2
Spain / Espagne	0.797 (0.126)			0.72	0.106	1.96	16/1
Group III/Groupe III	Results not significant/Résultats non significatifs						
Austria / Autriche	1.503 (0.142)			0.88	0.115	1.01	16/1
Finland/Finlande	Results not significant/Résultats non significatifs						
NorwayNorvège	Results not significant/Résultats non significatifs						
Sweden/Suède	Results not significant/Résultats non significatifs						

Plywood/Contreplaqué

	ENDR	WP	T	R^2	S_e	D/W	n/n
Group I / Groupe I	1.170 (0.205)	-0.241 (0.128)		0.95	0.223	0.64	80/6
France	0.707 (0.166)	-0.491 (0.174)	-0.0208 (0.0062)	0.77	0.067	2.48	16/3
Germany, Fed. Rep. of / Allemagne, Rép. féd. d'	1.125 (0.223)	-0.485 (0.435)		0.63	0.075	1.19	16/2
Netherlands / Pays-Bas	0.572 (0.291)	-0.871 (0.334)	0.066 (0.013)	0.96	0.096	2.15	16/3
Switzerland / Suisse	1.667 (0.471)			0.43	0.203	0.93	16/1
United Kingdom / Royaume-Uni	2.768 (0.579)	-1.442 (0.425)		0.58	0.112	2.89	16/2
Group II / Groupe II	Results not significant/Résultats non significatifs						
Italy / Italie	1.110 (0.478)			0.23	0.274	1.64	16/1
Spain / Espagne	1.956 (0.592)	-0.880 (0.489)	-0.095 (0.023)	0.76	0.185	1.67	16/3
Group III / Groupe III	0.683 (0.158)	-0.952 (0.212)		0.95	0.203	1.10	64/5
Austria/Autriche	Results not significant/Résultats non significatifs						
Finland/Finlande	Results not significant/Résultats non significatifs						
Norway / Norvège	1.256 (0.122)	-0.630 (0.238)		0.89	0.089	1.77	16/2
Sweden / Suède	0.955 (0.380)	-0.368 (0.122)	0.021 (0.005)	0.81	0.072	2.09	16/3

	ENDR	WP	T	R²	Se	D/W	n/n
			Particle board/Panneaux de particules				
Group I (C-O) Groupe I	1.120 (0.179)		0.0056 (0.0134)	0.99	0.074	2.29	75/6
France	1.137 (0.0995)		0.041 (0.0035)	0.98	0.048	2.50	16/2
Germany, Fed. Rep. of (C-O) Allemagne, Rép. féd. d'	0.951 (0.208)		-0.0092 (0.0129)	0.57	0.039	1.92	15/2
Netherlands Pays-Bas	1.722 (0.163)			0.88	0.094	0.94	16/1
Switzerland (C-O) Suisse	1.420 (0.212)		0.0176 (0.0055)	0.85	0.052	2.36	15/2
United Kingdom Royaume-Uni	2.275 (0.197)		0.1120 (0.0032)	0.99	0.058	1.47	16/2
Group II (C-O) Groupe II	0.641 (0.267)		0.0454 (0.0180)	0.92	0.119	1.89	30/3
Italy/Italie	Results not significant/Résultats non significatifs						
Spain Espagne	2.440 (0.36)		0.0295 (0.0160)	0.95	0.143	2.02	16/2
Group III Groupe III (C-O)	0.876 (0.313)		0.030 (0.0122)	0.97	0.102	2.15	60/5
Austria Autriche	1.735 (0.075)			0.97	0.061	1.89	16/1
Finland Finlande	2.041 (0.141)			0.93	0.117	2.29	16/1
Norway Norvège	0.889 (0.217)		0.0405 (0.0092)	0.98	0.055	2.01	16/2
Sweden Suède	2.717 (0.678)		0.0767 (0.0088)	0.90	0.148	1.46	16/2

	ENDR	WP	T	R^2	S_e	D/W	n/n
Fibreboard/Panneaux de fibres							
Group I (C-O) Groupe I	0.939 (0.275)		-0.0274 (0.0112)	0.98	0.118	2.16	75/6
France	0.771 (0.379)	-1.478 (0.373)	-0.1003 (0.0214)	0.72	0.132	2.74	16/3
Germany, Fed. Rep. of Allemagne, Rép. féd. d'	1.061 (0.249)	-0.906 (0.251)	-0.0262 (0.0046)	0.77	0.054	2.41	16/3
Netherlands/Pays-Bas	Results not significant/Résultats non significatifs						
Switzerland (C-O) Suisse	1.595 (0.326)		0.0960 (0.0455)	0.64	0.069	1.84	16/2
United Kingdom Royaume-Uni	1.778 (0.290)	-0.769 (0.271)		0.70	0.081	2.26	16/2
Group II/Groupe II	Results not significant/Résultats non significatifs						
Italy Italie	2.517 (0.564)			0.56	0.314	1.10	16/1
Spain/Espagne	Results not significant/Résultats non significatifs						
Group III Groupe III	0.844 (0.129)		-0.0324 (0.0048)	0.97	0.112	1.81	64/5
Austria Autriche	1.045 (0.319)		-0.0436 (0.0131)	0.39	0.104	1.94	16/2
Finland Finlande	0.841 (0.336)		-0.0284 (0.0155)	0.24	0.140	2.37	16/2
Norway Norvège	1.506 (0.389)		-0.0635 (0.0162)	0.47	0.086	1.97	16/2
Sweden Suède	1.072 (0.623)		-0.0275 (0.0064)	0.57	0.118	0.77	16/2

126

	ENDR	WP	T	R^2	S_e	D/W	n/n
		Wood-based panels/Panneaux dérivés du bois					
Group I (C-O) Groupe I	1.024 (0.134)		0.0207 (0.0038)	0.99	0.067	2.42	75/6
France	0.848 (0.082)		0.0173 (0.0032)	0.97	0.040	2.89	16/2
Germany, Fed. Rep. of (C-O) Allemagne, Rép. féd. d'	1.055 (0.221)			0.61	0.042	1.80	15/1
Netherlands Pays-Bas	0.735 (0.198)		0.0190 (0.0057)	0.91	0.056	1.96	16/2
Switzerland Suisse	1.610 (0.152)		0.0150 (0.0034)	0.95	0.052	1.93	16/2
United Kingdom Royaume-Uni	1.487 (0.280)		0.0464 (0.0044)	0.89	0.078	1.99	16/2
Group II (C-O) Groupe II	0.532 (0.282)		0.0382 (0.0171)	0.96	0.107	1.81	30/3
Italy/Italie		Results not significant/Résultats non significatifs					
Spain Espagne	1.736 (0.184)		0.0269 (0.0074)	0.98	0.070	2.44	16/2
Group III (C-O) Groupe III	0.961 (0.187)		0.0136 (0.0065)	0.94	0.090	2.12	60/5
Austria Autriche	1.436 (0.069)			0.97	0.056	2.24	16/1
Finland Finlande	1.223 (0.139)			0.84	0.116	2.08	16/1
Norway Norvège	1.151 (0.049)			0.97	0.038	2.57	16/1
Sweden Suède	1.441 (0.627)		0.0319 (0.0064)	0.67	0.118	1.06	16/2

Annex table 11.3
Sawn softwood: results of experimental regressions
(For explanation, see notes)

Tableau annexe 11.3
Sciages résineux : résultats des régressions
expérimentales (Pour explication, voir notes)

	ENDR	GDP	WP	T	S_e	D/W	n/n
All CD	0.364 (0.069)		0.004 (0.081)		0.126	1.35	192/13
All CD + PD	0.424 (0.068)		-0.403, + 0.393, -0.008 a/		0.119	1.39	192/15
All CD + EED	0.410, 0.329, 0.355 a/		-0.002 (0.083)		0.126	1.35	192/15
All no dummies	1.033 (0.012)		-0.277 (0.100)		0.198	0.69	192/2
4 major CD	0.661 (0.126)		-0.308 (0.118)		0.103	1.15	64/5
3 major CD	0.780 (0.112)		-0.604 (0.109)		0.079	1.87	48/4
Groups I and II CD	0.382 (0.087)		-0.021 (0.094)		0.123	1.17	128/9
Group I CD	0.692 (0.114)		-0.570 (0.127)		0.109	1.27	80/6
Group II CD	0.290 (0.115)		0.406 (0.117)		0.111	1.22	48/4
Group III CD	0.322 (0.121)		0.072 (0.161)		0.132	1.47	64/5
3 major ($\frac{AC}{\overline{ENDR}}$) CD			0.74 (0.08)		0.082	1.76	48/3
4 major ($\frac{AC}{\overline{ENDR}}$) CD			-0.502 (0.10)		0.108	1.02	64/4
3 major (GDP) CD		0.511 (0.154)	-0.526 (0.165)		0.103	1.45	48/4
4 major (GDP) CD		0.682 (0.142)	-0.532 (0.157)		0.105	1.27	64/5
Group III CD	0.354 (0.097)				0.131	1.48	64/4

a/ Estimates for groups I, II and III respectively.

a/ Estimations pour les groupes I, II et III.

Notes au tableau annexe 11.3

Notes to annex table 11.3

The equation used is the standard form except when indicated otherwise:

L'équation utilisée est de la forme normale, sauf indication contraire :

$$\left\{\frac{AC}{ENDR}\right\} : \quad \frac{AC^t}{ENDR_b} = a. \ WP^b . e^{CT}$$

$$(GDP) : \quad AC = a. \ GDP^b. \ WP^C. \ e^{dT}$$

For abbreviations, see annex table 11.2 and the following:
Pour les abbréviations, voir tableau annexe 11.2 ainsi que ce qui suit :

All   All countries (12) / Tous les pays (12)

3 major   France, Federal Republic of Germany, United Kingdom / France, République fédérale d'Allemagne, Royaume-Uni

4 major   " 3 major " + Italy / "3 major" + Italie

CD   With country dummies / Avec "dummy variable" par pays

PD   With price elasticity dummies / Avec "dummy variable" pour l'élasticité des prix

EED   With end-use elasticity dummies / Avec "dummy variable" pour l'élasticité de l'activité

Annex 11.4

Review of ETTS III econometric projections

(This annex was prepared by the Statistics and Economic Analysis Group
of the FAO Forestry Department) 1/

A comparison was made between the projected regional totals and per capita
consumption of the main product groups for 1980, as recorded in ETTS III with
projections made using the actual growth rates of GDP recorded for the period
1970-1980 and the actual population recorded for 1980.

In carrying out this exercise it was found that using the 1970 base data
and the growth assumptions recorded in ETTS III, it was not possible to
reconstruct the 1980 projections precisely as they were recorded. This may stem
from the fact that the original projections were formulated by the addition of
projections for individual countries whereas this calculation was carried out on
the regional average consumption and GDP per caput. The recalculated
projections for the region on the original GDP and population assumptions have
therefore also been shown (abbreviated as Rev. Calc.).

A revised projection of the 1980 per capita consumption was then
calculated, replacing the original estimate of growth of GDP per caput with the
actual growth of GDP per caput. Likewise a revised projection of the total
consumption was recalculated applying the actual 1980 population to the revised
projection of per capita consumption (abbreviated as Rev. Assumpt.).

Concerning the assumptions, the estimated development of population to 1980
is only slightly different to the actual, the difference exceeds 1% only in the
case of Central Europe, where the expectation was for a decrease and an increase
occurred (annex table 11.5.6). Growth in per caput GDP was generally
overestimated so that estimated 1980 GDP was 1-3% higher than actual (annex
table 11.5.5).

The estimates of consumption and actual consumption in 1980, are summarized
below:

Percentage difference of projections from actual results 1980

Europe	Sawnwood & panels	Sawnwood	Panels	Paper All	Cultural	Other
ETTS III	+ 5	− 3	+ 26	+ 22		
Rev. Calc.	+ 1	− 7	+ 24	+ 15	+ 13	+ 16
Rev. Assumpt.	− 2	− 12	+ 23	+ 7	+ 5	+ 9

1/ As this analysis was carried out in early 1984, subsequent revisions to the
data base were not taken into account and some of the data shown in this annex
may not coincide exactly with those used elsewhere in the study.

The published ETTS III projection for sawnwood and panels exceeds actual consumption in 1980 by 5%. The revised projection on the original assumption exceeds the actual result by 1%, while the simulated projection using actual GDP and population growth falls below actual consumption by 2%. Examination of the components sawnwood and wood-based panels individually shows that all projections of sawnwood for Europe as a whole fall below actual sawnwood consumption, while projections for wood-based panels exceed actual results by more than 20%.

The published ETTS III projection for paper exceeds actual consumption in 1980 by 22%. The revised projection using original GDP and population growth estimates exceeds actual results by 15%, while the simulated projection using actual GDP and population growth exceeds actual results by 7%. The differences are somewhat higher for other paper and paperboard than they are for cultural paper.

Examination of the divergence for individual regions between projection and actual results (annex tables 11.4.2 and 11.4.3) shows that they are spread over a wide range and are on average larger than those for Europe taken as a whole. In most cases the replacement of the estimated economic and population growth with actual growth leads to a closer estimate of the actual consumption.

Estimates of the error arising from variability not accounted for by the relationship with the explanatory variables were made in calculating the original relationships. These suggested that a range of one standard deviation would be for mechanical wood products 12% and for paper and paperboard 5%. These applied to estimates for single countries. The range for a group of countries would be expected to be somewhat less. It will be seen that deviations between projections and actual results are well within a range of a standard deviation in the case of mechanical wood products, and within two standard deviations in the case of paper. In the case of the individual components, the deviations tend to exceed one standard deviation.

It was recognized in establishing the original relationships that important explanatory variables had not been explicitly included. Among these was price. The assumption was made that price trends over the time span of the historic data used in calculating the relationship would continue over the period of the projections. It is manifest that the assumption broke down over the period between 1970 and 1980 when trends in prices, both of the products themselves and of other goods and services determining preferences, changed radically. It has to be accepted that the model used to project consumption was inadequate to handle these variations.

In conclusion it is emphasized that in future modelling of the outlook for forest products, an explanatory structure taking fuller account of the factors contributing to the determination of supply and demand is desirable. The limitations of availability of data in particular on the future development of those determining factors remains a limitation on the power of such exercises.

The attached tables provide the data underlying the above analysis "FAO/ETTS Pop. GDP" refers to the recalculated projections for 1980 using the original (ETTS III) assumptions regarding population and GDP. "FAO/Real Pop. GDP" refers to the projections using real data for population and GDP in 1980.

Annex table 11.4.1

Europe: Comparison of projections of consumption in 1980

	Sawnwood and panels	Sawnwood	Wood-based panels	Paper & paper-board	Cultural paper	Other paper & paperboard
	(million m3)			(million m.t.)		
WESTERN EUROPE						
ETTS III	112.10	75.39	36.71	50.90		
FAO/ETTS Pop. GDP	(109.63)	(73.75)	(35.88)	(48.30)	(20.30)	(28.00)
FAO/Real Pop. GDP	(107.83)	(72.23)	(35.60)	(46.19)	(19.48)	(26.71)
Actual	109.86	80.52	29.34	43.20	18.70	24.50
EASTERN EUROPE						
ETTS III	32.93	23.85	9.08	8.75		
FAO/ETTS Pop. GDP	(30.31)	(21.83)	(9.08)	(7.99)	(2.52)	(5.47)
FAO/Real Pop. GDP	(28.01)	(19.15)	(8.86)	(6.45)	(1.75)	(4.70)
Actual	28.56	21.66	6.90	5.80	1.42	4.38
TOTAL EUROPE						
ETTS III	145.03	99.24	45.79	59.65		
FAO/ETTS Pop. GDP	(139.94)	(95.58)	(44.96)	(56.29)	(22.82)	(33.47)
FAO/Real Pop. GDP	(135.84)	(91.38)	(44.46)	(52.64)	(21.23)	(31.41)
Actual	138.42	102.18	36.24	49.00	20.12	28.88

Annex table 11.4.2

Comparison of adjusted per capita consumption and
total consumption in 1980

	Sawnwood and panels	Sawnwood	Wood-based panels	Paper & paper-board	Cultural paper	Other paper & paperboard	Popula-tion 1979-81
	(m3/1000 inhabitants)			(m.t./1000 inhabitants)			(million)
NORDIC							
ETTS III	791	589	202	196			
FAO/ETTS Pop. GDP	(796)	(600)	(196)	(199)	(83)	(116)	
FAO/Real Pop. GDP	(806)	(608)	(198)	(206)	(86)	(120)	
Actual	835	668	167	192	84	108	17.40
EEC(9)							
ETTS III	290	186	104	135			
FAO/ETTS Pop. GDP	(290)	(186)	(104)	(141)	(61)	(80)	
FAO/Real Pop. GDP	(286)	(183)	(103)	(136)	(59)	(77)	
Actual	280	200	80	125	56	69	260.74
CENTRAL EUROPE							
ETTS III	435	295	140	140			
FAO/ETTS Pop. GDP	(409)	(275)	(134)	(153)	(64)	(89)	
FAO/Real Pop. GDP	(389)	(258)	(131)	(133)	(57)	(76)	
Actual	472	368	104	142	60	82	13.90
SOUTHERN EUROPE							
ETTS III	129	97	32	50			
FAO/ETTS Pop. GDP	(112)	(85)	(27)	(46)	(16)	(30)	
FAO/Real Pop. GDP	(107)	(80)	(27)	(41)	(14)	(27)	
Actual	130	98	32	41	14	27	129.30
EASTERN EUROPE							
ETTS III	301	218	83	80			
FAO/ETTS Pop. GDP	(277)	(194)	(83)	(73)	(23)	(50)	
FAO/Real Pop. GDP	(256)	(175)	(81)	(59)	(16)	(43)	
Actual	261	198	63	53	13	40	109.41

Annex table 11.4.3

Comparison of projections of consumption in 1980

	Sawnwood and panels	Sawnwood	Wood-based panels	Paper & paper-board	Cultural paper	Other paper & paperboard
	(million m3)			(million m.t.)		
NORDIC						
ETTS III	13.76	10.25	3.51	3.41		
FAO/ETTS Pop. GDP	(13.85)	(10.44)	(3.41)	(3.46)	(1.44)	(2.02)
FAO/Real Pop. GDP	(14.02)	(10.58)	(3.44)	(3.58)	(1.50)	(2.08)
Actual	14.53	11.62	2.91	3.34	1.46	1.88
EEC(9)						
ETTS III	75.62	48.50	27.12	35.20		
FAO/ETTS Pop. GDP	(75.62)	(48.50)	(27.12)	(36.76)	(15.90)	(20.86)
FAO/Real Pop. GDP	(74.57)	(47.72)	(26.85)	(35.46)	(15.38)	(20.08)
Actual	73.00	52.15	20.85	32.59	14.60	17.99
CENTRAL EUROPE						
ETTS III	6.05	4.10	1.95	1.95		
FAO/ETTS Pop. GDP	(5.68)	(3.82)	(1.86)	(2.13)	(0.89)	(1.24)
FAO/Real Pop. GDP	(5.41)	(3.59)	(1.82)	(1.85)	(0.79)	(1.06)
Actual	6.55	5.11	1.44	1.97	0.83	1.14
SOUTHERN EUROPE						
ETTS III	16.67	12.54	4.13	6.46		
FAO/ETTS Pop. GDP	(14.48)	(10.99)	(3.49)	(5.95)	(2.07)	(3.87)
FAO/Real Pop. GDP	(13.83)	(10.34)	(3.49)	(5.30)	(1.81)	(3.49)
Actual	15.78	11.64	4.14	5.30	1.81	3.49
EASTERN EUROPE						
ETTS III	32.93	23.85	9.08	8.75		
FAO/ETTS Pop. GDP	(30.31)	(21.83)	(9.08)	(7.99)	(2.52)	(5.47)
FAO/Real Pop. GDP	(28.01)	(19.15)	(8.86)	(6.45)	(1.75)	(4.70)
Actual	28.56	21.66	6.89	5.80	1.42	4.38

Annex table 11.4.4

Adjustments of per capita consumption

	Sawnwood and panels	Sawnwood	Wood-based panels	Paper & paper-board	Sawnwood and panels	Sawnwood	Wood-based panels	Paper & paper-board
	(m3/1000 inhabitants)			(m.t./1000 inhabitants)	(m3/1000 inhabitants)			(m.t./1000 inhabitants)
NORDIC								
ETTS III	791	589	202	196	791	589	202	196
FAO/ETTS Pop. GDP	(796)	(600)	(196)	(199)	(796)	(603)	(197)	(199)
FAO/Real Pop. GDP	(806)	(608)	(198)	(206)	(806)	(613)	(199)	(206)
EEC(9)								
ETTS III	290	186	104	135	290	186	104	135
FAO/ETTS Pop. GDP	(290)	(186)	(104)	(141)	(290)	(186)	(104)	(141)
FAO/Real Pop. GDP	(286)	(185)	(103)	(136)	(286)	(195)	(109)	(136)
CENTRAL EUROPE								
ETTS III	435	295	140	140	435	295	140	140
FAO/ETTS Pop. GDP	(409)	(275)	(134)	(153)	(409)	(293)	(142)	(409)
FAO/Real Pop. GDP	(389)	(258)	(131)	(133)	(389)	(273)	(138)	(389)
SOUTHERN EUROPE								
ETTS III	129	97	32	50	129	97	32	50
FAO/ETTS Pop. GDP	(112)	(85)	(27)	(46)	(112)	(90)	(29)	(46)
FAO/Real Pop. GDP	(107)	(80)	(27)	(41)	(107)	(84)	(29)	(41)
EASTERN EUROPE								
ETTS III	301	218	83	80	301	218	83	80
FAO/ETTS Pop. GDP	(277)	(194)	(83)	(73)	(277)	(212)	(91)	(73)
FAO/Real Pop. GDP	(256)	(175)	(81)	(59)	(256)	(190)	(87)	(59)

Annex table 11.4.5

Growth of GDP and Population 1970-1980

	GDP		Population		GDP/capita		
	Forecasted	Actual	Forecasted	Actual	Forecasted		Actual
	% annum		% annum		1970	1980	1980
					US$ (1970 prices)		
NORDIC	2.5	2.9	0.35	0.40	3 353	4 176	4 292
EEC(9)	3.0	2.85	0.40	0.35	2 486	3 277	3 182
CENTRAL EUROPE	2.4	2.3	0.03	0.20	2 430	3 242	2 888
SOUTHERN EUROPE	5.4	4.7	1.50	1.50	730	1 099	999
EASTERN EUROPE	5.6	2.35	0.60	0.60	1 576	2 570	2 157

Annex table 11.4.6

GDP and Population

	Forecast Total GDP (Billion US$)		Actual a/ Total GDP (Billion US$)		Forecast Total Population (Millions)		Actual
	1970	1980	1970	1980	1970	1980	1980
NORDIC	56	72	108	145	16.7	17.3	17.4
EEC(9)	627	859	1 204	1 596	252.2	262.1	260.8
CENTRAL EUROPE	33	44	82	103	13.6	13.5	13.9
SOUTHERN EUROPE	81	141	171	271	111.0	128.4	128.4
EASTERN EUROPE	163	281	188	237	103.4	109.3	109.3

a/   Data provided by ECE from which actual percentage growth of GDP in annex table 11.5.5 was calculated.

Annex table 11.5

Tableau annexe 11.5

NMP elasticities for wood-based panels in eastern Europe

Elasticités par rapport au Produit matériel net (PMN) pour les panneaux dérivés du bois en Europe orientale

	1964–1981	1980 – 1990		1990 – 2000	
	(real) (réel)	Low Bas	High Haut	Low Bas	High Haut
Bulgaria Bulgarie	0.72	0.55	0.65	0.45	0.55
Czechoslovakia Tchécoslovaquie	1.83	1.20	1.30	0.90	1.00
German. Dem. Rep. Rép. dém. allemande	1.20	0.95	1.05	0.85	0.95
Hungary Hongrie	1.49	1.05	1.15	0.95	1.05
Poland Pologne	2.43	1.40	1.50	1.10	1.20
Romania Roumanie	0.66	0.55	0.65	0.45	0.55

Annex table 12.1
Projections by GDP elasticities model:
sawnwood (including sleepers)

(base scenario)

Tableau annexe 12.1
Projections par le modèle "élasticités
par rapport au PIB":
sciages (y compris les traverses)
(scénario de base)

Country	1979-1981	1990	1995	2000	Av. annual change, 1980-2000 Taux de change-ment annuel moyen 1980-2000	Pays
	(1 000 000 m3)				(%)	
Finland	3.10	3.52	3.87	4.26	+ 1.6	Finlande
Iceland	0.07	0.07	0.08	0.08	+ 1.0	Islande
Norway	2.47	2.69	3.03	3.41	+ 1.6	Norvège
Sweden	5.11	4.72	5.13	5.58	+ 0.5	Suède
NORDIC COUNTRIES	10.75	11.00	12.11	13.33	+ 1.1	PAYS NORDIQUES
Belgium-Luxembourg	2.05	1.84	1.98	2.15	+ 0.2	Belgique-Luxembourg
Denmark	1.92	1.95	2.09	2.25	+ 0.8	Danemark
France	11.75	12.02	13.32	14.76	+ 1.1	France
Germany, Fed.Rep. of	14.44	14.32	15.34	16.50	+ 0.7	Allemagne, Rép. féd. d'
Ireland	0.64	0.59	0.62	0.65	-	Irlande
Italy	7.95	7.60	7.94	8.30	+ 0.2	Italie
Netherlands	3.15	2.92	3.11	3.32	+ 0.3	Pays-Bas
United Kingdom	8.55	8.66	8.79	8.93	+ 0.2	Royaume-Uni
EEC(9)	50.46	49.90	53.19	56.86	+ 0.6	CEE(9)
Austria	2.92	3.30	3.64	4.02	+ 1.6	Autriche
Switzerland	2.12	2.24	2.35	2.46	+ 0.8	Suisse
CENTRAL EUROPE	5.04	5.54	5.99	6.48	+ 1.3	EUROPE CENTRALE
Cyprus	0.13	0.12	0.13	0.14	+ 0.4	Chypre
Greece	1.05	0.79	0.86	0.93	- 0.6	Grèce
Israel	0.26	0.29	0.32	0.34	+ 1.3	Israel
Malta	0.02	0.03	0.04	0.04	+ 3.0	Malte
Portugal	1.25	1.45	1.56	1.70	+ 1.6	Portugal
Spain	3.44	3.17	3.32	3.47	+ -	Espagne
Turkey	4.64	5.46	6.78	8.41	+ 3.0	Turquie
Yugoslavia	3.52	4.10	4.57	5.09	+ 1.9	Yougoslavie
SOUTHERN EUROPE	14.32	15.41	17.58	20.12	+ 1.7	EUROPE MERIDIONALE
Bulgaria	1.67	1.88	2.17	2.50	+ 2.0	Bulgarie
Czechoslovakia	3.87	4.14	4.33	4.52	+ 0.7	Tchécoslovaquie
German Dem. Rep.	3.72	4.26	4.44	4.63	+ 1.1	Rép. dém. allemande
Hungary	2.07	2.11	2.28	2.47	+ 0.9	Hongrie
Poland	6.77	7.45	8.54	9.88	+ 1.9	Pologne
Romania	3.65	6.24	7.93	10.30	+ 5.3	Roumanie
EASTERN EUROPE	21.75	26.08	29.69	34.30	+ 2.3	EUROPE ORIENTALE
EUROPE	102.32	107.93	118.56	131.09	+ 1.3	EUROPE

Annex table 12.2
Projections by GDP elasticities model:
wood-based panels

(base scenario)

Tableau annexe 12.2
Projections par le modèle "élasticités
par rapport au PIB":
panneaux dérivés du bois
(scénario de base)

Country	1979–1981	1990	1995	2000	Av. annual change, 1980-2000 Taux de change-ment annuel moyen 1980-2000	Pays
	(1 000 000 m3)				(%)	
Finland	0.70	0.89	1.09	1.32	+ 3.3	Finlande
Iceland	0.03	0.03	0.04	0.05	+ 2.9	Islande
Norway	0.64	0.82	1.02	1.27	+ 3.5	Norvège
Sweden	1.21	1.56	1.88	2.28	+ 3.2	Suède
NORDIC COUNTRIES	2.57	3.30	4.03	4.92	+ 3.3	PAYS NORDIQUES
Belgium-Luxembourg	0.86	1.13	1.35	1.63	+ 3.3	Belgique-Luxembourg
Denmark	0.68	0.77	0.92	1.10	+ 2.4	Danemark
France	3.32	3.99	4.90	6.02	+ 3.0	France
Germany, Fed.Rep. of	8.04	9.66	11.51	13.76	+ 2.7	Allemagne, Rép. féd. d'
Ireland	0.18	0.22	0.28	0.36	+ 3.6	Irlande
Italy	3.12	3.87	4.95	6.33	+ 3.6	Italie
Netherlands	1.23	1.35	1.60	1.90	+ 2.2	Pays-Bas
United Kingdom	3.36	4.36	5.26	6.35	+ 3.2	Royaume-Uni
EEC(9)	20.78	25.35	30.77	37.45	+ 3.0	CEE(9)
Austria	0.70	0.92	1.13	1.38	+ 3.4	Autriche
Switzerland	0.66	0.80	0.93	1.08	+ 2.5	Suisse
CENTRAL EUROPE	1.35	1.72	2.06	2.46	+ 3.0	EUROPE CENTRALE
Cyprus	0.03	0.04	0.05	0.07	+ 4.3	Chypre
Greece	0.38	0.55	0.76	1.05	+ 5.2	Grèce
Israel	0.17	0.24	0.32	0.43	+ 4.8	Israël
Malta	0.01	0.01	0.02	0.02	+ 3.6	Malte
Portugal	0.33	0.41	0.55	0.78	+ 4.4	Portugal
Spain	1.42	2.00	2.56	3.28	+ 4.3	Espagne
Turkey	0.48	0.94	1.43	2.16	+ 7.8	Turquie
Yugoslavia	1.20	1.88	2.75	4.04	+ 6.2	Yougoslavie
SOUTHERN EUROPE	4.02	6.07	8.44	11.83	+ 5.5	EUROPE MERIDIONALE
Bulgaria	0.49	0.96	1.50	2.36	+ 8.2	Bulgarie
Czechoslovakia	1.21	1.76	2.25	2.88	+ 4.4	Tchécoslovaquie
German Dem. Rep.	1.43	2.44	3.10	3.95	+ 5.2	Rép. dém. allemande
Hungary	0.47	0.72	1.00	1.38	+ 5.5	Hongrie
Poland	2.08	3.09	4.22	5.82	+ 5.3	Pologne
Romania	1.20	3.19	4.98	8.02	+ 9.9	Roumanie
EASTERN EUROPE	6.88	12.16	17.05	24.41	+ 6.5	EUROPE ORIENTALE
EUROPE	35.60	48.60	62.35	81.07	+ 4.2	EUROPE

139

Annex table 12.3
Projections by GDP elasticities model:
newsprint

(base scenario)

Tableau annexe 12.3
Projections par le modèle "élasticités
par rapport au PIB":
papier journal
(scénario de base)

Country	1979– 1981	1990	1995	2000	Av. annual change, 1980–2000 Taux de change- ment annuel moyen 1980–2000	Pays
	(1 000 000 m3)				(%)	
Finland	0.14	0.20	0.23	0.27	+ 3.3	Finlande
Iceland	–	–	–	0.01	+ 1.9	Islande
Norway	0.10	0.14	0.17	0.20	+ 3.4	Norvège
Sweden	0.28	0.34	0.39	0.45	+ 2.4	Suède
NORDIC COUNTRIES	0.52	0.68	0.79	0.93	+ 3.0	PAYS NORDIQUES
Belgium–Luxembourg	0.20	0.23	0.26	0.30	+ 2.0	Belgique–Luxembourg
Denmark	0.16	0.19	0.22	0.24	+ 2.2	Danemark
France	0.63	0.70	0.83	0.97	+ 2.2	France
Germany, Fed.Rep. of	1.37	1.59	1.81	2.06	+ 2.1	Allemagne, Rép. féd. d'
Ireland	0.06	0.07	0.08	0.09	+ 1.9	Irlande
Italy	0.31	0.40	0.45	0.52	+ 2.6	Italie
Netherlands	0.44	0.48	0.54	0.61	+ 1.6	Pays–Bas
United Kingdom	1.41	1.59	1.76	1.94	+ 1.6	Royaume–Uni
EEC(9)	4.59	5.25	5.95	6.73	+ 1.9	CEE(9)
Austria	0.15	0.19	0.22	0.26	+ 2.8	Autriche
Switzerland	0.20	0.24	0.27	0.30	+ 2.1	Suisse
Central Europe	0.34	0.43	0.49	0.56	+ 2.5	EUROPE CENTRALE
Cyprus	–	–	–	–	–	Chypre
Greece	0.04	0.07	0.08	0.10	+ 4.7	Grèce
Israel	0.04	0.06	0.07	0.08	+ 3.6	Israël
Malta	–	–	–	–	–	Malte
Portugal	0.04	0.05	0.05	0.06	+ 2.1	Portugal
Spain	0.18	0.26	0.29	0.33	+ 3.1	Espagne
Turkey	0.14	0.24	0.31	0.41	+ 5.5	Turquie
Yugoslavia	0.05	0.03	0.04	0.05	+ –	Yougoslavie
SOUTHERN EUROPE	0.50	0.71	0.84	1.03	+ 3.7	EUROPE MERIDIONALE
Bulgaria	0.04	0.06	0.07	0.09	+ 4.5	Bulgarie
Czechoslovakia	0.07	0.08	0.10	0.11	+ 2.3	Tchécoslovaquie
German Dem. Rep.	0.14	0.18	0.21	0.24	+ 2.7	Rép. dém. allemande
Hungary	0.07	0.08	0.09	0.11	+ 2.3	Hongrie
Poland	0.13	0.16	0.20	0.24	+ 3.1	Pologne
Romania	0.06	0.11	0.14	0.19	+ 5.9	Roumanie
EASTERN EUROPE	0.51	0.67	0.81	0.98	+ 3.3	EUROPE ORIENTALE
EUROPE	6.47	7.74	8.88	10.23	+ 2.3	EUROPE

Annex table 12.4
Projections by GDP elasticities model:
printing and writing paper

(base scenario)

Tableau annexe 12.4
Projections par le modèle "élasticités
par rapport au PIB":
papiers impression et écriture
(scénario de base)

Country	1979–1981	1990	1995	2000	Av. annual change, 1980–2000 Taux de change- ment annuel moyen 1980–2000	Pays
	(1 000 000 m3)				(%)	
Finland	0.31	0.50	0.65	0.84	+ 5.1	Finlande
Iceland	–	–	0.01	0.01	+ 3.6	Islande
Norway	0.16	0.21	0.28	0.38	+ 4.4	Norvège
Sweden	0.50	0.76	0.97	1.23	+ 4.6	Suède
NORDIC COUNTRIES	0.98	1.47	1.91	2.46	+ 4.7	PAYS NORDIQUES
Belgium–Luxembourg	0.49	0.67	0.84	1.08	+ 4.1	Belgique–Luxembourg
Denmark	0.19	0.27	0.35	0.44	+ 4.3	Danemark
France	2.06	3.05	3.99	5.22	+ 4.8	France
Germany, Fed.Rep. of	3.24	4.70	5.89	7.43	+ 4.1	Allemagne, Rép. féd. d'
Ireland	0.04	0.05	0.06	0.08	+ 3.6	Irlande
Italy	1.62	2.13	2.65	3.29	+ 3.6	Italie
Netherlands	0.67	0.84	1.05	1.31	+ 3.4	Pays–Bas
United Kingdom	1.73	2.38	2.84	3.38	+ 3.4	Royaume–Uni
EEC(9)	10.05	14.09	17.67	22.23	+ 4.1	CEE(9)
Austria	0.10	0.15	0.20	0.26	+ 4.9	Autriche
Switzerland	0.37	0.46	0.57	0.70	+ 3.3	Suisse
CENTRAL EUROPE	0.47	0.61	0.77	0.96	+ 3.6	EUROPE CENTRALE
Cyprus	–	0.01	0.01	0.01	+ 5.6	Chypre
Greece	0.11	0.13	0.17	0.22	+ 3.5	Grèce
Israel	0.05	0.10	0.13	0.17	+ 6.3	Israël
Malta	–	–	–	0.01	+ 7.2	Malte
Portugal	0.08	0.13	0.16	0.21	+ 5.0	Portugal
Spain	0.74	0.95	1.18	1.47	+ 3.5	Espagne
Turkey	0.07	0.15	0.20	0.28	+ 7.2	Turquie
Yugoslavia	0.24	0.40	0.54	0.72	+ 5.7	Yougoslavie
SOUTHERN EUROPE	1.31	1.87	2.39	3.09	+ 4.4	EUROPE MERIDIONALE
Bulgaria	0.05	0.10	0.15	0.21	+ 7.5	Bulgarie
Czechoslovakia	0.15	0.21	0.26	0.32	+ 3.9	Tchécoslovaquie
German Dem. Rep.	0.20	0.32	0.40	0.50	+ 4.7	Rép. dém. allemande
Hungary	0.14	0.20	0.26	0.33	+ 4.4	Hongrie
Poland	0.22	0.30	0.39	0.50	+ 4.2	Pologne
Romania	0.09	0.18	0.26	0.37	+ 7.3	Roumanie
EASTERN EUROPE	0.85	1.31	1.72	2.23	+ 4.9	EUROPE ORIENTALE
EUROPE	13.66	19.35	24.46	30.97	+ 4.2	EUROPE

Annex table 12.5
Projections by GDP elasticities model:
other paper and paperboard

(base scenario)

Tableau annexe 12.5
Projections pour le modèle "élasticités
par rapport au PIB":
autres papiers et cartons
(scénario de base)

Country	1979– 1981	1990	1995	2000	Av. annual change, 1980–2000 Taux de change-ment annuel moyen 1980–2000	Pays
	(1 000 000 m3)				(%)	
Finland	0.68	0.73	0.78	0.84	+ 1.0	Finlande
Iceland	0.01	0.01	0.01	0.01	–	Islande
Norway	0.26	0.27	0.29	0.32	+ 1.0	Norvège
Sweden	0.87	0.91	0.96	1.02	+ 0.8	Suède
NORDIC COUNTRIES	1.84	1.92	2.04	2.19	+ 0.9	PAYS NORDIQUES
Belgium–Luxembourg	0.70	0.70	0.74	0.78	+ 0.5	Belgique–Luxembourg
Denmark	0.40	0.44	0.46	0.48	+ 0.9	Danemark
France	3.51	3.74	4.04	4.36	+ 1.1	France
Germany, Fed.Rep. of	5.00	5.14	5.37	5.62	+ 0.6	Allemagne, Rép. féd. d'
Ireland	0.18	0.19	0.22	0.27	+ 2.0	Irlande
Italy	3.27	3.62	4.28	5.06	+ 2.2	Italie
Netherlands	1.05	1.12	1.16	1.21	+ 0.7	Pays–Bas
United Kingdom	3.99	4.17	4.56	4.97	+ 1.1	Royaume–Uni
EEC(9)	18.11	19.12	20.83	22.75	+ 1.1	CEE(9)
Austria	0.66	0.73	0.79	0.85	+ 1.3	Autriche
Switzerland	0.46	0.45	0.46	0.47	+ 0.1	Suisse
CENTRAL EUROPE	1.12	1.18	1.25	1.32	+ 0.8	EUROPE CENTRALE
Cyprus	0.04	0.05	0.06	0.07	+ 2.8	Chypre
Greece	0.28	0.35	0.43	0.52	+ 3.2	Grèce
Israel	0.15	0.21	0.26	0.33	+ 4.0	Israël
Malta	0.02	0.02	0.03	0.04	+ 3.6	Malte
Portugal	0.28	0.36	0.44	0.55	+ 3.4	Portugal
Spain	1.74	2.14	2.52	2.99	+ 2.7	Espagne
Turkey	0.34	0.43	0.57	0.76	+ 4.1	Turquie
Yugoslavia	0.78	1.05	1.37	1.80	+ 4.3	Yougoslavie
SOUTHERN EUROPE	3.63	4.61	5.68	7.06	+ 3.4	EUROPE MERIDIONALE
Bulgaria	0.42	0.65	0.92	1.31	+ 5.9	Bulgarie
Czechoslovakia	0.85	0.96	1.13	1.34	+ 2.3	Tchécoslovaquie
German Dem. Rep.	1.04	1.54	1.80	2.12	+ 3.6	Rép. dém. allemande
Hungary	0.45	0.60	0.75	0.92	+ 3.7	Hongrie
Poland	1.03	1.17	1.39	1.68	+ 2.5	Pologne
Romania	0.55	1.13	1.56	2.21	+ 7.2	Roumanie
EASTERN EUROPE	4.35	6.05	7.55	9.58	+ 4.0	EUROPE ORIENTALE
EUROPE	29 05	32.88	37.35	42.90	+ 2.0	EUROPE

Annex table 12.6
Projections by GDP elasticities model:
total paper and paperboard

(base scenario)

Tableau annexe 12.6
Projections par le modèle "élasticités
par rapport au PIB":
papiers et cartons (total)
(scénario de base)

Country	1979–1981	1990	1995	2000	Av. annual change, 1980–2000 / Taux de change-ment annuel moyen 1980–2000	Pays
	(1 000 000 m3)				(%)	
Finland	1.14	1.43	1.66	1.95	+ 2.8	Finlande
Iceland	0.02	0.02	0.02	0.03	+ 1.6	Islande
Norway	0.52	0.62	0.74	0.90	+ 2.8	Norvège
Sweden	1.66	2.00	2.32	2.70	+ 2.5	Suède
NORDIC COUNTRIES	3.34	4.07	4.74	5.58	+ 2.6	PAYS NORDIQUES
Belgium–Luxembourg	1.39	1.60	1.84	2.15	+ 2.2	Belgique–Luxembourg
Denmark	0.76	0.90	1.02	1.17	+ 2.2	Danemark
France	6.20	7.50	8.86	10.55	+ 2.7	France
Germany, Fed.Rep. of	9.62	11.42	13.07	15.12	+ 2.3	Allemagne, Rép. féd. d'
Ireland	0.27	0.30	0.36	0.43	+ 2.4	Irlande
Italy	5.19	6.15	7.38	8.87	+ 2.7	Italie
Netherlands	2.17	2.43	2.75	3.13	+ 1.9	Pays–Bas
United Kingdom	7.14	8.15	9.15	10.29	+ 1.9	Royaume–Uni
EEC(9)	32.75	38.45	44.43	51.71	+ 2.3	CEE(9)
Austria	0.92	1.08	1.21	1.37	+ 2.0	Autriche
Switzerland	1.02	1.16	1.30	1.47	+ 1.8	Suisse
CENTRAL EUROPE	1.94	2.24	2.51	2.84	+ 1.9	EUROPE CENTRALE
Cyprus	0.05	0.06	0.07	0.08	+ 2.4	Chypre
Greece	0.44	0.55	0.68	0.84	+ 3.3	Grèce
Israel	0.24	0.37	0.46	0.58	+ 4.5	Israël
Malta	0.02	0.03	0.04	0.04	+ 3.6	Malte
Portugal	0.39	0.53	0.65	0.82	+ 3.8	Portugal
Spain	2.67	3.34	4.00	4.79	+ 3.0	Espagne
Turkey	0.55	0.82	1.08	1.44	+ 4.9	Turquie
Yugoslavia	1.07	1.48	1.95	2.57	+ 4.5	Yougoslavie
SOUTHERN EUROPE	5.44	7.18	8.93	11.16	+ 3.7	EUROPE MERIDIONALE
Bulgaria	0.52	0.81	1.14	1.60	+ 5.8	Bulgarie
Czechoslovakia	1.07	1.25	1.48	1.77	+ 2.5	Tchécoslovaquie
German Dem. Rep.	1.38	2.04	2.42	2.86	+ 3.7	Rép. dém. allemande
Hungary	0.66	0.89	1.10	1.36	+ 3.7	Hongrie
Poland	1.37	1.63	1.98	2.42	+ 2.9	Pologne
Romania	0.70	1.43	1.96	2.77	+ 7.1	Roumanie
EASTERN EUROPE	5.71	8.05	10.08	12.78	+ 4.1	EUROPE ORIENTALE
EUROPE	49.18	59.99	70.69	84.07	+ 2.7	EUROPE

Annex table 12.7/Tableau annexe 12.7

Projections by the end-use elasticities model, for sawn softwood
Projections du modèle "élasticités par rapport aux utilisations finales", pour les sciages résineux

Country/Pays	Actual/Réel 1979-83	Projections Low/Bas (1000 m3)			Projections High/Haut (1000 m3)			Average annual growth Croissance annuelle moyenne (%)	
		1990	1995	2000	1990	1995	2000	Low/Bas	High/Haut
Belgium-Luxembourg Belgique-Luxembourg	1189	1220	1263	1307	1279	1414	1560	+ 0.5	+ 1.4
Denmark/Danemark	1454	1515	1576	1637	1594	1728	1869	+ 0.6	+ 1.3
France	7596	7890	8293	8689	8337	9408	10431	+ 0.7	+ 1.7
Germany, Fed. Rep. of Allemagne, Rép. féd. d'	11393	11790	12269	12741	12419	13692	14921	+ 0.6	+ 1.4
Ireland/Irlande	526	548	573	598	597	687	793	+ 0.7	+ 2.2
Netherlands/Pays-Bas	2167	2208	2284	2360	2287	2530	2763	+ 0.5	+ 1.3
Switzerland/Suisse	1920	1962	2015	2067	2059	2248	2431	+ 0.4	+ 1.3
United Kingdom/Royaume-Uni	7482	7662	7859	8052	8088	8671	9238	+ 0.4	+ 1.1
Group I (West-Central) Groupe I (Ouest-Centre)	33728	34793	36133	37451	36660	40379	44007	+ 0.6	+ 1.4
Greece/Grèce	667	720	795	868	777	932	1127	+ 1.4	+ 2.8
Italy/Italie	5054	5369	5727	6077	5707	6475	7213	+ 1.0	+ 1.9
Portugal	1017	1171	1319	1461	1296	1615	1914	+ 1.9	+ 3.4
Spain/Espagne	2218	2381	2544	2704	2564	2950	3319	+ 1.0	+ 2.1
Turkey/Turquie	3567	4037	4518	4982	4678	5745	7115	+ 1.8	+ 3.7
Yugoslavia/Yougoslavie	2408	2655	3005	3342	2891	3496	4250	+ 1.7	+ 3.0
Group II (South) Groupe II (Sud)	14930	16333	17907	19433	17912	21213	24939	+ 1.4	+ 2.7
Austria/Autriche	2491	2557	2628	2696	2664	2845	3006	+ 0.4	+ 1.0
Finland/Finlande	2945	3077	3194	3305	3195	3439	3656	+ 0.6	+ 1.1
Norway/Norvège	2390	2406	2429	2452	2506	2578	2646	+ 0.1	+ 0.5
Sweden/Suède	4246	4337	4425	4510	4501	4771	5015	+ 0.3	+ 0.9
Group III (Exporters) Groupe III (Exportateurs)	12071	12377	12677	12963	12865	13633	14323	+ 0.4	+ 0.9
Total western Europe Total Europe occidentale	60729	63503	66718	69846	67437	75225	83269	+ 0.7	+ 1.7

Annex table 12.8/Tableau annexe 12.8
Projections by the end-use elasticities model, for sawn hardwood
Projections du modèle "élasticités par rapport aux utilisations finales", pour les sciages feuillus

Country/Pays	Actual/Réel	Projections						Average annual growth Croissance annuelle moyenne	
	1979-83	Low/Bas			High/Haut			Low/Bas	High/Haut
		1990	1995	2000	1990	1995	2000	(%)	
		(1000 m3)			(1000 m3)				
Belgium-Luxembourg Belgique-Luxembourg	632	697	792	890	737	890	1106	+ 1.8	+ 3.0
Denmark/Danemark	386	441	499	557	464	544	638	+ 2.0	+ 2.7
France	3643	4171	4920	5686	4421	5666	6947	+ 2.4	+ 3.5
Germany, Fed. Rep. of Allemagne, Rép. féd. d'	2275	2541	2871	3205	2708	3293	3893	+ 1.8	+ 2.9
Ireland/Irlande	76	88	103	118	98	131	176	+ 2.4	+ 4.5
Netherlands/Pays-Bas	757	811	914	1019	848	1064	1287	+ 1.6	+ 2.8
Switzerland/Suisse	193	208	229	249	223	268	313	+ 1.4	+ 2.6
United Kingdom/Royaume-Uni	1256	1365	1488	1611	1476	1705	1938	+ 1.3	+ 2.3
Group I (West-Central) Groupe I (Ouest-Centre)	9217	10323	11816	13335	10975	13571	16296	+ 2.0	+ 3.0
Greece/Grèce	286	335	402	467	356	459	596	+ 2.6	+ 3.9
Italy/Italie	2372	2664	2991	3310	2793	3297	3782	+ 1.8	+ 2.5
Portugal	236	306	371	433	331	444	550	+ 3.2	+ 4.6
Spain/Espagne	1080	1254	1425	1592	1330	1614	1886	+ 2.1	+ 3.0
Turkey/Turquie	964	1273	1581	1874	1408	1876	2514	+ 3.6	+ 5.2
Yugoslavia/Yougoslavie	1243	1556	1986	2394	1647	2173	2867	+ 3.5	+ 4.5
Group II (South) Groupe II (Sud)	6181	7387	8756	10069	7864	9862	12195	+ 2.6	+ 3.6
Austria/Autriche	395	468	553	639	518	676	836	+ 2.6	+ 4.0
Finland/Finlande	67	89	110	132	96	131	167	+ 3.6	+ 4.9
Norway/Norvège	55	62	73	83	69	79	89	+ 2.2	+ 2.6
Sweden/Suède	245	286	329	371	312	396	482	+ 2.2	+ 3.6
Group III (Exporters) Groupe III (Exportateurs)	762	906	1064	1226	995	1281	1574	+ 2.5	+ 3.9
Total western Europe Total Europe occidentale	16160	18615	21636	24630	19834	24714	30065	+ 2.2	+ 3.3

Annex table 12.9/Tableau annexe 12.9

Projections by the end-use elasticities model, for particle board

Projections du modèle "élasticités par rapport aux utilisations finales", pour les panneaux de particules

Country/Pays	Actual/Réel	Projections						Average annual growth	
		Low/Bas			High/Haut			Croissance annuelle moyenne	
	1979-83	1990	1995	2000	1990	1995	2000	Low/Bas	High/Haut
		(1000 m3)			(1000 m3)			(%)	
Belgium-Luxembourg / Belgique-Luxembourg	672	715	778	842	764	908	1080	+ 1.2	+ 2.5
Denmark/Danemark	435	489	545	602	517	600	696	+ 1.7	+ 2.5
France	2182	2376	2651	2929	2564	3160	3767	+ 1.6	+ 2.9
Germany, Fed. Rep. of / Allemagne, Rép. féd. d'	6119	6731	7486	8249	7190	8626	10092	+ 1.6	+ 2.7
Ireland/Irlande	127	147	170	193	164	216	286	+ 2.2	+ 4.3
Netherlands/Pays-Bas	500	528	582	636	554	677	804	+ 1.3	+ 2.5
Switzerland/Suisse	449	480	520	559	514	608	704	+ 1.2	+ 2.4
United Kingdom/Royaume-Uni	1933	2088	2261	2434	2255	2587	2923	+ 1.2	+ 2.2
Group I (West-Central) / Groupe I (Ouest-Centre)	12416	13553	14991	16444	14521	17381	20350	+ 1.5	+ 2.6
Greece/Grèce	283	334	401	463	348	440	556	+ 2.6	+ 3.6
Italy/Italie	2009	2257	2525	2779	2326	2695	3037	+ 1.7	+ 2.2
Portugal	276	347	411	469	368	472	565	+ 2.8	+ 3.8
Spain/Espagne	1051	1185	1314	1436	1243	1453	1648	+ 1.7	+ 2.4
Turkey/Turquie	350	437	520	596	475	598	758	+ 2.8	+ 4.1
Yugoslavia/Yougoslavie	790	943	1145	1329	988	1234	1541	+ 2.8	+ 3.6
Group II (South) / Groupe II (Sud)	4759	5504	6317	7072	5748	6892	8104	+ 2.1	+ 2.8
Austria/Autriche	601	683	774	864	739	904	1063	+ 1.9	+ 3.0
Finland/Finlande	422	466	507	548	512	611	708	+ 1.4	+ 2.8
Norway/Norvège	394	413	438	464	452	491	529	+ 0.9	+ 1.6
Sweden/Suède	669	731	792	851	789	931	1069	+ 1.3	+ 2.5
Group III (Exporters) / Groupe III (Exportateurs)	2086	2292	2511	2727	2493	2936	3369	+ 1.4	+ 2.6
Total western Europe / Total Europe occidentale	19261	21348	23818	26242	22761	27209	31823	+ 1.6	+ 2.7

Annex table 12.10/Tableau annexe 12.10
Projections by the end-use elasticities model, for plywood
Projections du modèle "élasticités par rapport aux utilisations finales", pour les contreplaqués

Country/Pays	Actual/Réel 1979-83	Projections						Average annual growth Croissance annuelle moyenne (%)	
		Low/Bas (1000 m3)			High/Haut (1000 m3)			Low/Bas	High/Haut
		1990	1995	2000	1990	1995	2000		
Belgium-Luxembourg Belgique-Luxembourg	156	165	177	190	177	211	250	+ 1.1	+ 2.5
Denmark/Danemark	117	127	137	147	138	158	180	+ 1.2	+ 2.3
France	635	687	761	835	747	922	1101	+ 1.5	+ 2.9
Germany, Fed. Rep. of Allemagne, Rép. féd. d'	796	850	916	983	922	1089	1261	+ 1.1	+ 2.5
Ireland/Irlande	36	40	45	50	46	59	77	+ 1.7	+ 4.0
Netherlands/Pays-Bas	459	462	468	474	494	559	624	+ 0.2	+ 1.6
Switzerland/Suisse	92	94	98	101	102	117	133	+ 0.5	+ 2.0
United Kingdom/Royaume-Uni	972	1031	1097	1163	1123	1276	1433	+ 0.9	+ 2.1
Group I (West-Central) Groupe I (Ouest-Centre)	3264	3456	3699	3944	3748	4391	5059	+ 1.0	+ 2.3
Greece/Grèce	44	54	69	84	60	84	121	+ 3.4	+ 5.4
Italy/Italie	414	483	565	649	520	657	798	+ 2.4	+ 3.5
Portugal	15	21	26	32	24	35	47	+ 4.1	+ 6.1
Spain/Espagne	94	111	130	148	123	159	196	+ 2.4	+ 3.9
Turkey/Turquie	40	54	70	86	65	94	138	+ 4.1	+ 6.8
Yugoslavia/Yougoslavie	116	144	185	228	161	225	317	+ 3.6	+ 5.4
Group II (South) Groupe II (Sud)	723	867	1045	1227	952	1254	1616	+ 2.8	+ 4.3
Austria/Autriche	18	20	21	23	21	24	27	+ 1.1	+ 2.1
Finland/Finlande	102	111	119	127	120	138	155	+ 1.2	+ 2.2
Norway/Norvège	73	76	81	85	82	88	93	+ 0.8	+ 1.3
Sweden/Suède	153	159	166	172	171	192	211	+ 0.6	+ 1.7
Group III (Exporters) Groupe III (Exportateurs)	346	366	387	407	394	441	486	+ 0.9	+ 1.8
Total western Europe Total Europe occidentale	4333	4690	5131	5579	5094	6086	7161	+ 1.3	+ 2.7

Annex table 12.11/Tableau annexe 12.11

Projections by the end-use elasticities model, for fibreboard

Projections du modèle 'élasticités par rapport aux utilisations finales', pour les panneaux de fibres

Country/Pays	Actual/Réel 1979-83	Low/Bas 1990	Low/Bas 1995	Low/Bas 2000	High/Haut 1990	High/Haut 1995	High/Haut 2000	Growth Low/Bas (%)	Growth High/Haut (%)
		(1000 m3)			(1000 m3)				
Belgium-Luxembourg / Belgique-Luxembourg	31	32	34	36	34	39	45	+ 0.8	+ 2.0
Denmark/Danemark	51	54	57	60	57	64	71	+ 0.9	+ 1.8
France	215	229	247	266	245	289	333	+ 1.1	+ 2.3
Germany, Fed. Rep. of / Allemagne, Rép. féd. d'	472	495	523	551	530	604	677	+ 0.8	+ 1.9
Ireland, Irlande	10	11	12	13	12	15	18	+ 1.1	+ 3.1
Netherlands/Pays-Bas	143	147	155	164	154	178	201	+ 0.7	+ 1.8
Switzerland/Suisse	77	79	83	86	85	96	106	+ 0.6	+ 1.7
United Kingdom/Royaume-Uni	374	389	404	420	417	459	501	+ 0.6	+ 1.5
Group I (West-Central) / Groupe I (Ouest-Centre)	1372	1435	1515	1595	1533	1743	1953	+ 0.8	+ 1.9
Greece/Grèce	26	29	33	37	32	40	51	+ 1.8	+ 3.6
Italy/Italie	261	283	309	335	305	359	413	+ 1.3	+ 2.4
Portugal	50	61	72	83	69	92	115	+ 2.7	+ 4.5
Spain/Espagne	225	250	275	300	273	328	383	+ 1.5	+ 2.9
Turkey/Turquie	70	84	99	113	100	131	174	+ 2.6	+ 4.9
Yugoslavia/Yougoslavie	64	74	88	103	82	105	136	+ 2.5	+ 4.1
Group II (South) / Groupe II (Sud)	695	781	876	971	860	1056	1272	+ 1.8	+ 3.2
Austria/Autriche	54	58	62	67	63	75	85	+ 1.2	+ 2.5
Finland/Finlande	152	171	189	206	186	223	259	+ 1.6	+ 2.8
Norway/Norvège	162	165	170	174	182	195	208	+ 0.4	+ 1.3
Sweden/Suède	300	318	337	355	346	400	453	+ 0.9	+ 2.2
Group III (Exporters) / Groupe III (Exportateurs)	667	712	757	802	778	893	1005	+ 1.0	+ 2.2
Total western Europe / Total Europe occidentale	2735	2928	3149	3368	3171	3691	4229	+ 1.1	+ 2.3

Annex table 12.12/Tableau annexe 12.12

Projections by the end-use elasticities model, for wood-based panels (projected as a group) a/

Projections du modèle 'élasticités par rapport aux utilisations finales', pour les panneaux dérivés du bois (projection par groupe) a/

Country/Pays	Actual/Réel	Projections						Average annual growth	
		Low/Bas			High/Haut			Croissance annuelle moyenne	
	1979-83	1990	1995	2000	1990	1995	2000	Low/Bas	High/Haut
		(1000 m3)			(1000 m3)			(%)	
Belgium-Luxembourg/Belgique-Luxembourg	858	906	977	1047	964	1128	1319	1.1	2.3
Denmark/Danemark	603	663	725	788	702	801	912	1.4	2.2
France	3032	3269	3600	3933	3509	4239	4972	1.4	2.6
Germany, Fed. Rep. of / Allemagne, Rép. féd. d'	7386	8015	8785	9555	8532	10042	11562	1.4	2.4
Ireland/Irlande	174	196	221	247	217	277	355	1.9	3.8
Netherlands/Pays-Bas	1102	1136	1198	1261	1194	1387	1580	0.7	1.9
Switzerland/Suisse	617	651	694	737	694	804	915	0.9	2.1
United Kingdom/Royaume-Uni	3280	3488	3720	3950	3749	4226	4704	1.0	1.9
Group I (West-Central) / Groupe I (Ouest-Centre)	17052	18323	19919	21518	19561	22903	26319	1.2	2.3
Greece/Grèce	353	401	463	518	416	501	606	2.0	2.9
Italy/Italie	2683	2937	3207	3459	3018	3398	3744	1.3	1.8
Protugal	341	409	467	520	430	526	609	2.2	3.1
Spain/Espagne	1370	1503	1629	1747	1567	1777	1967	1.3	1.9
Turkey/Turquie	460	545	624	696	587	707	856	2.2	3.3
Yugoslavia/Yougoslavie	970	1112	1297	1461	1160	1388	1661	2.2	2.9
Group II (South) / Groupe II (Sud)	6177	6906	7687	8400	7177	8296	9443	1.6	2.3
Austria/Autriche	673	770	879	988	841	1045	1246	2.0	3.3
Finland/Finlande	676	759	837	916	841	1024	1205	1.6	3.1
Norway/Norvège	630	658	698	738	728	796	862	0.8	1.7
Sweden/Suède	1121	1220	1318	1415	1334	1586	1836	1.2	2.6
Group III (Exporters) / Groupe III (Exportateurs)	3100	3406	3732	4057	3744	4450	5150	1.4	2.7
Total estern Europe / Total Europe occidentale	26328	28635	31338	33976	30482	35650	40911	1.4	2.3

a/ See text for explanation/Pour explication, voir texte.

Annex table 12.13
ETTS IV consumption scenarios:
sawnwood (total)

Tableau annexe 12.13
Scénarios ETTS IV pour la
consommation de sciages (total)

| Country | 1979–1981 | Low/Bas | | High/Haut | | Av. annual change, 1980–2000 / Taux de changement annuel moyen 1980–2000 | | Pays |
		1990	2000	1990	2000	Low Bas	High Haut	
		(1 000 000 m3)				(%)		
Finland	3.10	3.19	3.45	3.31	3.85	+ 0.5	+ 1.1	Finlande
Iceland	0.07	0.07	0.08	0.07	0.08	+ 0.7	+ 0.7	Islande
Norway	2.47	2.47	2.53	2.58	2.74	+ 0.1	+ 0.5	Norvège
Sweden	5.11	4.66	4.91	4.84	5.52	– 0.2	+ 0.4	Suède
NORDIC COUNTRIES	10.75	10.39	10.97	10.80	12.19	+ 0.1	+ 0.6	PAYS NORDIQUES
Belgium–Luxembourg	2.05	2.05	2.33	2.15	2.80	+ 0.6	+ 1.6	Belgique–Luxembourg
Denmark	1.92	1.97	2.22	2.07	2.53	+ 0.7	+ 1.4	Danemark
France	11.75	12.23	14.55	12.93	17.55	+ 1.1	+ 2.0	France
Germany, Fed.Rep. of	14.44	14.51	16.12	15.31	18.99	+ 0.5	+ 1.4	Allemagne, Rép.féd. d'
Ireland	0.64	0.64	0.72	0.70	0.97	+ 0.6	+ 2.1	Irlande
Italy	7.95	8.14	9.50	8.61	11.10	+ 0.9	+ 1.7	Italie
Netherlands	3.15	3.12	3.48	3.24	4.15	+ 0.5	+ 1.4	Pays–Bas
United Kingdom	8.55	9.04	9.68	9.59	11.20	+ 0.6	+ 1.4	Royaume–Uni
EEC(9)	50.46	51.70	58.60	54.60	69.29	+ 0.8	+ 1.6	CEE(9)
Austria	2.92	3.03	3.34	3.18	3.85	+ 0.7	+ 1.4	Autriche
Switzerland	2.12	2.23	2.38	2.34	2.80	+ 0.6	+ 1.4	Suisse
CENTRAL EUROPE	5.04	5.26	5.72	5.52	6.65	+ 0.7	+ 1.4	EUROPE CENTRALE
Cyprus	0.13	0.12	0.14	0.12	0.14	+ 0.4	+ 0.4	Chypre
Greece	1.05	1.05	1.34	1.14	1.73	+ 1.2	+ 2.5	Grèce
Israel	0.26	0.30	0.34	0.30	0.34	+ 1.3	+ 1.3	Israël
Malta	0.02	0.03	0.04	0.03	0.04	+ 3.5	+ 3.5	Malte
Portugal	1.25	1.55	1.96	1.70	2.53	+ 2.3	+ 3.6	Portugal
Spain	3.44	3.68	4.34	3.94	5.26	+ 1.2	+ 2.1	Espagne
Turkey	4.64	5.36	6.90	6.14	9.67	+ 2.0	+ 3.7	Turquie
Yugoslavia	3.52	4.21	5.73	4.54	7.12	+ 2.5	+ 3.6	Yougoslavie
SOUTHERN EUROPE	14.32	16.30	20.79	17.91	26.83	+ 1.9	+ 3.2	EUROPE MERIDIONALE
Bulgaria	1.67	1.67	1.67	1.79	1.93	–	+ 0.8	Bulgarie
Czechoslovakia	3.87	3.87	3.87	4.16	4.47	–	+ 0.8	Tchécoslovaquie
German Dem. Rep.	3.72	3.72	3.72	4.00	4.30	–	+ 0.8	Rép. dém. allemande
Hungary	2.06	2.07	2.07	2.23	2.40	–	+ 0.8	Hongrie
Poland	6.77	7.98	7.98	7.98	8.56	+ 0.8	+ 1.2	Pologne
Romania	3.65	3.64	3.64	3.92	4.22	–	+ 0.8	Roumanie
EASTERN EUROPE	21.75	22.95	22.95	24.08	25.88	+ 0.3	+ 0.9	EUROPE ORIENTALE
EUROPE	102.32	106.60	119.03	112.91	140.84	+ 0.8	+ 1.6	EUROPE

Annex table 12.14
ETTS IV consumption scenarios:
sawn softwood

Tableau annexe 12.14
Scénarios ETTS IV pour la
consommation des sciages résineux

Country	1979–1981	1990	2000	1990	2000	Av. annual change, 1980–2000 Taux de change- ment annuel moyen 1980–2000		Pays
		Low/Bas		High/Haut		Low Bas	High Haut	
	(1 000 000 m3)					(%)		
Finland	3.02	3.08	3.30	3.19	3.66	+ 0.4	+ 1.0	Finlande
Iceland	0.07	0.07	0.08	0.07	0.08	+ 0.7	+ 0.7	Islande
Norway	2.40	2.41	2.45	2.51	2.65	+ 0.1	+ 0.5	Norvège
Sweden	4.83	4.34	4.51	4.50	5.01	− 0.3	+ 0.2	Suède
NORDIC COUNTRIES	10.32	9.90	10.34	10.27	11.40	−	+ 0.5	PAYS NORDIQUES
Belgium–Luxembourg	1.24	1.22	1.31	1.28	1.56	+ 0.3	+ 1.1	Belgique–Luxembourg
Denmark	1.52	1.51	1.64	1.59	1.87	+ 0.4	+ 1.0	Danemark
France	7.77	7.89	8.69	8.34	10.43	+ 0.5	+ 1.5	France
Germany, Fed.Rep. of	11.79	11.79	12.74	12.42	14.92	+ 0.4	+ 1.2	Allemagne, Rép.féd. d'
Ireland	0.56	0.55	0.60	0.60	0.79	+ 0.3	+ 1.7	Irlande
Italy	5.29	5.37	6.08	5.71	7.21	+ 0.7	+ 1.6	Italie
Netherlands	2.26	2.21	2.36	2.29	2.76	+ 0.2	+ 1.0	Pays–Bas
United Kingdom	7.23	7.66	8.05	8.09	9.24	+ 0.6	+ 1.2	Royaume–Uni
EEC(9)	37.66	38.20	41.47	40.32	48.78	+ 0.5	+ 1.3	CEE(9)
Austria	2.47	2.56	2.70	2.66	3.01	+ 0.4	+ 1.0	Autriche
Switzerland	1.87	1.96	2.07	2.06	2.43	+ 0.5	+ 1.3	Suisse
CENTRAL EUROPE	4.34	4.52	4.77	4.72	5.44	+ 0.5	+ 1.1	EUROPE CENTRALE
Cyprus	0.12	0.11	0.13	0.11	0.13	+ 0.4	+ 0.4	Chypre
Greece	0.76	0.72	0.87	0.78	1.13	+ 0.7	+ 2.0	Grèce
Israel	0.23	0.27	0.31	0.27	0.31	+ 1.5	+ 1.5	Israël
Malta	0.02	0.02	0.03	0.02	0.03	+ 2.0	+ 2.0	Malte
Portugal	1.00	1.17	1.46	1.30	1.91	+ 1.9	+ 3.3	Portugal
Spain	2.26	2.38	2.70	2.56	3.32	+ 0.9	+ 1.9	Espagne
Turkey	3.63	4.04	4.98	4.68	7.11	+ 1.6	+ 3.4	Turquie
Yugoslavia	2.39	2.65	3.34	2.89	4.25	+ 1.7	+ 2.9	Yougoslavie
SOUTHERN EUROPE	10.42	11.36	13.82	12.61	18.19	+ 1.4	+ 2.8	EUROPE MERIDIONALE
Bulgaria	1.19	1.19	1.19	1.28	1.38	−	+ 0.8	Bulgarie
Czechoslovakia	3.06	3.06	3.06	3.30	3.55	−	+ 0.8	Tchécoslovaquie
German Dem. Rep.	3.10	3.10	3.10	3.34	3.60	−	+ 0.8	Rép. dém. allemande
Hungary	1.30	1.30	1.30	1.40	1.51	−	+ 0.8	Hongrie
Poland	5.38	6.40	6.40	6.40	6.89	+ 0.9	+ 1.3	Pologne
Romania	1.49	1.49	1.49	1.61	1.73	−	+ 0.8	Roumanie
EASTERN EUROPE	15.52	16.54	16.54	17.33	18.66	+ 0.3	+ 0.9	EUROPE ORIENTALE
EUROPE	78.15	80.52	86.94	85.25	102.47	+ 5.6	+ 1.4	EUROPE

Annex table 12.15
ETTS IV consumption scenarios:
sawn hardwood

Tableau annexe 12.15
Scénarios ETTS IV pour la
consommation des sciages feuillus

Country	1979–1981	1990	2000	1990	2000	Av. annual change, 1980–2000 Taux de change-ment annuel moyen 1980–2000		Pays
		Low/Bas		High/Haut		Low Bas	High Haut	
	(1 000 000 m3)					(%)		
Finland	0.07	0.09	0.13	0.10	0.17	+ 3.1	+ 4.5	Finlande
Iceland	–	–	–	–	–	–	–	Islande
Norway	0.06	0.06	0.08	0.07	0.09	+ 1.4	+ 2.0	Norvège
Sweden	0.25	0.29	0.37	0.31	0.48	+ 2.0	+ 3.3	Suède
NORDIC COUNTRIES	0.38	0.44	0.58	0.48	0.74	+ 2.1	+ 3.4	PAYS NORDIQUES
Belgium–Luxembourg	0.68	0.70	0.89	0.74	1.11	+ 1.3	+ 2.5	Belgique–Luxembourg
Denmark	0.39	0.44	0.56	0.46	0.64	+ 1.8	+ 2.5	Danemark
France	3.80	4.17	5.69	4.42	6.95	+ 2.0	+ 3.1	France
Germany, Fed.Rep. of	2.47	2.54	3.20	2.71	3.89	+ 1.3	+ 2.3	Allemagne, Rép.féd. d'
Ireland	0.08	0.09	0.12	0.10	0.18	+ 2.0	+ 4.1	Irlande
Italy	2.56	2.66	3.31	2.79	3.78	+ 1.3	+ 2.0	Italie
Netherlands	0.78	0.81	1.02	0.85	1.29	+ 1.3	+ 2.5	Pays–Bas
United Kingdom	1.31	1.36	1.61	1.48	1.94	+ 1.0	+ 1.9	Royaume–Uni
EEC(9)	12.06	12.77	16.40	13.55	19.78	+ 1.5	+ 2.5	CEE(9)
Austria	0.45	0.47	0.64	0.52	0.84	+ 1.8	+ 3.2	Autriche
Switzerland	0.19	0.21	0.25	0.22	0.31	+ 1.4	+ 2.5	Suisse
CENTRAL EUROPE	0.64	0.68	0.89	0.74	1.15	+ 1.7	+ 3.0	EUROPE CENTRALE
Cyprus	0.02	0.01	0.01	0.01	0.01	– 3.4	– 3.4	Chypre
Greece	0.29	0.33	0.47	0.36	0.60	+ 2.3	+ 3.5	Grèce
Israel	0.02	0.03	0.03	0.03	0.03	+ 2.0	+ 2.0	Israël
Malta	0.01	0.01	0.01	0.01	0.01	–	–	Malte
Portugal	0.18	0.31	0.43	0.33	0.55	+ 4.4	+ 5.7	Portugal
Spain	1.13	1.25	1.59	1.33	1.89	+ 1.7	+ 2.6	Espagne
Turkey	0.96	1.27	1.87	1.41	2.51	+ 3.4	+ 4.9	Turquie
Yugoslavia	1.14	1.56	2.39	1.65	2.87	+ 3.8	+ 4.7	Yougoslavie
SOUTHERN EUROPE	3.74	4.77	6.80	5.13	8.47	+ 3.0	+ 4.1	EUROPE MERIDIONALE
Bulgaria	0.45	0.45	0.45	0.48	0.52	–	+ 0.8	Bulgarie
Czechoslovakia	0.70	0.70	0.70	0.75	0.81	–	+ 0.8	Tchécoslovaquie
German Dem. Rep.	0.47	0.47	0.47	0.51	0.55	–	0.8	Réo. dém. allemande
Hungary	0.73	0.73	0.73	0.79	0.85	–	+ 0.8	Hongrie
Poland	1.01	1.20	1.20	1.20	1.29	+ 0.8	+ 1.2	Pologne
Romania	2.08	2.08	2.08	2.24	2.42	–	+ 0.8	Roumanie
EASTERN EUROPE	5.45	5.63	5.63	5.97	6.44	+ 0.1	+ 0.8	EUROPE ORIENTALE
EUROPE	22.37	24.29	30.30	25.87	35.58	+ 1.5	+ 2.5	EUROPE

Annex table 12.16
ETTS IV consumption scenarios:
wood-based panels

Tableau annexe 12.16
Scénarios ETTS IV pour la
consommation de panneaux dérivés du bois

Country	1979–1981	1990	2000	1990	2000	Av. annual change 1980–2000 Taux de change-ment annuel moyen 1980–2000		Pays
		Low/Bas		High/Haut		Low Bas	High Haut	
		(1 000 000 m3)				(%)		
Finland	0.70	0.75	0.90	0.83	1.13	+ 1.3	+ 2.4	Finlande
Iceland	0.03	0.03	0.05	0.03	0.05	+ 2.6	+ 2.6	Islande
Norway	0.64	0.66	0.73	0.72	0.84	+ 0.7	+ 1.4	Norvège
Sweden	1.21	1.24	1.40	1.34	1.76	+ 0.7	+ 1.9	Suède
NORDIC COUNTRIES	2.57	2.68	3.08	2.92	3.78	+ 0.9	+ 1.9	PAYS NORDIQUES
Belgium-Luxembourg	0.86	0.94	1.11	1.01	1.41	+ 1.3	+ 2.5	Belgique-Luxembourg
Denmark	0.68	0.70	0.84	0.75	0.98	+ 1.1	+ 1.8	Danemark
France	3.32	3.45	4.18	3.70	5.35	+ 1.2	+ 2.4	France
Germany, Fed.Rep. of	8.04	8.42	10.13	8.99	12.38	+ 1.2	+ 2.2	Allemagne, Rép.féd. d'
Ireland	0.18	0.20	0.26	0.22	0.38	+ 1.9	+ 3.8	Irlande
Italy	3.12	3.52	4.26	3.65	4.75	+ 1.6	+ 2.1	Italie
Netherlands·	1.23	1.19	1.32	1.24	1.67	+ 0.3	+ 1.5	Pays-Bas
United Kingdom	3.36	3.57	4.07	3.85	4.91	+ 1.0	+ 1.9	Royaume-Uni
EEC(9)	20.78	21.99	26.16	23.41	31.83	+ 1.2	+ 2.1	CEE(9)
Austria	0.70	0.78	0.96	0.84	1.19	+ 1.6	+ 2.7	Autriche
Switzerland	0.66	0.67	0.77	0.71	0.96	+ 0.8	+ 1.9	Suisse
CENTRAL EUROPE	1.35	1.45	1.73	1.55	2.15	+ 1.2	+ 2.3	EUROPE CENTRALE
Cyprus	0.03	0.04	0.07	0.04	0.07	+ 4.3	+ 4.3	Chypre
Greece	0.38	0.43	0.60	0.46	0.75	+ 2.3	+ 3.5	Grèce
Israel	0.17	0.24	0.43	0.24	0.43	+ 4.7	+ 4.7	Israël
Malta	-0.01	0.01	0.02	0.01	0.02	+ 3.5	+ 3.5	Malte
Portugal	0.33	0.43	0.58	0.46	0.72	+ 2.9	+ 4.0	Portugal
Spain	1.42	1.55	1.90	1.64	2.24	+ 1.5	+ 2.3	Espagne
Turkey	0.48	0.58	0.81	0.64	1.08	+ 2.6	+ 4.1	Turquie
Yugoslavia	1.20	1.36	1.87	1.44	2.21	+ 2.2	+ 3.1	Yougoslavie
SOUTHERN EUROPE	4.02	4.65	6.28	4.94	7.52	+ 2.3	+ 3.2	EUROPE MERIDIONALE
Bulgaria	0.49	0.71	0.92	0.75	1.02	+ 3.2	+ 3.7	Bulgarie
Czechoslovakia	1.21	1.63	2.05	1.68	2.15	+ 2.7	+ 2.9	Tchécoslovaquie
German Dem. Rep.	1.43	2.12	2.92	2.22	3.11	+ 3.6	+ 4.0	Rép. dém. allemande
Hungary	0.47	0.72	0.97	0.75	1.03	+ 3.7	+ 4.0	Hongrie
Poland	2.08	2.73	3.36	2.79	3.48	+ 2.4	+ 2.6	Pologne
Romania	1.20	1.70	2.20	1.82	2.44	+ 3.1	+ 3.6	Roumanie
EASTERN EUROPE	6.88	9.62	12.36	10.02	13.17	+ 3.0	3.3	EUROPE ORIENTALE
EUROPE	35.60	40.40	49.63	42.85	58.46	+ 1.7	+ 2.5	EUROPE

Annex table 12.17
ETTS IV consumption scenarios:
particle board

Tableau annexe 12.17
Scénarios ETTS IV pour la
consommation des panneaux de particules

Country	1979–1981	1990	2000	1990	2000	Av. annual change, 1980–2000 Taux de change-ment annuel moyen 1980–2000		Pays
		Low/Bas		High/Haut		Low Bas	High Haut	
		(1 000 000 m3)				(%)		
Finland	0.43	0.46	0.55	0.51	0.71	+ 1.2	+ 2.5	Finlande
Iceland	0.02	0.03	0.05	0.03	0.05	+ 4.7	+ 4.7	Islande
Norway	0.40	0.41	0.46	0.45	0.53	+ 0.7	+ 1.4	Norvège
Sweden	0.70	0.73	0.85	0.79	1.07	+ 1.0	+ 2.1	Suède
NORDIC COUNTRIES	1.55	1.63	1.91	1.78	2.36	+ 1.0	+ 2.1	PAYS NORDIQUES
Belgium–Luxembourg	0.63	0.71	0.84	0.76	1.08	+ 1.4	+ 2.7	Belgique–Luxembourg
Denmark	0.46	0.49	0.60	0.52	0.70	+ 1.3	+ 2.1	Danemark
France	2.25	2.38	2.93	2.56	3.77	+ 1.3	+ 2.6	France
Germany, Fed.Rep. of	6.36	6.73	8.25	7.19	10.09	+ 1.3	+ 2.3	Allemagne, Rép.féd. d'
Ireland	0.12	0.15	0.19	0.16	0.28	+ 2.3	+ 4.3	Irlande
Italy	1.89	2.26	2.78	2.33	3.04	+ 1.9	+ 2.4	Italie
Netherlands	0.54	0.53	0.64	0.55	0.80	+ 0.8	+ 2.0	Pays–Bas
United Kingdom	1.94	2.09	2.43	2.25	2.92	+ 1.1	+ 2.1	Royaume–Uni
EEC(9)	14.20	15.34	18.66	16.32	22.68	+ 1.4	+ 2.4	CEE(9)
Austria	0.60	0.68	0.86	0.74	1.06	+ 1.8	+ 2.9	Autriche
Switzerland	0.47	0.48	0.56	0.51	0.70	+ 0.9	+ 2.0	Suisse
CENTRAL EUROPE	1.07	1.16	1.42	1.25	1.76	+ 1.4	+ 2.5	EUROPE CENTRALE
Cyprus	0.01	0.02	0.03	0.02	0.03	+ 5.6	+ 5.6	Chypre
Greece	0.29	0.33	0.46	0.35	0.56	+ 2.3	+ 3.3	Grèce
Israel	0.05	0.09	0.19	0.09	0.19	+ 6.9	+ 6.9	Israël
Malta	-	-	-	-	-	-	-	Malte
Portugal	0.27	0.35	0.47	0.37	0.56	+ 2.8	+ 3.7	Portugal
Spain	1.08	1.18	1.44	1.24	1.65	+ 1.4	+ 2.1	Espagne
Turkey	0.36	0.44	0.60	0.47	0.76	+ 2.6	+ 3.8	Turquie
Yugoslavia	0.80	0.94	1.33	0.99	1.54	+ 2.6	+ 3.3	Yougoslavie
SOUTHERN EUROPE	2.86	3.35	4.52	3.53	5.29	+ 2.3	+ 3.1	EUROPE MERIDIONALE
Bulgaria	0.30	0.52	0.73	0.56	0.83	+ 4.6	+ 5.2	Bulgarie
Czechoslovakia	0.66	1.08	1.50	1.13	1.60	+ 4.2	+ 4.5	Tchécoslovaquie
German Dem. Rep.	0.93	1.64	2.36	1.74	2.55	+ 4.8	+ 5.2	Rép. dém. allemande
Hungary	0.31	0.56	0.81	0.59	0.87	+ 4.9	+ 5.3	Hongrie
Poland	1.22	1.86	2.50	1.92	2.62	+ 3.7	+ 3.9	Pologne
Romania	0.72	1.22	1.72	1.34	1.96	+ 4.4	+ 5.2	Roumanie
EASTERN EUROPE	4.14	6.88	9.62	7.28	10.43	+ 4.3	+ 4.7	EUROPE ORIENTALE
EUROPE	23.82	28.36	36.13	30.16	42.52	+ 2.1	+ 2.9	EUROPE

Annex table 12.18
ETTS IV consumption scenarios:
plywood

Tableau annexe 12.18
Scénarios ETTS IV pour la
consommation de contreplaqués

Country	Av. annual change, 1979-1981	1990	2000	1990	2000	Av. annual change, 1980-2000 Taux de changement annuel moyen 1980-2000		Pays
		Low/Bas		High/Haut		Low Bas	High Haut	
		(1 000 000 m3)				(%)		
Finland	0.10	0.11	0.13	0.12	0.15	+ 1.3	+ 2.0	Finlande
Iceland			0.04	0.04	0.05	0.03	0.02	Islande
Norway	0.07	0.08	0.09	0.08	0.09	+ 1.3	+ 1.3	Norvège
Sweden	0.16	0.16	0.17	0.17	0.21	+ 0.3	+ 1.4	Suède
NORDIC COUNTRIES	0.34	0.35	0.39	0.37	0.45	+ 0.7	+ 1.4	PAYS NORDIQUES
Belgium-Luxembourg	0.16	0.16	0.19	0.18	0.25	+ 0.9	+ 2.3	Belgique-Luxembourg
Denmark	0.13	0.13	0.15	0.14	0.18	+ 0.7	+ 1.6	Danemark
France	0.67	0.69	0.83	0.75	1.10	+ 1.1	+ 2.5	France
Germany, Fed.Rep. of	0.84	0.85	0.98	0.92	1.26	+ 0.8	+ 2.0	Allemagne, Rép.féd.d'
Ireland	0.04	0.04	0.05	0.05	0.08	+ 1.1	+ 3.5	Irlande
Italy	0.44	0.48	0.65	0.52	0.80	+ 2.0	+ 3.0	Italie
Netherlands	0.50	0.46	0.47	0.49	0.62	- 0.3	+ 1.1	Pays-Bas
United Kingdom	0.96	1.03	1.16	1.12	1.43	+ 1.0	+ 2.0	Royaume-Uni
EEC(9)	3.74	3.84	4.48	4.17	5.72	+ 0.9	+ 2.1	CEE(9)
Austria	0.02	0.02	0.02	0.02	0.03		+ 2.0	Autriche
Switzerland	0.09	0.09	0.10	0.10	0.13	+ 0.5	+ 1.9	Suisse
CENTRAL EUROPE	0.11	0.11	0.12	0.12	0.16	+ 0.4	+ 1.9	EUROPE CENTRALE
Cyprus	0.01	0.02	0.03	0.02	0.03	+ 5.6	+ 5.6	Chypre
Greece	0.05	0.05	0.08	0.06	0.12	+ 2.4	+ 4.5	Grèce
Israel	0.05	0.07	0.15	0.07	0.15	+ 5.6	+ 5.6	Israël
Malta	0.01	0.01	0.02	0.01	0.02	+ 3.5	+ 3.5	Malte
Portugal	0.01	0.02	0.03	0.02	0.05	+ 5.6	+ 8.4	Portugal
Spain	0.10	0.11	0.15	0.12	0.20	+ 2.0	+ 3.5	Espagne
Turkey	0.04	0.05	0.09	0.06	0.14	+ 4.1	+ 6.5	Turquie
Yugoslavia	0.14	0.14	0.23	0.16	0.32	+ 2.5	+ 4.2	Yougoslavie
SOUTHERN EUROPE	0.40	0.47	0.78	0.52	1.03	+ 3.4	+ 4.8	EUROPE MERIDIONALE
Bulgaria	0.05	0.05	0.05	0.05	0.05			Bulgarie
Czechoslovakia	0.27	0.27	0.27	0.27	0.27			Tchécoslovaquie
German Dem. Rep.	0.09	0.09	0.09	0.09	0.09			Rép. dém. allemande
Hungary	0.04	0.04	0.04	0.04	0.04			Hongrie
Poland	0.22	0.22	0.22	0.22	0.22			Pologne
Romania	0.16	0.16	0.16	0.16	0.16			Roumanie
EASTERN EUROPE	0.84	0.84	0.84	0.84	0.84			EUROPE ORIENTALE
EUROPE	5.44	5.61	6.61	6.02	8.20	+ 1.0	+ 2.1	EUROPE

Annex table 12.19
ETTS IV consumption scenarios:
fibreboard

Tableau annexe 12.19
Scénarios ETTS IV pour la
consommation de panneaux de fibres

Country	Av. annual change 1979-1981	1990 Low/Bas	2000 Low/Bas	1990 High/Haut	2000 High/Haut	Av. annual change 1980-2000 Taux de change-ment annuel moyen 1980-2000 Low Bas	High Haut	Pays
	(1 000 000 m3)					(%)		
Finland	0.15	0.17	0.21	0.19	0.26	+ 1.7	+ 2.8	Finlande
Iceland		–	0.02	0.02	0.02	0.0	–	Islande
Norway	0.15	0.16	0.17	0.18	0.21	+ 0.6	+ 1.7	Norvège
Sweden	0.32	0.32	0.35	0.35	0.45	+ 0.4	+ 1.7	Suède
NORDIC COUNTRIES	0.63	0.65	0.73	0.72	0.92	+ 0.7	+ 1.9	PAYS NORDIQUES
Belgium-Luxembourg	0.03	0.03	0.04	0.03	0.04	+ 1.4	+ 1.4	Belgique-Luxembourg
Denmark	0.06	0.05	0.06	0.06	0.07	0.0	+ 0.8	Danemark
France	0.24	0.23	0.27	0.24	0.33	+ 0.6	+ 1.6	France
Germany, Fed.Rep. of	0.49	0.49	0.55	0.53	0.68	+ 0.6	+ 1.6	Allemagne, Rép.féd. d'
Ireland	0.01	0.01	0.01	0.01	0.02	0.0	+ 3.5	Irlande
Italy	0.28	0.28	0.33	0.30	0.41	+ 0.8	+ 1.9	Italie
Netherlands	0.14	0.15	0.16	0.15	0.20	+ 0.7	+ 1.8	Pays-Bas
United Kingdom	0.40	0.39	0.42	0.42	0.50	+ 0.2	+ 1.1	Royaume-Uni
EEC(9)	1.65	1.63	1.84	1.74	2.25	+ 0.5	+ 1.6	CEE(9)
Austria	0.06	0.06	0.07	0.07	0.08	+ 0.8	+ 1.4	Autriche
Switzerland	0.07	0.08	0.09	0.08	0.11	+ 1.3	+ 2.3	Suisse
CENTRAL EUROPE	0.13	0.14	0.16	0.14	0.19	+ 1.0	+ 1.9	EUROPE CENTRALE
Cyprus			0.10	0.01	0.01	–	–	Chypre
Greece	0.02	0.03	0.04	0.03	0.05	+ 3.5	+ 4.7	Grèce
Israel	0.01	0.02	0.03	0.02	0.03	+ 5.6	+ 5.6	Israël
Malta			0.00	0.01	0.01	–	–	Malte
Portugal	0.05	0.06	0.08	0.07	0.11	+ 2.4	+ 4.0	Portugal
Spain	0.23	0.25	0.30	0.27	0.38	+ 1.3	+ 2.5	Espagne
Turkey	0.07	0.08	0.11	0.10	0.17	+ 2.3	+ 4.5	Turquie
Yugoslavia	0.05	0.07	0.10	0.08	0.14	+ 3.5	+ 5.3	Yougoslavie
SOUTHERN EUROPE	0.44	0.51	0.66	0.57	0.88	+ 2.0	+ 3.5	EUROPE MERIDIONALE
Bulgaria	0.08	0.08	0.08	0.08	0.08	–	–	Bulgarie
Czechoslovakia	0.21	0.21	0.21	0.21	0.21	–	–	Tchécoslovaquie
German Dem. Rep.	0.36	0.36	0.36	0.37	0.36	–	–	Rép. dém. allemande
Hungary	0.10	0.10	0.10	0.10	0.10	–	–	Hongrie
Poland	0.60	0.60	0.60	0.60	0.60	–	–	Pologne
Romania	0.25	0.25	0.25	0.25	0.25	–	–	Roumanie
EASTERN EUROPE	1.60	1.60	1.60	1.60	1.60	–	–	EUROPE ORIENTALE
EUROPE	4.45	4.53	4.99	4.77	5.84	+ 0.6	+ 1.4	EUROPE

Annex table 12.20
ETTS IV consumption scenarios:
paper and paperboard

Tableau annexe 12.20
Scénarios ETTS IV pour la
consommation de papiers et cartons

Country	1979– 1981	1990	2000	1990	2000	Av. annual change, 1980–2000 Taux de change- ment annuel moyen 1980–2000		Pays
		Low/Bas		High/Haut		Low Bas	High Haut	
	(1 000 000 m3)					(%)		
Finland	1.14	1.34	1.68	1.52	2.26	+ 2.0	+ 3.5	Finlande
Iceland	0.02	0.02	0.02	0.02	0.03	–	+ 2.1	Islande
Norway	0.52	0.58	0.78	0.66	1.04	+ 2.1	+ 3.5	Norvège
Sweden	1.66	1.88	2.34	2.14	3.13	+ 1.8	+ 3.3	Suède
NORDIC COUNTRIES	3.34	3.82	4.82	4.34	6.46	+ 1.9	+ 3.4	PAYS NORDIQUES
Belgium–Luxembourg	1.39	1.50	1.86	1.70	2.50	+ 1.5	+ 3.0	Belgique–Luxembourg
Denmark	0.76	0.85	1.01	0.96	1.35	+ 1.5	+ 3.0	Danemark
France	6.20	7.02	9.10	8.00	12.24	+ 1.9	+ 3.5	France
Germany, Fed.Rep. of	9.62	10.71	13.04	12.19	17.53	+ 1.5	+ 3.1	Allemagne, Rép.féd. d'
Ireland	0.27	0.28	0.36	0.33	0.53	+ 1.3	+ 3.3	Irlande
Italy	5.19	5.65	7.35	6.70	10.74	+ 1.8	+ 3.7	Italie
Netherlands	2.17	2.28	2.71	2.59	3.62	+ 1.1	+ 2.6	Pays–Bas
United Kingdom	7.14	7.52	8.64	8.83	12.31	+ 1.0	+ 2.8	Royaume–Uni
EEC(9)	32.75	35.81	44.07	41.30	69.82	+ 1.5	+ 3.1	CEE(9)
Austria	0.92	1.01	1.19	1.14	1.57	+ 1.3	+ 2.8	Autriche
Switzerland	1.02	1.09	1.27	1.24	1.70	+ 1.1	+ 2.5	Suisse
CENTRAL EUROPE	1.94	2.10	2.46	2.38	3.27	+ 1.2	+ 2.6	EUROPE CENTRALE
Cyprus	0.05	0.05	0.06	0.06	0.10	+ 2.5	+ 4.7	Chypre
Greece	0.44	0.50	0.70	0.60	1.02	+ 2.5	+ 4.4	Grèce
Israel	0.24	0.31	0.44	0.40	0.70	+ 3.1	+ 5.5	Israël
Malta	0.02	0.02	0.04	0.03	0.06	+ 2.9	+ 5.7	Malte
Portugal	0.39	0.49	0.67	0.58	1.00	+ 2.7	+ 4.8	Portugal
Spain	2.67	3.06	3.94	3.65	5.83	+ 2.0	+ 4.0	Espagne
Turkey	0.55	0.75	1.19	0.89	1.75	+ 3.9	+ 6.0	Turquie
Yugoslavia	1.07	1.36	2.11	1.62	3.14	+ 3.5	+ 5.5	Yougoslavie
SOUTHERN EUROPE	5.44	6.54	9.15	7.83	13.60	+ 2.6	+ 4.7	EUROPE MERIDIONALE
Bulgaria	0.52	0.55	0.60	0.60	0.73	+ 0.7	+ 1.7	Bulgarie
Czechoslovakia	1.07	1.16	1.25	1.30	1.51	+ 0.7	+ 1.7	Tchécoslovaquie
German Dem. Rep.	1.38	1.50	1.61	1.65	1.92	+ 0.8	+ 1.7	Rép. dém. allemande
Hungary	0.66	0.71	0.76	0.77	0.89	+ 0.7	+ 1.5	Hongrie
Poland	1.37	1.55	1.67	1.56	1.83	+ 1.0	+ 1.4	Pologne
Romania	0.70	0.76	0.82	0.84	0.98	+ 0.8	+ 1.7	Roumanie
EASTERN EUROPE	5.71	6.23	6.71	6.72	7.86	+ 0.8	+ 1.6	EUROPE ORIENTALE
EUROPE	49.18	54.50	67.21	62.57	92.01	+ 1.6	+ 3.2	EUROPE

Annex table 12.21
ETTS IV consumption scenarios:
newsprint

Tableau annexe 12.21
Scénarios ETTS IV pour la
consommation de papier journal

Country	1979-1981	1990	2000	1990	2000	Av. annual change, 1980-2000 Taux de changement annuel moyen 1980-2000		Pays
		Low/Bas		High/Haut		Low Bas	High Haut	
	(1 000 000 m.t./t.m.)					(%)		
Finland	0.14	0.19	0.24	0.21	0.31	+ 2.7	+ 4.1	Finlande
Iceland	–	–	0.01	0.01	0.01	+ 1.9	+ 2.8	Islande
Norway	0.10	0.13	0.17	0.15	0.22	+ 2 7	+ 4.0	Norvège
Sweden	0.28	0.32	0.40	0.36	0.51	+ 1.8	+ 3.1	Suède
NORDIC COUNTRIES	0.52	0.64	0.82	0.73	1.05	+ 2.3	+ 3.6	PAYS NORDIQUES
Belgium–Luxembourg	0.20	0.21	0.26	0.24	0.34	+ 1.3	+ 2.7	Belgique–Luxembourg
Denmark	0.16	0.18	0.22	0.20	0.28	+ 1.4	+ 2.8	Danemark
France	0.63	0.67	0.86	0.74	1.10	+ 1.5	+ 2.8	France
Germany, Fed.Rep. of	1.37	1.50	1.82	1.68	2.33	+ 1.4	+ 2.7	Allemagne, Rép.féd. d'
Ireland	0.06	0.06	0.08	0.07	0.10	+ 1.6	+ 2.6	Irlande
Italy	0.31	0.38	0.46	0.42	0.58	+ 2.0	+ 3.2	Italie
Netherlands	0.44	0.45	0.54	0.50	0.69	+ 1.0	+ 2.3	Pays–Bas
United Kingdom	1.41	1.52	1.74	1.67	2.16	+ 1.0	+ 2.2	Royaume–Uni
EEC(9)	4.59	4.97	5.98	5.52	7.58	+ 1.3	+ 2.6	CEE(9)
Austria	0.15	0.18	0.23	0.20	0.30	+ 2.2	+ 3.5	Autriche
Switzerland	0.20	0.23	0.27	0.26	0.34	+ 1.5	+ 2.7	Suisse
CENTRAL EUROPE	0.34	0.41	0.50	0.46	0.64	+ 1.9	+ 3.2	EUROPE CENTRALE
Cyprus	–	–	–	–	–	–	–	Chypre
Greece	0.04	0.07	0.09	0.08	0.11	+ 4.1	+ 5.2	Grèce
Israel	0.04	0.06	0.07	0.06	0.09	+ 2.8	+ 4.1	Israël
Malta	–	–	–	–	–	–	–	Malte
Portugal	0.04	0.04	0.06	0.05	0.07	+ 2.0	+ 2.8	Portugal
Spain	0.18	0.24	0.30	0.27	0.37	+ 2.6	+ 3.7	Espagne
Turkey	0.14	0.23	0.35	0.26	0.47	+ 4.7	+ 6.2	Turquie
Yugoslavia	0.05	0.03	0.04	0.03	0.05	– 1.1	–	Yougoslavie
SOUTHERN EUROPE	0.50	0.70	0.91	0.75	1.16	+ 3.0	+ 4.3	EUROPE MERIDIONALE
Bulgaria	0.04	0.04	0.04	0.04	0.05	–	+ 0.5	Bulgarie
Czechoslovakia	0.07	0.07	0.07	0.08	0.08	–	+ 0.5	Tchécoslovaquie
German Dem. Rep.	0.14	0.14	0.14	0.15	0.16	–	+ 0.5	Rép. dém. allemande
Hungary	0.07	0.07	0.08	0.07	0.07	–	+ 0.5	Hongrie
Poland	0.13	0.13	0.13	0.14	0.14	–	+ 0.5	Pologne
Romania	0.06	0.06	0.06	0.06	0.07	–	+ 0.5	Roumanie
EASTERN EUROPE	0.51	0.51	0.51	0.54	0.57	–	+ 0.5	EUROPE ORIENTALE
EUROPE	6.47	7.23	8.72	8.00	11.00	+ 1.5	+ 2.7	EUROPE

Annex table 12.22
ETTS IV consumption scenarios:
printing and writing paper

Tableau annexe 12.22
Scénarios ETTS IV pour la consommation
des papiers impression et écriture

Country	1979-1981	Low/Bas		High/Haut		Av. annual change, 1980-2000 / Taux de changement annuel moyen 1980-2000		Pays
		1990	2000	1990	2000	Low/Bas	High/Haut	
	(1 000 000 m.t./t.m.)					(%)		
Finland	0.31	0.46	0.71	0.54	0.99	+ 4.2	+ 6.0	Finlande
Iceland			0.01		0.01	+ 3.6	+ 3.6	Islande
Norway	0.16	0.20	0.32	0.23	0.45	+ 3.6	+ 5.3	Norvège
Sweden	0.50	0.70	1.05	0.81	1.46	+ 3.8	+ 5.5	Suède
NORDIC COUNTRIES	0.98	1.36	2.09	1.58	2.91	+ 3.9	+ 5.6	PAYS NORDIQUES
Belgium-Luxembourg	0.49	0.62	0.91	0.72	1.27	+ 3.1	+ 4.7	Belgique-Luxembourg
Denmark	0.19	0.26	0.37	0.30	0.52	+ 3.4	+ 5.2	Danemark
France	2.06	2.84	4.43	3.29	6.16	+ 3.9	+ 5.7	France
Germany, Fed.Rep. of	3.24	4.37	6.30	5.05	8.76	+ 3.4	+ 5.1	Allemagne, Rép.féd. d'
Ireland	0.04	0.04	0.07	0.05	0.08	+ 2.8	+ 3.6	Irlande
Italy	1.62	2.02	2.90	2.25	3.73	+ 3.0	+ 4.3	Italie
Netherlands	0.67	0.78	1.11	0.90	1.54	+ 2.5	+ 4.3	Pays-Bas
United Kingdom	1.73	2.25	2.98	2.52	3.83	+ 2.8	+ 4.1	Royaume-Uni
EEC(9)	10.05	13.18	19.07	15.08	25.89	+ 3.3	+ 4.9	CEE(9)
Austria	0.10	0.14	0.22	0.17	0.31	+ 4.0	+ 5.8	Autriche
Switzerland	0.37	0.43	0.59	0.50	0.82	+ 2.4	+ 4.1	Suisse
CENTRAL EUROPE	0.47	0.57	0.81	0.67	1.13	+ 2.8	+ 4.5	EUROPE CENTRALE
Cyprus		0.01	0.01	0.01	0.01	+ 5.6	+ 5.6	Chypre
Greece	0.11	0.12	0.19	0.14	0.25	+ 2.8	+ 4.2	Grèce
Israel	0.05	0.10	0.15	0.11	0.19	+ 5.7	+ 6.9	Israël
Malta			0.01		0.01	+ 7.2	+ 7.2	Malte
Portugal	0.08	0.12	0.18	0.13	0.24	+ 4.1	+ 5.6	Portugal
Spain	0.74	0.90	1.30	1.01	1.67	+ 2.9	+ 4.1	Espagne
Turkey	0.07	0.14	0.24	0.16	0.32	+ 6.4	+ 7.9	Turquie
Yugoslavia	0.24	0.38	0.63	0.43	0.83	+ 4.9	+ 6.4	Yougoslavie
SOUTHERN EUROPE	1.31	1.77	2.71	1.99	3.52	+ 3.7	+ 5.1	EUROPE MERIDIONALE
Bulgaria	0.05	0.05	0.05	0.05	0.06	-	+ 0.5	Bulgarie
Czechoslovakia	0.15	0.15	0.15	0.16	0.17	-	+ 0.5	Tchécoslovaquie
German. Dem. Rep.	0.20	0.20	0.20	0.21	0.22	-	+ 0.5	Rép. dém. allemande
Hungary	0.14	0.14	0.14	0.14	0.15	-	+ 0.5	Hongrie
Poland	0.22	0.27	0.27	0.27	0.28	+ 1.0	+ 1.2	Pologne
Romania	0.09	0.09	0.09	0.10	0.10	-	+ 0.5	Roumanie
EASTERN EUROPE	0.85	0.90	0.90	0.93	0.98	+ 0.3	+ 0.7	EUROPE ORIENTALE
EUROPE	13.66	17.78	25.58	20.25	34.43	+ 3.2	+ 4.7	EUROPE

Annex table 12.23
ETTS IV consumption scenarios:
other paper and paperboard

Tableau annexe 12.23
Scénarios ETTS IV pour la consommation
des autres papiers et cartons

Country	1979–1981	1990	2000	1990	2000	Av. annual change, 1980–2000 Taux de change- ment annuel moyen 1980–2000		Pays
		Low/Bas		High/Haut		Low Bas	High Haut	
		(1 000 000 m.t./t.m.)				(%)		
Finland	0.68	0.69	0.73	0.77	0.96	+ 0.4	+ 1.7	Finlande
Iceland	0.01	0.01	0.01	0.01.	0.01	–	+ 1.1	Islande
Norway	0.26	0.25	0.28	0.28	0.37	+ 0.4	+ 1.8	Norvège
Sweden	0.87	0.85	0.89	0.96	1.16	+ 0.1	+ 1.5	Suède
NORDIC COUNTRIES	1.84	1.80	1.91	2.02	2.50	+ 0.2	+ 1.5	PAYS NORDIQUES
Belgium–Luxembourg	0.70	0.66	0.68	0.75	0.89	– 0.1	+ 1.2	Belgique–Luxembourg
Denmark	0.40	0.41	0.42	0.47	0.55	+ 0.2	+ 1.6	Danemark
France	3.51	3.52	3.81	3.97	4.99	+ 0.4	+ 1.8	France
Germany, Fed.Rep. of	5.00	4.84	4.91	5.46	6.44	– 0.1	+ 1.3	Allemagne, Rép.féd. d'
Ireland	0.18	0.17	0.21	0.21	0.34	+ 0.8	+ 3.2	Irlande
Italy	3.27	3.25	3.98	4.03	6.44	+ 1.0	+ 3.4	Italie
Netherlands	1.05	1.05	1.06	1.18	1.38	–	+ 1.4	Pays-Bas
United Kingdom	3.99	3.75	3.91	4.64	6.32	– 0.1	+ 2.3	Royaume–Uni
EEC(9)	18.11	17.65	18.98	20.71	27.35	+ 0.2	+ 2.1	CEE(9)
Austria	0.66	0.69	0.74	0.78	0.97	+ 0.6	+ 1.9	Autriche
Switzerland	0.46	0.42	0.41	0.48	0.54	– 0.5	+ 0.8	Suisse
CENTRAL EUROPE	1.12	1.11	1.15	1.26	1.51	+ 0.1	+ 1.5	EUROPE CENTRALE
Cyprus	0.04	0.04	0.05	0.05	0.08	+ 1.1	+ 3.5	Chypre
Greece	0.28	0.31	0.42	0.38	0.66	+ 2.1	+ 4.4	Grèce
Israel	0.15	0.19	0.26	0.23	0.42	+ 2.8	+ 5.3	Israël
Malta	0.02	0.02	0.03	0.03	0.05	+ 2.6	+ 4.7	Malte
Portugal	0.28	0.33	0.44	0.40	0.69	+ 2.3	+ 4.6	Portugal
Spain	1.74	1.92	2.35	2.38	3.80	+ 1.5	+ 4.0	Espagne
Turkey	0.34	0.39	0.60	0.48	0.95	+ 2.9	+ 5.3	Turquie
Yugoslavia	0.78	0.94	1.43	1.16	2.27	+ 3.1	+ 5.5	Yougoslavie
SOUTHERN EUROPE	3.63	4.14	5.58	5.11	8.92	+ 2.2	+ 4.6	EUROPE MERIDIONALE
Bulgaria	0.42	0.46	0.51	0.51	0.62	+ 1.0	+ 2.0	Bulgarie
Czechoslovakia	0.85	0.94	1.03	1.06	1.26	+ 1.0	+ 2.0	Tchécoslovaquie
German Dem. Rep.	1.04	1.16	1.27	1.29	1.54	+ 1.0	+ 2.0	Rép. dém. allemande
Hungary	0.45	0.50	0.55	0.56	0.67	+ 1.0	+ 2.0	Hongrie
Poland	1.03	1.15	1.27	1.15	1.41	+ 1.0	+ 1.5	Pologne
Romania	0.55	0.61	0.67	0.68	0.81	+ 1.0	+ 2.0	Roumanie
EASTERN EUROPE	4.35	4.82	5.30	5.25	6.31	+ 1.0	+ 1.9	EUROPE ORIENTALE
EUROPE	29.05	29.52	32.92	34.35	46.59	+ 0.6	+ 2.4	EUROPE

Annex table 12.24
ETTS IV consumption scenarios:
sleepers, veneer sheets and
dissolving pulp

Tableau annexe 12.24
Scénarios ETTS IV pour la consommation future de traverses, de
feuilles de placage et de pâte de
à dissoudre

Country	Sleepers Traver-ses 2000 a/	Veneer sheets Placa-ges 2000 a/	Dissolving pulp Pâte à dissoudre 1979-81	1990	2000	Pays
	(1 000 000 m3)		(1 000 000 m.t./t.m.)			
Finland	0.02	0.01	0.05	0.04	0.04	Finlande
Iceland	–	–	–	–	–	Islande
Norway	–	0.01	–	–	–	Norvège
Sweden	0.03	0.03	0.03	0.03	0.03	Suède
NORDIC COUNTRIES	0.05	0.06	0.08	0.07	0.07	PAYS NORDIQUES
Belgium–Luxembourg	0.13	0.04	0.04	0.03	0.03	Belgique–Luxembourg
Denmark	0.02	0.03	–	–	–	Danemark
France	0.17	0.15	0.15	0.14	0.12	France
Germany, Fed.Rep. of	0.18	0.35	0.24	0.21	0.19	Allemagne, Rép.féd. d'
Ireland	–	–	–	–	–	Irlande
Italy	0.11	0.50	0.04	0.03	0.03	Italie
Netherlands	0.10	0.05	0.03	0.03	0.0	Pays–Bas
United Kingdom	0.02	0.06	0.27	0.24	0.22	Royaume–Uni
EEC(9)	0.74	1.18	0.76	0.68	0.62	CEE(9)
Austria	0.05	0.02	0.12	0.11	0.10	Autriche
Switzerland	0.06	0.02	–	–	–	Suisse
CENTRAL EUROPE	0.06	0.05	0.12	0.11	0.10	EUROPE CENTRALE
Cyprus	–	–	–	–	–	Chypre
Greece	–	0.02	–	–	–	Grèce
Israel	–	0.06	–	–	–	Israël
Malta	–	–	–	–	–	Malte
Portugal	0.07	–	–	–	–	Portugal
Spain	0.05	0.01	0.07	0.06	0.06	Espagne
Turkey	0.05	0.01	0.03	0.03	0.02	Turquie
Yugoslavia	–	0.21	0.09	0.08	0.07	Yougoslavie
SOUTHERN EUROPE	0.16	0.32	0.20	0.17	0.15	EUROPE MERIDIONALE
Bulgaria	0.03	0.06	–	–	–	Bulgarie
Czechoslovakia	0.11	0.07	0.08	0.07	0.06	Tchécoslovaquie
German Dem. Rep.	0.15	0.05	0.24	0.22	0.20	Rép. dém. allemande
Hungary	0.04	0.02	0.01	0.01	0.01	Hongrie
Poland	0.38	0.04	0.07	0.07	0.06	Pologne
Romania	0.07	0.07	0.07	0.06	0.06	Roumanie
EASTERN EUROPE	0.78	0.30	0.48	0.43	0.39	EUROPE ORIENTALE
EUROPE	1.80	1.90	1.65	1.46	1.33	EUROPE

a/ Same levels as 1979–81.

a/ Même niveau qu'en 1979–81.

Annex table 12.25
ETTS IV consumption scenarios:
pitprops and other industrial wood

Tableau annexe 12.25
Scénarios ETTS IV pour la consommation des
bois de mine et des autres bois ronds industriels

Country	Pitprops/Bois de mine			OIW a/	Total			Pays
	1979–81	1990	2000	2000	1979–81	1990	2000	
Finland	–	–	–	1.13	1.13	1.13	1.13	Finlande
Iceland	–	–	–	0.02	0.02	0.02	0.02	Islande
Norway	–	–	–	0.17	0.17	0.17	0.17	Norvège
Sweden	–	–	–	0.23	0.23	0.23	0.23	Suède
NORDIC COUNTRIES	–	–	–	1.55	1.55	1.55	1.55	PAYS NORDIQUES
Belgium–Luxembourg	0.15	0.12	0.09	0.09	0.23	0.21	0.18	Belgique–Luxembourg
Denmark	–	–	–	0.37	0.37	0.37	0.37	Danemark
France	0.33	0.26	0.21	0.60	0.93	0.86	0.81	France
Germany, Fed.Rep. of	0.40	0.32	0.26	0.81	1.21	1.13	1.07	Allemagne,Rép.féd. d'
Ireland	–	–	–	–	–	–	–	Irlande
Italy	0.03	0.02	0.02	1.28	1.31	1.30	1.30	Italie
Netherlands	–	–	–	0.31	0.31	0.31	0.31	Pays–Bas
United Kingdom	0.25	0.20	0.16	0.34	0.59	0.54	0.50	Royaume–Uni
EEC(9)	1.16	0.92	0.74	3.77	4.93	4.69	4.51	CEE(9)
Austria	0.05	0.04	0.03	0.57	0.62	0.61	0.60	Autriche
Switzerland	–	–	–	0.10	0.10	0.10	0.10	Suisse
CENTRAL EUROPE	0.05	0.04	0.03	0.67	0.72	0.71	0.70	EUROPE CENTRALE
Cyprus	–	–	–	–	–	–	–	Chypre
Greece	0.02	0.01	0.01	0.13	0.15	0.14	0.14	Grèce
Israel	–	–	–	0.03	0.03	0.03	0.03	Israël
Malta	–	–	–	–	–	–	–	Malte
Portugal	–	–	–	0.09	0.09	0.09	0.09	Portugal
Spain	0.64	0.51	0.41	0.47	1.12	0.98	0.88	Espagne
Turkey	0.61	0.49	0.39	0.33	0.94	0.82	0.72	Turquie
Yugoslavia	0.48	0.38	0.31	0.74	1.22	1.12	1.05	Yougoslavie
SOUTHERN EUROPE	1.74	1.39	1.12	1.80	3.54	3.19	2.92	EUROPE MERIDIONALE
Bulgaria	0.20	0.16	0.13	0.56	0.76	0.72	0.69	Bulgarie
Czechoslovakia	0.48	0.38	0.30	1.78	2.26	2.16	2.08	Tchécoslovaquie
German Dem. Rep.	0.20	0.16	0.13	2.49	2.69	2.65	2.62	Rép. dém. allemande
Hungary	0.60	0.48	0.38	0.52	1.12	1.00	0.90	Hongrie
Poland	1.45	1.16	0.93	1.67	3.12	2.83	2.60	Pologne
Romania	0.64	0.51	0.41	2.01	2.65	2.52	2.42	Roumanie
EASTERN EUROPE	3.56	2.85	2.28	9.03	12.60	11.88	11.31	EUROPE ORIENTALE
EUROPE	6.50	5.20	4.17	16.82	23.33	22.02	20.99	EUROPE

a/ Other industrial wood.
    2000 at same level as 1979–81.

a/ Autres bois ronds industriels.
    2000 au même niveau qu'en 1979–81.

Annex table 12.26
ETTS IV consumption scenarios:
fuelwood

Tableau annexe 12.26
Scénarios ETTS IV pour la consommation
de bois de chauffage

Country	1979-1981 a/	1990	2000	1990	2000	Av. annual change, 1980-2000 Taux de change-ment annuel moyen 1980-2000		Pays
		Low/Bas		High/Haut		Low Bas	High Haut	
	(1 000 000 m3)					(%)		
Finland	3.70	4.00	4.30	4.20	4.70	+ 0.8	+ 1.2	Finlande
Iceland	-	-	-	-	-	-	-	Islande
Norway	1.75	2.00	2.20	2.20	2.40	+ 1.1	+ 1.6	Norvège
Sweden	4.50	7.30	8.90	11.60	13.00	+ 3.5	+ 5.5	Suède
NORDIC COUNTRIES	9.95	13.30	15.40	18.00	20.10	+ 2.2	+ 3.6	PAYS NORDIQUES
Belgium-Luxembourg*	0.37	0.43	0.48	0.48	0.61	+ 1.3	+ 2.5	Belgique-Luxembourg*
Denmark	0.25	0.25	0.20	0.35	0.45	- 1.1	+ 3.0	Danemark
France	10.00	10.50	11.20	13.50	18.50	+ 0.6	+ 3.1	France
Germany, Fed.Rep. of	3.55	4.50	5.00	4.50	5.00	+ 1.7	+ 1.7	Allemagne, Rép.féd. d'
Ireland	0.24	0.25	0.28	0.28	0.30	+ 0.8	+ 1.1	Irlande
Italy	3.46	3.97	4.50	4.50	5.53	+ 1.3	+ 2.4	Italie
Netherlands*	0.08	0.09	0.10	0.10	0.13	+ 1.3	+ 2.5	Pays-Bas*
United Kingdom	0.15	0.16	0.30	0.38	0.72	+ 3.6	+ 8.2	Royaume-Uni
EEC(9)	18.12	20.15	22.06	24.09	31.24	+ 1.0	+ 2.8	CEE(9)
Austria*	1.41	1.50	1.57	2.23	3.05	+ 0.5	+ 3.9	Autriche*
Switzerland	0.81	0.86	0.90	1.28	1.75	+ 0.5	+ 3.9	Suisse
CENTRAL EUROPE	2.22	2.36	2.47	3.51	4.80	+ 0.5	+ 3.9	EUROPE CENTRALE
Cyprus	0.02	0.03	0.03	0.03	0.03	+ 2.1	+ 2.1	Chypre
Greece	0.72	0.75	0.75	0.79	0.79	+ 0.2	+ 0.5	Grèce
Israel	0.01	0.03	0.04	0.04	0.05	+ 6.7	+ 7.3	Israël
Malta	-	-	-	-	-	-	-	Malte
Portugal	5.45	6.77	8.56	7.20	9.00	+ 2.3	+ 2.5	Portugal
Spain	2.00	2.30	3.10	2.60	3.60	+ 2.2	+ 3.0	Espagne
Turkey*	15.70	14.50	13.00	17.00	18.00	- 0.9	+ 0.7	Turquie*
Yugoslavia	6.10	7.40	7.60	7.80	8.00	+ 1.1	+ 1.4	Yougoslavie
SOUTHERN EUROPE	30.00	31.78	33.08	35.46	39.47	+ 0.5	+ 1.4	EUROPE MERIDIONALE
Bulgaria	1.93	1.68	1.61	1.68	1.61	- 0.9	- 0.9	Bulgarie
Czechoslovakia	1.36	1.00	1.90	1.00	1.90	+ 1.7	+ 1.7	Tchécoslovaquie
German Dem. Rep.*	0.68	0.67	0.76	0.69	0.70	+ 0.5	+ 0.6	Rép. dém. allemande*
Hungary	2.38	2.85	2.89	2.90	2.96	+ 1.0	+ 1.1	Hongrie
Poland	1.92	1.94	1.97	2.16	2.16	+ 0.1	+ 0.6	Pologne
Romania*	3.40	3.37	3.78	3.44	3.78	+ 0.5	+ 0.6	Roumanie*
EASTERN EUROPE	11.67	11.51	12.91	11.87	13.11	+ 0.5	+ 0.6	EUROPE ORIENTALE
EUROPE	71.96	79.10	85.92	92.93	108.72	+ 0.9	+ 2.1	EUROPE

a/ National estimates, where available.
These differ from recorded consumption
(see annex table 19.1).

a/ Estimations nationales, si disponibles.
Celles-ci sont différentes des données
officielles (voir tableau annexe 19.1).

Annex table 12.27/Tableau annexe 12.27

Group I (west-central Europe), sawnwood and wood-based panels: alternative scenarios
Groupe I (Europe ouest-centre), sciages et panneaux dérivés du bois : scénarios alternatifs

		Scenarios 2000: / Scénarios 2000:						
		Base			Alternative			
	1979-1983	Low Bas	High Haut	User sector recession Récession secteur utilisateurs	Marketing and product development Commercialisation et développement des produits		Prices Prix	
					Success Réussite	Failure Echec	Rise Hausse	Fall Baisse
	(million m3)							
Sawn softwood/Sciages résineux	33.7	37.4	44.0	29.6	41.2	34.0	33.5	41.7
Sawn hardwood/Sciages feuillus	9.2	13.3	16.3	8.5	14.7	12.1	12.1	14.5
Sawnwood/Sciages a/	42.9	50.8	60.3	38.1	55.8	46.2	c/	c/
Plywood/Contreplaqués	3.3	3.9	5.1	2.7	4.3	3.6	3.8	4.1
Particle board/Panneaux de particules	12.4	16.4	20.4	11.0	18.9	14.9	c/	c/
Fibreboard/Panneaux de fibres	1.4	1.6	2.0	1.2	1.8	1.4	c/	c/
Panels/Panneaux, total b/ A	17.1	22.0	27.4	14.9	24.2	20.0	c/	c/
B	17.1	21.5	26.3	15.0	23.7	19.6	c/	c/
	(average annual percent change, 1979-83 to 2000) (changement annuel moyen en pourcentage, 1979-83 à 2000)							
Sawn softwood/Sciage résineux	..	+ 0.6	+ 1.4	- 0.7	+ 1.1	-	-	+ 1.1
Sawn hardwood/Sciages feuillus	..	+ 2.0	+ 3.0	- 0.4	+ 2.5	+ 1.4	+ 1.5	+ 2.4
Sawnwood/Sciages a/	..	+ 0.9	+ 1.8	- 0.6	+ 1.4	+ 0.4	c/	c/
Plywood/Contreplaqués	..	+ 1.0	+ 2.3	- 1.0	+ 1.5	+ 0.5	+ 0.7	+ 1.2
Particle board/Panneaux de particules	..	+ 1.5	+ 2.6	- 0.6	+ 2.0	+ 1.0	c/	c/
Fibreboard/Panneaux de fibres	..	+ 0.8	+ 1.9	- 0.8	+ 1.3	+ 0.3	c/	c/
Panels/Panneaux, total b/ A	..	+ 1.4	+ 2.5	- 0.7	+ 1.8	+ 0.8	c/	c/
B	..	+ 1.2	+ 2.3	- 0.7	+ 1.7	+ 0.7	c/	c/

Note: For explanation see text / Pour explication, voir texte.
      Projections for individual countries available on request from the secretariat / Projections par
      pays disponibles auprès du secrétariat.

a/ Excluding sleepers. Sum of projections for sawn softwood and sawn hardwood / Traverses non
comprises. Somme des projections pour sciages résineux et feuillus.

b/ Excluding veneer sheets. Total A is the sum of projections for individual panels. Total B is the
independently calculated projection for wood-based panels as a group, used as a cross check / Feuilles de
placage non comprises. Le total A est la somme des projections pour les différents panneaux . Le total
B est la projection, calculée indépendamment, pour tous les panneaux dérivés du bois pris en tant que
groupe, et utilisée comme vérification.

c/ Same as in base low scenario, as no price elasticities found for this product / Identique au scénario
de base, bas, aucune élasticité par rapport au prix n'ayant été constatée pour ce produit.

Annex table 12.28/Tableau annexe 12.28

Group II (South), sawnwood and wood-based panels: alternative scenarios
Groupe II (Sud), sciages et panneaux dérivés du bois : scénarios alternatifs

	1979-1983	Scenarios 2000: / Scénarios 2000:						
		Base		Alternative				
		Low Bas	High Haut	User sector recession Récession secteur utilisateurs	Marketing and product development Commercialisation et développement des produits		Prices Prix	
					Success Réussite	Failure Echec	Rise Hausse	Fall Baisse
				(million m3)				
Sawn softwood/Sciages résineux	14.9	19.4	24.9	13.1	21.4	17.7	c/	c/
Sawn hardwood/Sciages feuillus	6.2	10.1	12.2	5.9	11.1	9.2	8.4	11.8
Sawnwood/Sciages a/	21.1	29.5	37.1	19.0	32.4	26.8	c/	c/
Plywood/Contreplaqués	0.7	1.2	1.6	0.6	1.3	1.1	c/	c/
Particle board/Panneaux de particules	4.8	7.1	8.1	4.6	7.8	6.4	c/	c/
Fibreboard/Panneaux de fibres	0.7	1.0	1.3	0.6	1.1	0.8	c/	c/
Panels/Panneaux, total b/ A	6.2	9.3	11.0	5.9	10.2	8.4	c/	c/
B	6.2	8.4	9.4	6.0	9.2	7.6	c/	c/
		(average annual percent change, 1979-83 to 2000) (changement annuel moyen en pourcentage, 1979-83 à 2000)						
Sawn softwood/Sciage résineux	..	+ 1.4	+ 2.7	- 0.7	+ 1.9	+ 0.9	c/	c/
Sawn hardwood/Sciages feuillus	..	+ 2.6	+ 3.6	- 0.3	+ 3.1	+ 2.1	+ 1.7	+ 3.5
Sawnwood/Sciages a/	..	+ 1.8	+ 3.0	- 0.6	+ 2.3	+ 1.3	c/	c/
Plywood/Contreplaqués	..	+ 2.8	+ 4.3	- 0.6	+ 3.3	+ 2.3	c/	c/
Particle board/Panneaux de particules	..	+ 2.1	+ 2.8	- 0.1	+ 2.6	+ 1.6	c/	c/
Fibreboard/Panneaux de fibres	..	+ 1.8	+ 3.2	- 0.8	+ 2.3	+ 1.3	c/	c/
Panels/Panneaux, total b/ A	..	+ 2.2	+ 3.1	- 0.2	+ 2.7	+ 1.6	c/	c/
B	..	+ 1.6	+ 2.3	- 0.2	+ 2.1	+ 1.1	c/	c/

Note: For explanation see text / Pour explication, voir texte.
Projections for individual countries available on request from the secretariat / Projections par pays disponibles auprès du secrétariat.

a/ Excluding sleepers. Sum of projections for sawn softwood and sawn hardwood / Traverses non comprises. Somme des projections pour sciages résineux et feuillus.

b/ Excluding veneer sheets. Total A is the sum of projections for individual panels. Total B is the independently calculated projection for wood-based panels as a group, used as a cross check / Feuilles de placage non comprises. Le total A est la somme des projections pour les différents panneaux . Le total B est la projection, calculée indépendamment, pour tous les panneaux dérivés du bois pris en tant que groupe, et utilisée comme vérification.

c/ Same as in base low scenario, as no price elasticties found for this product / Identique au scénario de base, bas, aucune élasticité par rapport au prix n'ayant été constatée pour ce produit.

Annex table 12.29/Tableau annexe 12.29

Group III (exporters), sawnwood and wood-based panels: alternative scenarios
Groupe III (exportateurs), sciages et panneaux dérivés du bois : scénarios alternatifs

		Scenarios 2000: / Scénarios 2000:						
		Base		Alternative				
	1979-1983	Low Bas	High Haut	User sector recession Récession secteur utilisateurs	Marketing and product development Commercialisation et développement des produits		Prices Prix	
					Success Réussite	Failure Echec	Rise Hausse	Fall Baisse
				(million m3)				
Sawn softwood/Sciages résineux	12.1	13.0	14.3	11.3	14.3	11.8	c/	c/
Sawn hardwood/Sciages feuillus	0.8	1.2	1.6	0.7	1.3	1.1	c/	c/
Sawnwood/Sciages a/	12.8	14.2	15.9	12.0	15.6	12.9	c/	c/
Plywood/Contreplaqués	0.3	0.4	0.5	0.3	0.4	0.4	0.3	0.5
Particle board/Panneaux de particules	2.1	2.7	3.4	1.9	3.0	2.5	c/	c/
Fibreboard/Panneaux de fibres	0.7	0.8	1.0	0.6	0.9	0.7	c/	c/
Panels/Panneaux, total b/ A	3.1	3.9	4.9	2.7	4.3	3.6	c/	c/
B	3.1	4.1	5.1	2.7	4.5	3.7	c/	c/
		(average annual percent change, 1979-83 to 2000) (changement annuel moyen en pourcentage, 1979-83 à 2000)						
Sawn softwood/Sciage résineux	..	+ 0.4	+ 0.9	- 0.4	+ 0.9	- 0.1	c/	c/
Sawn hardwood/Sciages feuillus	..	+ 2.5	+ 3.9	- 0.4	+ 3.0	+ 2.0	c/	c/
Sawnwood/Sciages	..	+ 0.5	+ 1.1	- 0.4	+ 1.0	-	c/	c/
Plywood/Contreplaqués	..	+ 0.8	+ 1.8	- 0.6	+ 1.4	+ 0.4	- 0.1	+ 1.8
Particle board/Panneaux de particules	..	+ 1.4	+ 2.6	- 0.6	+ 1.9	+ 0.9	c/	c/
Fibreboard/Panneaux de fibres	..	+ 1.0	+ 2.2	- 0.8	+ 1.5	+ 0.5	c/	c/
Panels/Panneaux, total b/ A	..	+ 1.3	+ 2.4	- 0.7	+ 1.8	+ 0.8	c/	c/
B	..	+ 1.4	+ 2.7	- 0.8	+ 1.9	+ 0.9	c/	c/

Note: For explanation see text / Pour explication, voir texte.
Projections for individual countries available on request from the secretariat / Projections par pays disponibles auprès du secrétariat.

a/ Excluding sleepers. Sum of projections for sawn softwood and sawn hardwood / Traverses non comprises. Somme des projections pour sciages résineux et feuillus.

b/ Excluding veneer sheets. Total A is the sum of projections for individual panels. Total B is the independently calculated projection for wood-based panels as a group, used as a cross check / Feuilles de placage non comprises. Le total A est la somme des projections pour les différents panneaux . Le total B est la projection, calculée indépendamment, pour tous les panneaux dérivés du bois pris en tant que groupe, et utilisée comme vérification.

c/ Same as in base low scenario, as no price elasticities found for this product / Identique au scénario de base, bas, aucune élasticité par rapport au prix n'ayant été constatée pour ce produit.

Annex table 14.1 / Tableau annexe 14.1

| Average number of employees in the Forest industries (ISIC 331, 332 and 341) | Nombre moyen d'employés des industries forestières (CITI 331, 332 and 341) |

	thousands of employees/milliers d'employés										
	1970	1971	1972	1973	1974	1975	1976	1977	1978	1979	1980
Finland/Finlande	100	104	103	108	112	101	103	103	100	106	119
Norway/Norvège	53	53	53	53	53	52	52	51	51	49	49
Sweden/Suède	138	134	134	138	140	137	138	136	133	133	131
Belgium/Belgique	80	79	79	80	81	74	71	68	64	62	58
Denmark/Danemark	38	33	34	36	33	29	31	31	31	31	30
France	446	447	456	475	487	351	348	344	334	332	327
Germany,Fed.Rep.of/ Allemagne,Rép.féd.d'	499	497	496	500	481	446	437	445	446	448	..
Ireland/Irlande	13	13	14	14	14	13	..	13	13	..	..
Italy/Italie	215	217	213	219	218	211	207	202	197	191	187
Netherlands/Pays-Bas	92	89	87	86	86	82	79	76	75	74	72
United Kingdom/ Royaume-Uni	498	473	466	481	469	463	453	448	433	425	422
Austria/Autriche	57	59	61	62	61	59	61	61	60	59	59
Greece/Grèce	17	18	19	20	34	36	..	..	..	..	..
Portugal	..	58	69	70	71	73	74	75	74	75	73
Spain/Espagne	168	193	202	212	220	223	227	..	..	..	..
Turkey/Turquie	21	..	..	..	30	32	32	31	34	31	35
Yugoslavia/ Yougoslavie	161	172	179	181	195	197	203	213	214	225	233
Czechoslovakia/ Tchécoslovaquie	169	168	169	171	174	176	174	175	176	177	175
Hungary/Hongrie	76	70	68	69	73	71	69	68	68	65	63
Poland/Pologne	250	255	264	270	275	278	279	283	280	270	263
Canada	250	253	259	273	288	274	283	276	291	300	297
United States/ Etats-Unies	1495	1467	1560	1605	1571	1402	1486	1571	1619	1660	1594

Annex table 14.2 / Tableau annexe 14.2

Average number of employees in Wood, and wood and cork products, excluding furniture (ISIC 331)	Nombre moyen d'employés dans l'industrie du bois et la fabrication d'ouvrages en bois et en liège à l'exclusion des meubles (CITI 331)

	thousands of employees/milliers d'employés										
	1970	1971	1972	1973	1974	1975	1976	1977	1978	1979	1980
Finland/Finlande	44	44	43	45	47	39	39	40	40	45	47
Norway/Norvège	19	20	21	21	22	22	22	22	23	22	22
Sweden/Suède	63	61	61	63	63	60	59	57	56	56	54
Belgium/Belgique	(52)	(52)	(52)	(53)	(54)	(49)	(48)	(47)	(44)	(41)	(39)
Denmark/Danemark	13	11	11	11	10	9	9	9	9	10	10
France	(316)	(318)	(325)	(336)	(347)	110	105	103	99	98	98
Germany,Fed.Rep.of/ Allemagne,Rép.féd.d'	(287)	(292)	(299)	(306)	(293)	(271)	(268)	(280)	(283)	(285)	..
Ireland/Irlande	4	4	4	5	5	5	..	5	5	..	..
Italy/Italie	72	61	58	61	59	57	55	54	52	51	49
Netherlands/Pays-Bas	(58)	(57)	(56)	(55)	(54)	(51)	(49)	(47)	(47)	(46)	(45)
United Kingdom/ Royaume-Uni	127	122	120	126	125	122	121	119	110	105	105
Austria/Autriche	5	5	6	6	5	9	10	10	9	9	9
Greece/Grèce	7	7	8	9	17	16	..	..	..	..	..
Portugal	19	37	43	44	45	43	42	41	41	43	42
Spain/Espagne	34	36	37	39	39	36	37	..	..	..	..
Turkey/Turquie	9	..	..	..	12	13	12	12	13	12	13
Yugoslavia/ Yougoslavie	74	76	78	81	88	84	89	93	94	98	100
Czechoslovakia/ Tchécoslovaquie	66	65	66	67	68	68	65	66	66	66	66
Hungary/Hongrie	28	20	19	19	23	22	21	20	20	18	17
Poland/Pologne	103	106	109	111	113	114	113	113	110	106	97
Canada	88	92	103	112	107	98	106	109	119	122	117
United States/ Etats-Unies	482	478	515	532	522	460	491	524	539	555	512

Note: Figures in parentheses include also ISIC Group 332.
Les données entre parenthèses comprennent la classe CITI 332.

Annex table 14.3 / Tableau annexe 14.3

Average number of employees in Paper and paper products (ISIC 341)

Nombre moyen d'employés dans la fabrication de papier et d'articles en papier (CITI 341)

	thousands of employees/milliers d'employés										
	1970	1971	1972	1973	1974	1975	1976	1977	1978	1979	1980
Finland/Finlande	46	49	48	50	52	52	51	51	48	49	59
Norway/Norvège	24	23	22	21	21	20	20	19	18	18	17
Sweden/Suède	59	58	58	59	60	60	63	62	60	61	61
Belgium/Belgique	28	27	27	27	27	25	23	21	20	21	19
Denmark/Danemark	11	10	10	11	10	9	9	9	9	9	9
France	130	129	131	137	140	136	134	131	126	123	120
Germany,Fed.Rep.of/ Allemagne,Rép.féd.d'	212	205	197	194	188	175	169	165	163	163	..
Ireland/Irlande	6	6	6	6	6	5	..	5	5	..	..
Italy/Italie	75	75	76	76	75	77	75	72	70	67	66
Netherlands/Pays-Bas	34	32	31	31	32	31	30	29	28	28	27
United Kingdom/ Royaume-Uni	248	238	232	234	240	220	212	213	208	206	199
Austria/Autriche	27	27	26	26	26	25	24	24	22	22	22
Greece/Grèce	6	7	7	7	8	8	..	..	..	..	..
Portugal	13	13	14	14	14	16	16	17	17	18	18
Spain/Espagne	48	49	50	52	54	55	55	..	..	..	..
Turkey/Turquie	11	..	..	..	15	16	16	16	17	15	18
Yugoslavia/ Yougoslavie	30	31	31	32	35	33	39	40	40	40	42
Czechoslovakia/ Tchécoslovaquie	43	43	44	44	45	46	46	46	47	47	47
Hungary/Hongrie	17	17	17	17	17	16	16	16	16	16	15
Poland/Pologne	54	55	58	58	59	60	58	57	56	55	54
Canada	120	118	119	122	130	126	129	123	125	127	129
United States/ Etats-Unies	659	632	633	645	646	589	615	627	639	658	647

169

Annex table 14.4 / Tableau annexe 14.4

Average number of employees in Pulp, paper and paperboard (ISIC 3411)

Nombre moyen d'employés dans la fabrication de pâtes, papiers et cartons (CITI 3411)

	thousands of employees/milliers d'employés										
	1970	1971	1972	1973	1974	1975	1976	1977	1978	1979	1980
Finland/Finlande	39	41	40	42	37	43	41	41	39	39	41
Norway/Norvège	18	17	16	16	16	16	15	14	14	13	13
Sweden/Suède	48	47	48	48	49	50	52	52	50	50	49
Belgium/Belgique	..	..	..	..	..	..	..	..	..	..	..
Denmark/Danemark	3	3	3	3	3	3	3	2	2	2	2
France	..	..	..	..	..	..	..	..	..	..	..
Germany,Fed.Rep.of/ Allemagne,Rép.féd.d'	76	71	66	63	60	57	55	54	54	53	..
Ireland/Irlande	..	..	..	1	1	1	..	1	1	..	..
Italy/Italie	..	..	..	..	..	..	..	..	..	..	..
Netherlands/Pays-Pas	..	..	..	..	..	..	..	..	..	..	..
United Kingdom/ Royaume-Uni	78	69	63	61	63	59	57	55	55	54	48
Austria/Autriche	19	18	17	17	17	17	16	16	15	15	14
Greece/Grèce	4	4	4	4	4	5	..	..	..	..	..
Portugal	9	9	9	9	9	11	12	12	12	12	12
Spain/Espagne	24	24	23	23	25	24	24	..	..	..	..
Turkey/Turquie	..	..	..	..	..	..	..	11	11	10	11
Yugoslavia/ Yougoslavie	23	24	24	23	25	19	23	23	23	20	22
Czechoslovakia/ Tchécoslovaquie	..	..	..	..	..	..	..	..	..	..	..
Hungary/Hongrie	..	..	..	..	..	..	..	..	..	..	..
Poland/Pologne	30	31	32	32	31	31	30	30	30	29	36
Canada	80	79	79	80	86	84	87	85	86	87	87
United States/ Etats-Unies	272	263	258	262	266	245	257	259	257	268	263

Annex 15.1
Domestic supply of wood residues,
chips and particles

Tableau annexe 15.1
Approvisionnement domestique des
déchets de bois, plaquettes et
particules

Countries / Pays	Domestic supply Approvisionnement domestique (1000 m3)		Estimated recovery ratio/Taux de récupération estimé (%)					
			Real/Réal		Scenarios/Scénarios			
					Low/Bas		High/Haut	
	1969-71	1979-81	1969-71	1979-81	1990	2000	1990	2000
Finland / Finlande	4290	7830	58.5	86.8	87.8	88.8	91.2	96.8
Iceland / Islande	-	-	-	-				
Norway / Norvège	880	1867	51.3	89.0	90.0	91.0	94.0	99.0
Sweden / Suède	7867	9968	78.1	90.3	91.3	92.3	95.3	100.3
NORDIC COUNTRIES / PAYS NORDIQUES	13037	19664	68.2	88.7	..	..	..	..
Belgium-Luxembourg / Belgique/Luxembourg	258	273	38.1	45.0	55.0	65.0	65.0	75.0
Denmark / Danemark	425	193	64.0	20.7	30.7	40.7	50.7	60.7
France	2640	4077	37.5	54.5	64.5	59.5	74.5	84.5
Germany, Fed. Rep. of / Allemagne, Rép. féd. d'	2798	5019	48.7	104.1	104.1	104.1	106.1	108.1
Ireland / Irlande	20	71	42.6	61.2	66.2	71.2	71.2	81.2
Italy / Italie	1533	1405	63.2	49.3	59.3	69.3	69.3	79.3
Netherlands / Pays-Bas	231	166	77.0	56.0	66.0	71.0	76.0	86.0
UK / Royaume-Uni	258	567	27.6	49.3	59.3	69.3	69.3	79.3
EEC(9) / CEE(9)	8163	11772	45.8	64.4	..	..	..	..
Austria / Autriche	1429	2812	54.8	89.6	90.6	91.6	94.6	99.6
Switzerland / Suisse	354	568	53.3	74.9	79.9	84.9	84.9	89.9
CENTRAL EUROPE / EUROPE CENTRALE	1783	3381	54.5	86.8	..	..	..	..
Cyprus / Chypre	-	-	-	-	-	-	-	-
Greece / Grèce	16	150	5.6	38.2	48.2	58.2	68.2	78.2
Israel / Israël	-	-	-	-	-	-	-	-
Malta / Malte	-	-	-	-	-	-	-	-
Portugal	27	765	2.2	40.6	50.6	60.6	60.6	70.6
Spain / Espagne	319	220	16.2	11.0	21.0	31.0	41.0	61.0
Turkey / Turquie	-	453	-	13.6	23.6	33.6	43.6	63.6
Yugoslavia / Yougoslavie	386	847	13.4	23.9	33.9	43.9	53.9	73.9
SOUTHERN EUROPE / EUROPE MERIDIONALE	748	2435	9.3	21.5	..	..	..	..
Bulgaria / Bulgarie	-	342	-	34.4	44.4	54.4	74.4	84.4
Czechoslovakia / Tchécoslovaquie	424	1037	19.5	36.0	46.0	56.0	66.0	76.0
German Dem. Rep. / Rép. dém. allemande	1183	1313	77.3	74.4	79.4	84.4	84.4	89.4
Hungary / Hongrie	100	129	15.2	11.1	21.1	31.1	41.1	61.6
Poland / Pologne	1176	1683	31.8	46.2	56.2	66.2	66.2	76.2
Romania / Roumanie	943	1383	24.6	40.5	50.5	60.5	60.5	70.5
EASTERN EUROPE / EUROPE ORIENTALE	3826	5888	28.9	42.5	..	..	..	..
EUROPE	27557	43140	44.8	62.0	..	..	..	..

Annex table 15.2
Recovery of waste paper

Tableau annexe 15.2
Récupération de vieux papiers

Country	Recovery/Récupération (1000 m.t./t.m.)				Recovery rate/Taux de récupération (%) Real/Réel				Scenarios/Scénarios Low/bas		High/Haut		Pays
	1959-61	1969-71	1972-74	1979-81	1959-61	1969-71	1972-74	1979-81	1990	2000	1990	2000	
Finland	76	137	153	235	19.9	19.0	16.3	20.6	24.6	28.6	26.6	31.6	Finlande
Iceland	-	-	-	-	-	-	-	-	-	-	-	-	Islande
Norway	48	82	94	118	15.9	17.5	19.3	22.7	26.7	29.7	28.7	33.7	Norvège
Sweden	233	343	437	547	26.5	22.8	27.5	32.9	34.9	36.9	36.9	40.9	Suède
NORDIC COUNTRIES	357	562	684	900	22.7	20.8	22.6	27.0	..	..	..	..	PAYS NORDIQUES
Belgium-Luxembourg	141	295	386	379	24.4	27.8	36.3	27.3	30.3	32.3	32.3	36.3	Belgique-Luxembourg
Denmark	81	118	182	214	19.4	16.4	24.6	28.1	31.1	33.1	33.1	37.1	Danemark
France	697	1300	1614	1983	27.1	27.0	28.2	32.0	34.0	36.0	36.0	40.0	France
Germany, Fed. Rep. of	1141	2248	2502	3340	26.7	29.7	30.5	34.8	36.8	38.8	38.8	42.8	Allemagne, Rép. féd. d'
Ireland	10	22	57	48	8.8	10.3	19.8	17.5	21.5	25.5	23.5	29.5	Irlande
Italy	264	730	993	1491	16.9	20.6	23.8	28.7	31.7	33.7	33.7	37.7	Italie
Netherlands	329	711	837	966	34.4	40.3	43.9	42.7	43.7	44.7	45.7	48.7	Pays-Bas
UK	1436	1975	2083	2166	28.1	28.1	27.7	30.4	32.4	34.4	34.4	38.4	Royaume-Uni
EEC(9)	4099	7399	8654	10587	26.3	27.7	29.0	32.3	..	..	..	..	CEE(9)
Austria	70	183	205	274	23.5	29.7	26.6	29.9	32.9	34.9	34.9	38.9	Autriche
Switzerland	157	303	363	386	31.6	33.2	38.0	37.8	39.8	41.8	41.8	44.8	Suisse
CENTRAL EUROPE	227	486	568	660	28.6	28.1	32.9	34.1	..	..	..	..	EUROPE CENTRALE
Cyprus	..	..	..	-	..	..	..	-	4.0	8.0	6.0	12.0	Chypre
Greece	23	35	35*	40	18.3	16.5	11.2	9.0	13.0	17.0	15.0	21.0	Grèce
Israel	..	..	..	41	..	..	..	15.5	19.5	23.5	21.5	27.5	Israël
Malta	..	..	..	-	..	..	..	-	4.0	8.0	6.0	12.0	Malte
Portugal	18	9	22	172	17.0	3.2	7.2	39.7	41.7	42.7	43.7	46.7	Portugal
Spain	94	368	644	991	25.2	25.4	31.1	37.1	39.1	41.1	41.1	44.1	Espagne
Turkey	7	..	..	218	7.6	..	..	39.4	41.4	42.4	43.4	47.4	Turquie
Yugoslavia	..	..	177	353	..	..	27.1	33.0	35.0	37.0	37.0	41.0	Yougoslavie
SOUTHERN EUROPE	..	..	..	1815	..	..	..	33.1	..	..	..	..	EUROPE MERIDIONALE
Bulgaria	..	..	91	154	..	..	17.4	31.8	33.8	35.8	35.8	39.8	Bulgarie
Czechoslovakia	..	..	..	377	..	..	..	35.3	37.3	39.3	39.3	43.3	Tchécoslovaquie
German Dem. Rep.	..	..	..	596	..	..	..	43.1	44.1	45.1	46.1	49.1	Rép. dém. allemande
Hungary	..	..	154	251	..	..	29.8	38.2	40.2	41.2	42.2	45.2	Hongrie
Poland	..	..	385	443	..	..	29.5	32.2	34.3	36.3	36.3	40.3	Pologne
Romania	..	..	140	188	..	..	24.2	26.9	29.9	32.9	31.9	35.9	Roumanie
EASTERN EUROPE	..	..	1395*	2009	..	..	27.1	35.5	..	..	..	..	EUROPE ORIENTALE
EUROPE	(4825)a/	(8859)a/	12179	15971	..	..	27.9	32.4	..	..	..	..	EUROPE

a/ Total of countries with data.

a/ Total des pays ayant des données.

Annex table 16.1

Europe's trade in coniferous logs by country groups and main countries 1964-66 to 1979-81

Tableau annexe 16.1

Commerce européen de grumes de conifères, par groupes de pays et pays principaux, 1964-66 à 1979-81

Country	Volume (1000 m3)				Percentage share/Part en pourcentage (%)				Pays
	1964-66	1969-71	1974-76	1979-81	1964-66	1969-71	1974-76	1979-81	
Exports/Exportations									
Europe	1351.2	1548.3	2723.6	3438.8	100.0	100.0	100.0	100.0	Europe
Nordic countries	452.7	465.0	390.4	1056.6	33.5	30.0	14.3	30.7	Pays nordiques
EEC(9)	479.1	536.2	1346.8	958.8	35.4	34.6	49.4	27.9	CEE(9)
Central Europe	209.7	324.7	590.1	719.2	15.5	21.0	21.7	20.9	Europe centrale
Southern Europe	6.2	19.3	13.1	27.4	.5	1.5	.5	.8	Europe méridionale
Eastern Europe	203.5	203.1	383.2	676.8	15.1	13.1	14.1	19.7	Europe orientale
Sweden	348.5	395.9	228.9	189.8	25.8	25.6	8.4	5.5	Suède
France	227.4	172.3	119.2	121.4	16.8	11.1	4.4	3.5	France
Czechoslovakia	194.8	182.1	267.7	633.7	14.4	11.8	9.8	18.4	Tchécoslovaquie
Belgium-Luxembourg	138.1	92.4	182.5	337.3	10.2	6.0	6.7	9.8	Belgique-Luxembourg
Austria	109.9	144.6	236.6	290.4	8.1	9.3	8.7	8.5	Autriche
Switzerland	99.8	180.1	353.5	428.8	7.4	11.6	13.0	12.5	Suisse
Germany, Fed. Rep. of	87.2	184.2	1001.8	450.3	6.5	11.9	36.8	13.1	Allemagne, Rép. féd. d'
Finland	58.6	31.0	129.0	622.8	4.3	2.0	4.7	18.1	Finlande
Norway	45.6	38.1	31.6	244.0	3.4	2.5	1.2	7.1	Norvège
Other	41.3	127.6	172.8	120.3	3.1	8.2	6.3	3.5	Autre
Imports/Importations									
Europe	2852.7	3176.4	4906.0	5556.7	100.0	100.0	100.0	100.0	Europe
Nordic countries	779.2	746.9	1347.8	1284.7	27.3	23.5	27.5	23.1	Pays nordiques
EEC(9)	1334.7	1388.8	1939.9	2237.5	46.8	43.7	39.5	40.3	CEE(9)
Central Europe	106.4	157.4	736.7	1085.5	3.7	5.0	15.0	19.5	Europe centrale
Southern Europe	114.4	108.7	191.1	129.1	4.0	3.4	3.9	2.3	Europe méridionale
Eastern Europe	518.0	774.6	690.5	819.9	18.2	24.4	14.1	14.8	Europe orientale
Italy	646.2	744.4	1128.0	1348.0	22.7	23.5	23.0	24.3	Italie
Finland	621.6	506.3	725.9	509.8	21.8	15.9	14.8	9.2	Finlande
Germany, Fed. Rep. of	524.4	467.5	346.9	703.0	18.4	14.7	7.0	12.6	Allemagne, Rép. féd. d'
Hungary	440.7	426.5	366.9	511.8	15.4	13.4	7.5	9.2	Hongrie
Norway	120.5	208.0	239.9	186.8	4.2	6.5	4.9	3.4	Norvège
Bulgaria	60.6	347.8	323.6	227.9	2.1	11.0	6.6	4.1	Bulgarie
Sweden	36.9	32.5	382.0	508.1	1.3	1.0	7.8	9.1	Suède
Austria	26.3	82.0	690.1	1025.8	0.9	2.6	14.1	18.5	Autriche
Other	375.5	361.4	701.8	535.5	13.2	11.4	14.3	9.6	Autre

Annex table 16.2

Europe's trade in non-coniferous logs by country groups and main countries 1964-66 to 1979-81

Tableau annexe 16.2

Commerce européen de grumes de feuillus, par groupes de pays et pays principaux, 1964-66 à 1979-81

Country	Volume (1000 m3)				Percentage share/Part en pourcentage (%)				Pays
	1964-66	1969-71	1974-76	1979-81	1964-66	1969-71	1974-76	1979-81	
Exports/Exportations									
Europe	1055.8	1556.1	2038.4	2448.7	100.0	100.0	100.0	100.0	Europe
Nordic countries	52.7	46.8	29.6	29.7	5.0	3.0	1.4	1.2	Pays nordiques
EEC(9)	891.9	1082.6	1126.6	1264.3	84.5	69.6	55.3	51.6	CEE(9)
Central Europe	92.2	160.2	382.7	558.0	8.7	10.3	18.8	22.8	Europe centrale
Southern Europe	11.6	81.6	278.0	323.6	1.1	5.2	13.6	13.2	Europe méridionale
Eastern Europe	7.4	184.9	221.5	273.1	.7	11.9	10.9	11.2	Europe orientale
France	670.6	759.6	718.0	785.6	63.5	48.8	35.2	32.1	France
Belgium-Luxembourg	93.4	145.4	120.7	173.9	8.9	9.4	5.9	7.1	Belgique-Luxembourg
Germany, Fed. Rep. of	84.9	107.8	246.0	196.6	8.1	6.9	12.1	8.0	Allemagne, Rép. féd. d'
Switzerland	50.7	99.1	246.2	347.3	4.8	6.4	12.1	14.2	Suisse
Austria	41.5	61.1	136.5	210.7	3.9	3.9	6.7	8.6	Autriche
Czechoslovakia	4.1	160.7	162.3	247.3	0.4	10.3	7.9	10.1	Tchécoslovaquie
Yugoslavia	0.4	66.5	254.4	288.5	-	4.3	12.5	11.8	Yougoslavie
Other	110.2	155.9	154.3	198.8	10.4	10.0	7.6	8.1	Autre
Imports/Importations									
Europe	6698.5	8633.4	8708.3	9153.3	100.0	100.0	100.0	100.0	Europe
Nordic countries	64.0	49.5	56.9	35.2	1.0	.6	.6	.4	Pays nordiques
EEC(9)	5836.5	6690.7	6441.7	6232.7	87.1	77.5	74.0	68.1	CEE(9)
Central Europe	269.4	266.1	222.8	1141.6	4.0	3.1	2.6	12.5	Europe centrale
Southern Europe	327.2	1209.6	1640.5	1495.4	4.9	14.0	18.8	16.3	Europe méridionale
Eastern Europe	201.4	417.5	346.4	248.4	3.0	4.8	4.0	2.7	Europe orientale
Germany, Fed. Rep. of	1855.1	1773.5	1226.8	1115.3	27.7	20.5	14.1	12.2	Allemagne, Rép. féd. d'
France	1349.8	1658.4	1810.4	1669.7	20.2	19.2	20.8	18.2	France
Italy	1257.2	1955.3	2277.0	2626.0	18.8	22.7	26.1	28.7	Italie
United Kingdom	468.1	325.3	228.1	156.3	7.0	3.8	2.6	1.7	Royaume-Uni
Belgium-Luxembourg	373.6	451.2	406.4	334.8	5.6	5.2	4.7	3.6	Belgique-Luxembourg
Netherlands	363.5	396.3	394.6	284.0	5.4	4.6	4.5	3.1	Pays-Bas
Switzerland	239.4	220.0	158.1	165.0	3.6	2.6	1.8	1.8	Suisse
Spain	127.0	736.1	931.8	630.7	1.9	8.5	10.7	6.9	Espagne
Portugal	53.2	190.4	310.3	308.0	0.8	2.2	3.6	3.4	Portugal
Austria	30.0	46.1	64.7	976.6	0.4	0.5	0.8	10.7	Autriche
Greece	29.7	103.0	203.3	281.0	0.4	1.2	2.3	3.1	Grèce
Other	551.9	777.8	696.8	605.5	8.2	9.0	8.0	6.6	Autre

Annex table 16.3

Europe's trade in coniferous pulpwood (round and split)
by country groups and main countries, 1964-66 to 1979-81

Tableau annexe 16.3

Commerce européen de bois de trituration (rondins et quartiers)
résineux, par groupes de pays et pays principaux, 1964-66 à 1979-81

Country	Volume (1000 m3)				Percentage share/Part en pourcentage (%)				Pays
	1964-66	1969-71	1974-76	1979-81	1964-66	1969-71	1974-76	1979-81	
Exports/Exportations									
Europe	2937.2	5054.0	6494.7	7043.4	100.0	100.0	100.0	100.0	Europe
Nordic countries	1951.0	3498.8	1559.9	1629.5	66.5	69.3	24.0	23.1	Pays nordiques
EEC(9)	337.5	669.2	2433.3	2154.9	11.5	13.2	37.5	30.6	CEE(9)
Central Europe	9.8	59.9	99.5	158.5	.3	1.2	1.5	2.3	Europe centrale
Southern Europe	24.2	117.1	67.4	189.3	.8	2.3	1.0	2.7	Europe méridionale
Eastern Europe	614.7	709.0	2334.6	2911.2	20.9	14.0	36.0	41.3	Europe orientale
Sweden	1346.3	3067.5	1412.8	466.2	45.9	60.7	21.8	6.6	Suède
Finland	466.2	286.7	27.8	739.9	15.9	5.7	0.4	10.5	Finlande
Poland	356.3	151.3	683.7	747.8	12.1	3.0	10.5	10.6	Pologne
France	238.4	302.3	212.2	122.2	8.1	6.0	3.3	1.7	France
Czechoslovakia	202.5	475.2	1189.7	1756.3	6.9	9.4	18.3	25.0	Tchécoslovaquie
Norway	138.5	144.6	119.3	423.4	4.7	2.8	1.8	6.0	Norvège
Germany, Fed. Rep. of	71.2	261.6	1693.9	1140.3	2.4	5.2	26.1	16.2	Allemagne, Rép. féd. d'
Other	117.8	364.8	1154.9	1647.3	4.0	7.2	17.8	23.4	Autre
Imports/Importations									
Europe	7882.2	9706.3	10090.7	10348.1	100.0	100.0	100.0	100.0	Europe
Nordic countries	3437.3	3905.1	4674.3	4639.5	43.6	40.2	46.3	44.9	Pays nordiques
EEC(9)	2857.1	3281.5	2434.6	2194.2	36.2	33.8	24.1	21.2	CEE(9)
Central Europe	341.4	702.1	711.1	976.1	4.3	7.2	7.1	9.4	Europe centrale
Southern Europe	216.9	504.1	851.3	1256.3	2.8	5.2	8.4	12.1	Europe méridionale
Eastern Europe	1029.5	1313.5	1419.4	1282.0	13.1	13.6	14.1	12.4	Europe orientale
Norway	1989.1	2715.4	1375.3	622.2	25.2	28.0	13.6	6.0	Norvège
Finland	1231.4	1031.0	1429.2	1315.7	15.6	10.6	14.1	12.7	Finlande
France	897.9	960.6	560.7	426.1	11.4	9.9	5.6	4.1	France
German Dem. Rep.	773.1	970.0	824.1	673.1	9.8	10.0	8.2	6.5	Rép. dém. allemande
Germany, Fed. Rep. of	536.3	779.2	430.1	478.6	6.8	8.0	4.3	4.6	Allemagne, Rép. féd. d'
Italy	462.7	582.3	662.0	605.9	5.9	6.0	6.6	5.9	Italie
Netherlands	379.3	344.8	218.3	272.2	4.8	3.5	2.2	2.6	Pays-Bas
Belgium-Luxembourg	320.3	412.9	497.4	410.5	4.1	4.3	4.9	4.0	Belgique-Luxembourg
United Kingdom	260.6	201.7	66.0	0.8	3.3	2.1	0.6	-	Royaume-Uni
Austria	222.2	396.4	546.6	763.1	2.8	4.1	5.4	7.4	Autriche
Sweden	216.8	158.7	1869.8	2701.6	2.7	1.6	18.5	26.1	Suède
Spain	120.2	134.5	175.9	414.2	1.5	1.4	1.7	4.0	Espagne
Yugoslavia	52.9	347.1	666.1	842.0	0.7	3.6	6.6	8.1	Yougoslavie
Other	419.4	671.7	769.2	822.1	5.4	6.9	7.7	8.0	Autre

Annex table 16.4   Tableau annexe 16.4

Europe's trade in non-coniferous pulpwood (round and split), by country groups and main countries, 1964-66 to 1979-81

Commerce européen de bois de triturations (rondins et quartiers) feuillus, par groupes de pays et pays principaux, 1964-66 à 1979-81

Country	Volume (1000 m3)				Percentage share/Part en pourcentage (%)				Pays
	1964-66	1969-71	1974-76	1979-81	1964-66	1969-71	1974-76	1979-81	
Exports/Exportations									
Europe	2479.7	3303.2	3279.4	3820.6	100.0	100.0	100.0	100.0	Europe
Nordic countries	238.4	434.1	263.4	228.7	9.6	13.2	8.0	6.0	Pays nordiques
EEC(9)	479.0	899.1	1304.7	1777.3	19.3	27.2	39.8	46.5	CEE(9)
Central Europe	4.4	17.5	8.3	10.3	.2	.5	.3	.3	Europe centrale
Southern Europe	371.3	360.3	249.6	761.4	15.0	10.9	7.6	19.9	Europe méridionale
Eastern Europe	1386.6	1592.2	1453.4	1042.9	55.9	48.2	44.3	27.3	Europe orientale
Romania	653.7	577.1	307.5	167.3	26.4	17.5	9.4	4.4	Roumanie
France	469.2	880.8	1255.9	1558.7	18.9	26.7	38.3	40.8	France
Czechoslovakia	339.2	424.6	248.0	-	13.7	12.8	7.5	-	Tchécoslovaquie
Yugoslavia	303.7	310.7	200.0	306.8	12.3	9.4	6.1	8.0	Yougoslavie
Poland	261.5	188.0	363.7	387.2	10.5	5.7	11.1	10.1	Pologne
Sweden	132.7	129.3	153.3	105.6	5.4	3.9	4.7	2.8	Suède
Hungary	128.5	378.5	526.9	472.7	5.2	11.5	16.1	12.4	Hongrie
Other	190.7	414.2	224.1	822.3	7.7	12.5	6.8	21.5	Autre
Imports/Importations									
Europe	2521.4	4338.0	5453.2	6697.0	100.0	100.0	100.0	100.0	Europe
Nordic countries	316.6	1273.9	2065.9	2031.8	12.6	29.4	37.9	30.3	Pays nordiques
EEC(9)	1655.9	2236.7	2226.9	3294.8	65.7	51.5	40.8	49.2	CEE(9)
Central Europe	363.1	509.6	529.4	637.7	14.4	11.7	9.7	9.5	Europe centrale
Southern Europe	49.2	202.3	449.8	648.1	1.9	4.7	8.3	9.7	Europe méridionale
Eastern Europe	136.6	115.5	181.2	84.9	5.4	2.7	3.3	1.3	Europe orientale
Italy	658.7	846.8	602.2	705.0	26.1	19.5	11.0	10.5	Italie
Germany, Fed. Rep. of	621.4	673.5	415.9	523.5	24.6	15.5	7.6	7.8	Allemagne, Rép. féd. d'
Austria	326.8	478.0	507.7	531.2	13.0	11.0	9.3	7.9	Autriche
Belgium-Luxembourg	211.9	652.7	1205.1	1994.4	8.4	15.1	22.1	29.8	Belgique-Luxembourg
Norway	140.3	191.4	120.8	87.3	5.6	4.4	2.2	1.3	Norvège
Finland	105.9	713.3	1807.7	1698.7	4.2	16.5	33.2	25.4	Finlande
Other	456.4	782.3	793.8	1156.9	18.1	18.0	14.6	17.3	Autre

Annex table 16.5

Europe's trade in chips and particles, by country groups and main countries, 1964-66 to 1979-81

Tableau annexe 16.5

Commerce européen de plaquettes et particules, par groupes de pays et pays principaux, 1964-66 à 1979-81

Country	Volume (1000 m3)				Percentage share/Part en pourcentage (%)				Pays
	1964-66	1969-71	1974-76	1979-81	1964-66	1969-71	1974-76	1979-81	
Exports/Exportations									
Europe	342.2	591.3	740.9	1082.4	100.0	100.0	100.0	100.0	Europe
Nordic countries	220.6	410.7	475.8	730.4	64.5	69.5	64.2	67.5	Pays nordiques
EEC(9)	104.7	111.1	179.7	300.5	30.6	18.8	24.2	27.7	CEE(9)
Central Europe	15.4	24.0	13.0	2.9	4.5	4.1	1.8	.3	Europe centrale
Southern Europe	0.0	0.0	0.0	34.8	0.0	0.0	0.0	3.2	Europe méridionale
Eastern Europe	1.5	45.5	72.4	13.8	.4	7.6	9.8	1.3	Europe orientale
Finland	140.0	147.2	109.3	196.4	40.9	24.9	14.8	18.1	Finlande
France	95.7	102.8	117.9	140.7	28.0	17.4	15.9	13.0	France
Norway	73.9	62.2	37.7	249.9	21.6	10.5	5.1	23.1	Norvège
Austria	15.4	24.0	10.7	-	4.5	4.1	1.4	-	Autriche
Sweden	6.7	201.3	328.8	284.1	2.0	34.0	44.4	26.3	Suède
Denmark	-	-	42.9	108.9	-	-	5.8	10.1	Danemark
Other	10.5	53.8	93.6	102.4	3.0	9.1	12.6	9.4	Autre
Imports/Importations									
Europe	242.1	549.6	861.9	2107.0	100.0	100.0	100.0	100.0	Europe
Nordic countries	193.3	389.9	684.9	1597.6	79.8	71.0	79.5	75.8	Pays nordiques
EEC(9)	42.1	131.3	155.5	450.7	17.4	23.9	18.0	21.4	CEE(9)
Central Europe	6.7	24.4	20.5	58.4	2.8	4.4	2.4	2.8	Europe centrale
Southern Europe	0.0	4.0	1.0	.3	0.0	.7	.1	0.0	Europe méridionale
Eastern Europe	0.0	0.0	0.0	0.0	0.0	0.0	0.0	0.0	Europe orientale
Sweden	166.0	205.2	270.0	1210.0	68.6	37.3	31.3	57.4	Suède
Belgium-Luxembourg	30.6	104.8	77.1	120.3	12.6	19.1	9.0	5.7	Belgique-Luxembourg
Norway	19.7	165.8	307.0	266.8	8.1	30.2	35.6	12.7	Norvège
France	7.0	13.7	68.6	293.6	2.9	2.5	8.0	14.0	France
Finland	7.6	18.9	107.9	120.8	3.2	3.4	12.5	5.7	Finlande
Other	11.2	41.2	31.3	95.5	4.6	7.5	3.6	4.5	Autre

Annex table 16.6

Europe's trade in wood residues, by country groups and main countries, 1964-66 to 1979-81

Tableau annexe 16.6

Commerce européen de déchets de bois, par groupes de pays et pays principaux, 1964-66 à 1979-81

Country	Volume (1000 m3)				Percentage share/Part en pourcentage (%)				Pays
	1964-66	1969-71	1974-76	1979-81	1964-66	1969-71	1974-76	1979-81	
Exports/Exportations									
Europe	629.8	755.4	1798.0	2429.5	100.0	100.0	100.0	100.0	Europe
Nordic countries	150.0	138.5	173.3	245.1	23.8	18.3	9.7	10.1	Pays nordiques
EEC(9)	305.0	416.2	1190.7	1413.4	48.4	55.1	66.2	58.2	CEE(9)
Central Europe	105.6	175.9	274.0	152.2	16.8	23.3	15.2	6.3	Europe centrale
Southern Europe	0.0	4.4	.8	228.9	0.0	.6	0.0	9.4	Europe méridionale
Eastern Europe	69.2	20.4	159.2	389.9	11.0	2.7	8.9	16.0	Europe orientale
France	170.1	182.8	421.6	425.1	27.0	24.2	23.5	17.5	France
Sweden	118.3	109.3	48.3	76.0	18.8	14.4	2.7	3.1	Suède
Germany, Fed. Rep. of	70.3	133.6	569.8	682.4	11.2	17.7	31.7	28.1	Allemagne, Rép. féd. d'
Switzerland	69.6	120.0	183.9	41.5	11.1	15.9	10.2	1.7	Suisse
Netherlands	60.0	94.4	187.3	220.4	9.5	12.5	10.4	9.1	Pays-Bas
Czechoslovakia	50.7	4.3	135.6	349.0	8.0	0.6	7.5	14.4	Tchécoslovaquie
Other	90.8	111.0	251.5	635.1	14.4	14.7	14.0	26.1	Autre
Imports/Importations									
Europe	1366.8	1799.0	3079.3	2583.1	100.0	100.0	100.0	100.0	Europe
Nordic countries	798.3	801.6	1458.0	617.1	58.4	44.6	47.4	23.9	Pays nordiques
EEC(9)	512.3	810.2	1127.9	1157.9	37.5	45.0	36.6	44.8	CEE(9)
Central Europe	56.2	166.6	477.0	643.0	4.1	9.3	15.5	24.9	Europe centrale
Southern Europe	0.0	20.6	16.4	47.5	0.0	0.0	0.0	4.6	Europe méridionale
Eastern Europe	0.0	0.0	0.0	117.6	0.0	0.0	0.0	4.6	Europe orientale
Finland	760.9	687.7	1261.0	252.9	55.7	38.2	41.0	9.8	Finlande
Germany, Fed. Rep. of	290.2	508.1	312.8	473.1	21.2	28.2	10.2	18.3	Allemagne, Rép. féd. d'
Italy	171.9	155.6	437.7	123.2	12.6	8.7	14.2	4.8	Italie
Switzerland	40.5	109.7	294.1	379.5	3.0	6.1	9.6	14.7	Suisse
Belgium-Luxembourg	29.1	106.9	216.6	345.0	2.1	5.9	7.0	13.3	Belgique-Luxembourg
Norway	26.6	98.5	141.8	204.2	2.0	5.5	4.6	7.9	Norvège
Austria	15.7	56.9	182.9	263.5	1.1	3.2	5.9	10.2	Autriche
Other	31.9	75.6	232.4	541.7	2.3	4.2	7.5	21.0	Autre

Annex table 16.7

Europe's trade in coniferous sawnwood, by country groups and main countries, 1964-66 to 1979-81

Tableau annexe 16.7

Commerce européen de sciages résineux, par groupes de pays et pays principaux, 1964-66 à 1979-81

Country	Volume (1000 m3)				Percentage share/Part en pourcentage (%)				Pays
	1964-66	1969-71	1974-76	1979-81	1964-66	1969-71	1974-76	1979-81	
Exports/Exportations									
Europe	16939.8	19274.5	18027.8	21516.0	100.0	100.0	100.0	100.0	**Europe**
Nordic countries	9553.0	11745.8	10445.8	12814.3	56.4	61.0	58.0	59.6	Pays nordiques
EEC(9)	619.2	524.3	976.2	918.0	3.7	2.7	5.4	4.3	CEE(9)
Central Europe	2763.3	3378.9	3304.6	4139.4	16.3	17.5	18.3	19.2	Europe centrale
Southern Europe	704.1	701.3	945.5	1273.4	4.1	3.6	5.2	5.9	Europe méridionale
Eastern Europe	3300.2	2924.2	2355.7	2370.9	19.5	15.2	13.1	11.0	Europe orientale
Sweden	5294.3	7042.1	6431.4	6143.9	31.3	36.5	35.7	28.6	Suède
Finland	4139.8	4593.4	3646.7	6290.5	24.4	23.8	20.2	29.2	Finlande
Austria	2760.7	3359.2	3231.4	4088.4	16.3	17.4	17.9	19.0	Autriche
Romania	1492.2	1302.2	780.8	610.0	8.8	6.8	4.3	2.8	Romania
Poland	895.7	739.3	624.0	662.5	5.3	3.8	3.5	3.1	Pologne
Czechoslovakia	751.2	610.1	706.3	1020.7	4.4	3.2	3.9	4.8	Tchécoslovaquie
Portugal	391.3	434.7	483.0	886.2	2.3	2.3	2.7	4.1	Portugal
France	337.4	190.1	102.6	149.6	2.0	1.0	0.6	0.7	France
Germany, Fed. Rep. of	171.5	199.4	692.6	427.1	1.0	1.0	3.8	2.0	Allemagne, Rép. féd. d'
Other	705.7	804.0	1329.0	1237.1	4.2	4.2	7.4	5.7	Autre
Imports/Importations									
Europe	26019.5	27306.0	24950.4	27778.0	100.0	100.0	100.0	100.0	**Europe**
Nordic countries	354.3	412.0	370.9	638.3	1.4	1.5	1.5	2.3	Pays nordiques
EEC(9)	21487.2	21785.6	19357.0	21897.0	82.6	79.8	77.6	78.8	CEE(9)
Central Europe	328.5	347.3	310.5	839.3	1.3	1.3	1.2	3.0	Europe centrale
Southern Europe	1314.4	1734.7	1735.0	1786.7	5.0	6.3	7.0	6.5	Europe méridionale
Eastern Europe	2535.1	3026.4	3177.0	2616.7	9.7	11.1	12.7	9.4	Europe orientale
United Kingdom	8801.4	7979.4	7013.2	6222.4	33.8	29.2	28.1	22.4	Royaume-Uni
Germany, Fed. Rep. of	3771.3	3812.8	2593.7	4188.8	14.5	14.0	10.4	15.1	Allemagne, Rép. féd. d'
Netherlands	2627.2	2656.7	2262.3	2237.7	10.1	9.7	9.1	8.1	Pays-Bas
Italy	2616.9	3203.6	3316.3	4217.4	10.1	11.7	13.3	15.2	Italie
France	1440.8	1677.2	1750.3	2366.1	5.5	6.1	7.0	8.5	France
Germany, Fed. Rep of	1349.0	1383.8	1535.2	1314.5	5.2	5.1	6.1	4.7	Allemagne, Rép. féd. d'
Denmark	1080.2	1182.9	1177.8	1202.5	4.1	4.3	4.7	4.3	Danemark
Belgium-Luxembourg	882.6	891.6	897.1	1036.3	3.4	3.3	3.6	3.7	Belgique-Luxembourg
Hungary	779.1	1046.5	1024.4	820.7	3.0	3.8	4.1	3.0	Hongrie
Spain	601.7	777.7	812.6	671.0	2.3	2.9	3.3	2.4	Espagne
Other	2069.3	2693.8	2567.5	3500.6	8.0	9.9	10.3	12.6	Autre

Annex table 16.8

Europe's trade in non-coniferous sawnwood, by country groups and main countries, 1964-66 to 1979-81

Tableau annexe 16.8

Commerce européen de sciages feuillus, par groupes de pays et pays principaux, 1964-66 à 1979-81

Country	Volume (1000 m3)				Percentage share/Part en pourcentage (%)				Pays
	1964-66	1969-71	1974-76	1979-81	1964-66	1969-71	1974-76	1979-81	
Exports/Exportations									
Europe	1937.9	2419.5	2627.0	2933.0	100.0	100.0	100.0	100.0	Europe
Nordic countries	69.4	80.6	69.9	68.2	3.6	3.3	2.7	2.3	Pays nordiques
EEC(9)	568.7	725.2	1077.6	1283.5	29.3	30.0	41.0	43.8	CEE(9)
Central Europe	68.2	82.5	138.2	180.3	3.5	3.4	5.3	6.1	Europe centrale
Southern Europe	499.4	605.7	608.3	803.3	25.8	25.0	23.1	27.4	Europe méridionale
Eastern Europe	732.2	925.5	733.0	597.7	37.8	38.3	27.9	20.4	Europe orientale
Romania	659.6	752.5	527.0	422.3	34.0	31.1	20.1	14.4	Roumanie
Yugoslavia	491.5	583.1	591.2	779.6	25.4	24.1	22.5	26.6	Yougoslavie
France	325.1	373.7	487.9	499.8	16.8	15.5	18.5	17.0	France
Germany, Fed. Rep. of	104.7	155.2	338.0	362.6	5.4	6.4	12.9	12.4	Allemagne, Rép. féd. d'
Denmark	79.6	96.9	62.0	85.2	4.1	4.0	2.4	2.9	Danemark
Finland	56.7	54.7	32.9	34.3	2.9	2.3	1.2	1.2	Finlande
Austria	55.6	56.7	81.0	110.6	2.8	2.3	3.1	3.8	Autriche
Other	165.1	346.7	507.0	638.6	8.5	14.3	19.3	21.7	Autre
Imports/Importations									
Europe	2754.2	3637.3	4553.6	6020.3	100.0	100.0	100.0	100.0	Europe
Nordic countries	167.5	176.2	180.8	159.1	6.1	4.8	4.0	2.6	Pays nordiques
EEC(9)	2156.0	2819.9	3467.2	4884.8	78.3	77.5	76.1	81.1	CEE(9)
Central Europe	110.0	137.2	178.6	223.6	4.0	3.8	3.9	3.7	Europe centrale
Southern Europe	171.5	351.3	571.9	701.9	6.2	9.7	12.6	11.7	Europe méridionale
Eastern Europe	149.2	152.7	155.1	50.9	5.4	4.2	3.4	.9	Europe orientale
United Kingdom	904.3	773.0	610.5	682.3	32.9	21.3	13.4	11.3	Royaume-Uni
Italy	413.6	712.8	803.3	1240.5	15.0	19.6	17.6	20.6	Italie
Germany, Fed. Rep. of	240.6	393.3	560.8	906.8	8.7	10.8	12.3	15.1	Allemagne, Rép. féd. d'
Netherlands	189.8	295.1	555.0	722.1	6.9	8.1	12.2	12.0	Pays-Bas
France	163.1	278.9	427.0	700.4	5.9	7.7	9.4	11.6	France
Belgium-Luxembourg	143.2	241.8	375.7	485.0	5.2	6.6	8.2	8.1	Belgique-Luxembourg
Sweden	93.8	89.1	85.7	80.5	3.4	2.5	1.9	1.3	Suède
Spain	90.0	183.2	422.2	494.7	3.3	5.0	9.3	8.2	Espagne
Other	515.8	670.1	713.4	708.0	18.7	18.4	15.7	11.8	Autre

Annex table 16.9

Europe's trade in plywood, by country groups and main countries, 1964-66 to 1979-81

Tableau annexe 16.9

Commerce européen de contreplaqués, par groupes de pays et pays principaux, 1964-66 à 1979-81

Country	Volume (1000 m3)				Percentage share/Part en pourcentage (%)				Pays
	1964-66	1969-71	1974-76	1979-81	1964-66	1969-71	1974-76	1979-81	
Exports/Exportations									
Europe	895.1	1201.4	1141.8	1488.5	100.0	100.0	100.0	100.0	Europe
Nordic countries	460.5	607.2	415.1	552.1	51.5	50.5	36.4	37.1	Pays nordiques
EEC(9)	202.5	339.4	440.8	590.6	22.6	28.3	38.6	39.7	CEE(9)
Central Europe	4.4	3.1	7.7	16.1	.5	.3	.7	1.1	Europe centrale
Southern Europe	79.9	105.8	129.6	131.3	8.9	8.8	11.3	8.8	Europe méridionale
Eastern Europe	147.8	145.9	148.6	198.4	16.5	12.1	13.0	13.3	Europe orientale
Finland	451.4	594.0	383.2	529.3	50.4	49.4	33.6	35.6	Finlande
Romania	100.5	102.6	107.2	125.3	11.2	8.6	9.4	8.4	Roumanie
France	62.1	97.0	167.2	170.6	6.9	8.1	14.6	11.4	France
Germany, Fed. Rep. of	52.4	60.1	49.2	55.1	5.9	5.0	4.3	3.7	Allemagne, Rép. féd. d'
Italy	42.6	98.3	89.9	96.3	4.8	8.2	7.9	6.5	Italie
Yugoslavia	28.3	25.5	45.0	38.4	3.2	2.1	3.9	2.6	Yougoslavie
Belgium-Luxembourg	27.6	44.9	69.6	113.7	3.1	3.7	6.1	7.6	Belgique-Luxembourg
Other	130.2	179.0	230.5	359.8	14.5	14.9	20.2	24.2	Autre
Imports/Importations									
Europe	1246.1	1971.5	2402.0	3253.0	100.0	100.0	100.0	100.0	Europe
Nordic countries	56.1	92.9	123.4	177.2	4.5	4.7	5.1	5.4	Pays nordiques
EEC(9)	1092.4	1683.6	2044.9	2812.3	87.6	85.4	85.1	86.5	CEE(9)
Central Europe	32.2	59.1	49.5	96.2	2.6	3.0	2.1	3.0	Europe centrale
Southern Europe	32.0	36.6	39.5	27.6	2.6	1.9	1.7	.8	Europe méridionale
Eastern Europe	33.4	99.3	144.7	139.7	2.7	5.0	6.0	4.3	Europe orientale
United Kingdom	831.8	1015.7	936.4	1002.6	66.8	51.5	39.0	30.8	Royaume-Uni
Netherlands	71.8	141.2	262.5	490.6	5.8	7.2	10.9	15.1	Pays-Bas
Germany, Fed. Rep. of	67.9	216.8	302.6	491.5	5.4	11.0	12.6	15.1	Allemagne, Rép. féd. d'
Sweden	42.8	63.4	78.2	100.0	3.4	3.2	3.2	3.1	Suède
Belgium-Luxembourg	35.9	58.9	124.6	217.7	2.9	3.0	5.2	6.7	Belgique-Luxembourg
Denmark	33.5	85.0	146.4	155.5	2.7	4.3	6.1	4.8	Danemark
France	31.6	132.0	211.2	309.6	2.5	6.7	8.8	9.5	France
Switzerland	31.2	54.7	41.0	74.4	2.5	2.8	1.7	2.3	Suisse
Other	99.6	203.8	299.1	411.1	8.0	10.3	12.5	12.6	Autre

Annex table 16.10

Europe's trade in particle board, by country groups and main countries, 1964-66 to 1979-81

Tableau annexe 16.10

Commerce européen de panneaux de particules, par groupes de pays et pays principaux, 1964-66 à 1979-81

Country	Volume (1000 m3)				Percentage share/Part en pourcentage (%)				Pays
	1964-66	1969-71	1974-76	1979-81	1964-66	1969-71	1974-76	1979-81	
Exports/Exportations									
Europe	885.1	1905.7	3725.4	4805.9	100.0	100.0	100.0	100.0	Europe
Nordic countries	105.4	322.3	751.2	921.7	11.9	16.9	20.2	19.2	Pays nordiques
EEC(9)	614.9	1192.2	2110.9	2377.5	69.5	62.6	56.6	49.4	CEE(9)
Central Europe	19.9	166.1	479.6	883.4	2.2	8.7	12.9	18.4	Europe centrale
Southern Europe	10.7	65.4	59.8	355.1	1.2	3.4	1.6	7.4	Europe méridionale
Eastern Europe	134.2	159.7	323.9	268.2	15.2	8.4	8.7	5.6	Europe orientale
Belgium-Luxembourg	310.3	698.9	1144.6	1133.4	34.6	36.7	30.7	23.6	Belgique-Luxembourg
Germany, Fed. Rep. of	100.3	227.1	545.0	674.6	11.2	11.9	14.6	14.1	Allemagne, Rép. féd. d'
France	87.4	84.2	222.1	376.7	9.8	4.4	6.0	7.8	France
Italy	84.1	99.1	41.8	35.4	9.4	5.2	1.1	0.7	Italie
Romania	60.8	72.6	265.0	231.3	6.8	3.8	7.1	4.8	Roumanie
Finland	51.7	170.5	283.5	344.5	5.8	9.0	7.6	7.2	Finlande
Sweden	41.0	100.9	364.9	523.0	4.6	5.3	9.8	10.9	Suède
Other	41.0	100.9	364.9	523.0	4.6	5.3	9.8	10.9	Autre
Imports/Importations									
Europe	869.4	1940.4	3735.8	4820.5	100.0	100.0	100.0	100.0	Europe
Nordic countries	39.3	82.3	126.5	154.9	4.5	4.2	3.4	3.2	Pays nordiques
EEC(9)	723.4	1460.3	2779.1	4029.9	83.2	75.3	74.4	83.6	CEE(9)
Central Europe	12.3	65.7	74.7	111.2	1.4	3.4	2.0	2.3	Europe centrale
Southern Europe	18.2	120.5	119.8	35.8	2.1	6.2	3.2	.8	Europe méridionale
Eastern Europe	76.2	211.6	635.7	488.7	8.8	10.9	17.0	10.1	Europe orientale
Germany, Fed. Rep. of	295.9	325.2	420.8	911.0	34.0	16.8	11.3	18.9	Allemagne, Rép. féd. d'
Netherlands	170.2	319.2	559.8	520.9	19.6	16.5	15.0	10.8	Pays-Bas
United Kingdom	128.2	452.9	991.1	1406.3	14.7	23.3	26.5	29.2	Royaume-Uni
German Dem. Rep.	57.3	79.9	248.8	195.0	6.6	4.1	6.7	4.1	Rép. dém. allemande
Denmark	48.3	65.2	155.9	156.7	5.5	3.4	4.2	3.2	Danemark
Belgium-Luxembourg	43.0	81.5	90.7	107.3	5.0	4.2	2.4	2.2	Belgique-Luxembourg
France	33.5	194.6	277.2	414.4	3.9	10.0	7.4	8.6	France
Other	93.0	421.9	991.5	1108.9	10.7	21.7	26.5	23.0	Autre

Annex table 16.11

Europe's trade in chemical pulp (paper), by country groups and main countries, 1964-66 to 1979-81

Tableau annexe 16.11

Commerce européen de pâte de bois chimique (pour la fabrication du papier) par groupes de pays et pays principaux, 1964-66 à 1979-81

Country	Volume (1000 m3)				Percentage share/Part en pourcentage (%)				Pays
	1964-66	1969-71	1974-76	1979-81	1964-66	1969-71	1974-76	1979-81	
Exports/Exportations									
Europe	5231.9	5704.4	5312.8	5864.6	100.0	100.0	100.0	100.0	Europe
Nordic countries	4644.3	4835.7	4358.4	4477.0	88.8	84.8	82.0	76.3	Pays nordiques
EEC(9)	180.9	292.8	334.7	414.1	3.5	5.1	6.3	7.1	CEE(9)
Central Europe	148.4	142.2	136.9	273.1	2.8	2.5	2.6	4.7	Europe centrale
Southern Europe	157.0	332.5	386.2	638.7	3.0	5.8	7.3	10.9	Europe méridionale
Eastern Europe	101.3	101.2	96.6	61.7	1.9	1.8	1.8	1.0	Europe orientale
Sweden	2669.6	2956.0	3148.4	2656.3	51.1	51.8	59.3	45.3	Suède
Finland	1768.3	1647.6	951.2	1635.8	33.8	28.9	17.9	27.9	Finlande
Norway	206.4	232.1	258.8	185.7	3.9	4.1	4.9	3.2	Norvège
Austria	143.7	128.6	125.0	255.2	2.7	2.2	2.4	4.2	Autriche
Portugal	138.3	300.2	330.5	411.8	2.6	5.3	6.2	7.0	Portugal
France	111.5	138.8	136.4	169.4	2.2	2.4	2.5	2.9	France
Other	194.1	301.1	362.5	561.4	3.7	5.3	6.8	9.5	Autre
Imports/Importations									
Europe	5648.1	7644.6	8248.1	9867.9	100.0	100.0	100.0	100.0	Europe
Nordic countries	53.0	166.8	273.2	384.5	.9	2.2	3.3	3.9	Pays nordiques
EEC(9)	4895.4	6303.7	6452.5	7717.6	86.7	82.5	78.2	78.2	CEE(9)
Central Europe	148.4	248.0	359.5	433.2	2.6	3.2	4.4	4.4	Europe centrale
Southern Europe	293.0	420.9	434.6	565.6	5.2	5.5	5.3	5.7	Europe méridionale
Eastern Europe	258.3	505.2	728.3	767.0	4.6	6.6	8.8	7.8	Europe orientale
United Kingdom	1811.2	1867.1	1628.2	1441.2	32.1	24.4	19.7	14.6	Royaume-Uni
Germany, Fed. Rep. of	892.5	1347.1	1577.8	2171.0	15.8	17.6	19.1	22.0	Allemagne, Rép. Féd. d'
Italy	775.1	1094.7	1113.0	1567.6	13.7	14.4	13.5	15.9	Italie
France	723.2	1025.4	1194.0	1527.7	12.8	13.4	14.5	15.5	France
Netherlands	397.7	546.5	541.4	556.4	7.1	7.1	6.6	5.6	Pays-Bas
Belgium-Luxembourg	187.3	299.1	314.1	332.5	3.3	3.9	3.8	3.4	Belgique-Luxembourg
Spain	181.8	268.8	271.0	256.1	3.2	3.5	3.3	2.6	Espagne
Switzerland	115.5	182.4	196.6	264.5	2.0	2.4	2.4	2.7	Suisse
Other	563.8	1013.5	1412.0	1750.9	10.0	13.3	17.1	17.7	Autre

Annex table 16.12

Europe's trade in newsprint, by country groups and main countries, 1964-66 to 1979-81

Tableau annexe 16.12

Commerce européen de papier journal, par groupes de pays et pays principaux, 1964-66 à 1979-81

Country	Volume (1000 m3)				Percentage share/Part en pourcentage (%)				Pays
	1964-66	1969-71	1974-76	1979-81	1964-66	1969-71	1974-76	1979-81	
	Exports/Exportations								
Europe	2040.4	2577.1	2479.7	3620.2	100.0	100.0	100.0	100.0	Europe
Nordic countries	1781.5	2298.9	2242.2	3234.7	87.3	89.2	90.4	89.4	Pays nordiques
EEC(9)	124.0	128.6	145.4	218.0	6.1	5.0	5.9	6.0	CEE(9)
Central Europe	63.0	62.7	42.4	63.0	3.1	2.4	1.7	1.7	Europe centrale
Southern Europe	4.5	12.0	10.1	29.5	.2	.5	.4	.8	Europe méridionale
Eastern Europe	67.4	74.9	39.6	75.0	3.3	2.9	1.6	2.1	Europe orientale
Finland	1093.9	1170.4	926.6	1455.7	53.6	45.4	37.4	40.2	Finlande
Sweden	437.5	672.8	908.9	1259.7	21.4	26.1	36.7	34.8	Suède
Norway	250.1	455.7	406.7	519.3	12.3	17.7	16.4	14.3	Norvège
Austria	63.0	62.6	31.7	31.5	3.1	2.4	1.3	0.9	Autriche
Netherlands	41.0	35.0	26.8	37.9	2.0	1.4	1.0	1.1	Pays-Bas
Other	154.9	180.6	179.0	316.1	7.6	7.0	7.2	8.7	Autre
	Imports/Importations								
Europe	1839.8	2577.8	3125.9	3597.8	100.0	100.0	100.0	100.0	Europe
Nordic countries	3.1	3.2	3.4	3.9	.2	.1	.1	.1	Pays nordiques
EEC(9)	1617.8	2184.0	2684.4	3101.2	87.9	84.7	85.9	86.2	CEE(9)
Central Europe	2.3	17.0	11.1	24.7	.1	.7	.4	.7	Europe centrale
Southern Europe	149.6	224.4	222.1	249.5	8.1	8.7	7.1	6.9	Europe méridionale
Eastern Europe	67.0	149.2	204.9	218.0	3.7	5.8	6.5	6.1	Europe orientale
United Kingdom	642.5	788.2	1117.8	1169.4	34.9	30.6	35.8	32.5	Royaume-Uni
Germany, Fed. Rep. of	520.5	699.2	684.3	827.2	28.3	27.1	21.9	23.0	Allemagne, Rép. féd. d'
Denmark	120.3	141.9	135.3	165.4	6.5	5.5	4.3	4.6	Danemark
Netherlands	106.7	219.9	262.4	322.8	5.8	8.5	8.4	9.0	Pays-Bas
France	103.6	172.5	306.6	363.3	5.6	6.7	9.8	10.1	France
Belgium-Luxembourg	74.8	106.8	121.0	130.7	4.1	4.1	3.9	3.6	Belgique-Luxembourg
Spain	50.2	61.9	97.0	77.3	2.8	2.4	3.1	2.1	Espagne
Other	221.2	387.4	401.5	541.7	12.0	15.1	12.8	15.1	Autre

Annex table 16.13

Europe's trade in printing and writing paper, by country groups and main countries, 1964-66 to 1979-81

Tableau annexe 16.13

Commerce européen de papiers impression-écriture, par groupes de pays et pays principaux, 1964-66 à 1979-81

Country	Volume (1000 m3)				Percentage share/Part en pourcentage (%)				Pays
	1964-66	1969-71	1974-76	1979-81	1964-66	1969-71	1974-76	1979-81	
Exports/Exportations									
Europe	1373.5	2820.6	4093.2	5704.2	100.0	100.0	100.0	100.0	Europe
Nordic countries	719.3	1275.0	1873.7	2425.3	52.4	45.2	45.8	42.5	Pays nordiques
EEC(9)	406.8	1078.1	1471.5	2360.7	29.6	38.2	35.9	41.4	CEE(9)
Central Europe	166.4	295.9	498.6	655.6	12.1	10.5	12.2	11.5	Europe centrale
Southern Europe	29.3	56.8	111.6	136.0	2.1	2.0	2.7	2.4	Europe méridionale
Eastern Europe	51.7	114.8	138.8	126.0	3.8	4.1	3.4	2.2	Europe orientale
Finland	392.4	779.2	1334.4	1688.8	28.6	27.6	32.6	29.6	Finlande
Sweden	174.9	271.2	345.4	523.8	12.7	9.6	8.4	9.2	Suède
Norway	152.0	224.6	193.9	212.7	11.1	8.0	4.7	3.7	Norvège
Austria	163.9	285.3	459.5	581.5	11.9	10.1	11.2	10.2	Autriche
Netherlands	128.0	179.9	260.3	267.9	9.3	6.4	6.4	4.7	Pays-Bas
Germany, Fed. Rep. of	86.4	287.2	425.4	743.5	6.3	10.2	10.4	13.1	Allemagne, Rép. féd. d'
France	62.7	166.9	301.4	526.5	4.6	5.9	7.4	9.2	France
Other	213.2	626.3	773.9	1159.5	15.5	22.2	18.9	20.3	Autre
Imports/Importations									
Europe	955.8	1887.5	2786.9	4252.6	100.0	100.0	100.0	100.0	Europe
Nordic countries	14.3	26.0	43.7	91.9	1.5	1.4	1.6	2.2	Pays nordiques
EEC(9)	856.3	1634.5	2319.0	3630.3	89.6	86.6	83.2	85.3	CEE(9)
Central Europe	31.6	81.4	124.6	216.2	3.3	4.3	4.5	5.1	Europe centrale
Southern Europe	18.7	56.6	128.6	174.0	2.0	3.0	4.6	4.1	Europe méridionale
Eastern Europe	34.9	89.0	171.0	140.2	3.6	4.7	6.1	3.3	Europe orientale
Germany, Fed. Rep. of	308.5	546.7	711.5	1126.2	32.2	29.0	25.5	26.5	Allemagne, Rép. féd. d'
Belgium-Luxembourg	135.4	204.3	232.5	341.7	14.2	10.8	8.3	8.0	Belgique-Luxembourg
Netherlands	122.3	207.9	285.2	393.6	12.8	11.0	10.2	9.3	Pays-Bas
United Kingdom	104.3	306.0	517.7	890.7	10.9	16.2	18.6	20.9	Royaume-Uni
France	93.5	254.6	388.4	575.5	9.8	13.5	13.9	13.5	France
Denmark	56.6	81.9	109.4	135.4	5.9	4.3	3.9	3.2	Danemark
Italy	31.4	26.0	54.9	143.6	3.3	1.4	2.0	3.4	Italie
Switzerland	28.4	73.3	110.3	159.8	3.0	3.9	4.1	3.8	Suisse
Other	75.4	186.8	377.0	486.1	7.9	9.9	13.5	11.4	Autre

Annex table 16.14

Europe's trade in other paper and paperboard, by country groups and main countries, 1964-66 to 1979-81

Tableau annexe 16.14

Commerce européen d'autres papiers et cartons, par groupes de pays et pays principaux, 1964-66 à 1979-81

Country	Volume (1000 m3)				Percentage share/Part en pourcentage (%)				Pays
	1964-66	1969-71	1974-76	1979-81	1964-66	1969-71	1974-76	1979-81	
Exports/Exportations									
Europe	3871.9	5539.2	6696.0	9054.7	100.0	100.0	100.0	100.0	Europe
Nordic countries	2796.3	3830.3	4026.6	4929.1	72.2	69.2	60.1	54.4	Pays nordiques
EEC(9)	832.2	1290.9	1938.5	2818.4	21.5	23.3	29.0	31.1	CEE(9)
Central Europe	120.7	189.5	314.6	436.5	3.1	3.4	4.7	4.8	Europe centrale
Southern Europe	49.6	82.8	178.9	375.7	1.3	1.5	2.7	4.2	Europe méridionale
Eastern Europe	73.1	145.7	237.4	495.0	1.9	2.6	3.5	5.5	Europe orientale
Sweden	1328.4	1928.2	2176.2	2929.6	34.3	34.8	32.5	32.4	Suède
Finland	1206.7	1596.4	1551.4	1695.1	31.2	28.8	23.2	18.7	Finlande
Netherlands	326.8	367.5	334.3	637.5	8.4	6.6	5.0	7.0	Pays-Bas
Norway	261.2	305.7	299.0	304.4	6.7	5.5	4.5	3.4	Norvège
France	130.0	206.5	310.9	518.2	3.4	3.7	4.6	5.7	France
Austria	108.9	167.6	243.9	304.3	2.8	3.0	3.6	3.4	Autriche
Belgium-Luxembourg	105.5	104.4	116.6	165.9	2.7	1.9	1.7	1.8	Belgique-Luxembourg
United Kingdom	103.0	158.3	247.1	264.8	2.7	2.9	3.7	2.9	Royaume-Uni
Germany, Fed. Rep. of	87.7	259.1	612.9	948.2	2.3	4.7	9.2	10.5	Allemagne, Rép. féd. d'
Other	213.7	445.5	803.7	1286.7	5.5	8.1	12.0	14.2	Autre
Imports/Importations									
Europe	4047.2	6424.6	7066.0	8909.3	100.0	100.0	100.0	100.0	Europe
Nordic countries	78.1	187.6	267.2	320.1	1.9	2.9	3.8	3.6	Pays nordiques
EEC(9)	3435.1	4988.9	5366.4	7000.7	84.9	77.7	75.9	78.6	CEE(9)
Central Europe	111.2	259.6	252.9	326.5	2.7	4.0	3.6	3.6	Europe centrale
Southern Europe	172.8	394.4	489.5	532.5	4.3	6.1	6.9	6.0	Europe méridionale
Eastern Europe	250.0	594.1	690.0	729.5	6.2	9.3	9.8	8.2	Europe orientale
United Kingdom	1078.5	1429.1	1527.2	1696.6	26.6	22.2	21.6	19.0	Royaume-Uni
Germany, Fed. Rep. of	997.9	1474.9	1410.4	1828.2	24.7	23.0	20.0	20.6	Allemagne, Rép. féd. d'
France	395.1	638.3	753.5	1126.6	9.8	9.9	10.6	12.6	France
Netherlands	264.9	351.1	382.3	729.3	6.5	5.5	5.4	8.2	Pays-Bas
Belgium-Luxembourg	253.6	352.9	468.9	525.4	6.3	5.5	6.6	5.9	Belgique-Luxembourg
Italy	224.2	397.3	406.6	602.8	5.5	6.2	5.8	6.8	Italie
Denmark	173.9	256.3	302.3	342.5	4.3	4.0	4.3	3.8	Danemark
German Dem. Rep.	81.9	179.5	222.6	233.0	2.0	2.8	3.2	2.6	Rép. dém. allemande
Hungary	71.7	124.5	175.1	183.9	1.8	1.9	2.1	2.1	Hongrie
Other	505.5	1220.7	1417.1	1641.0	12.5	19.0	20.0	18.4	Autre

Annex table 16.15/Tableau annexe 16.15

Sawnwood: ratio of exports to production, and imports to consumption
in selected countries 1965 to 1980

Sciages : rapport des exportations à la production et des importations
à la consommation, dans certains pays, 1965 à 1980

	1965	1970	1975	1980	
		(%)			
Coniferous sawnwood		EX/PR			**Sciages résineux**
Finland	60.9	65.1	58.0	68.0	Finlande
Sweden	52.5	57.3	49.9	53.2	Suède
Austria	65.5	65.5	61.9	67.4	Autriche
Portugal	30.6	29.6	23.6	50.5	Portugal
Romania	48.6	46.1	45.1	26.9	Roumanie
		IM/CO			
France	22.6	23.4	20.9	32.3	France
Germany, Fed. Rep. of	36.5	35.2	26.5	36.3	Allemagne, Rép. féd. d'
Italy	78.2	84.3	80.5	79.6	Italie
United Kingdom	95.9	93.4	89.5	85.1	Royaume-Uni
Spain	35.9	34.4	28.8	33.1	Espagne
Broadleaved sawnwood		EX/PR			**Sciages feuillus**
Belgium-Luxembourg	9.8	12.8	20.4	33.6	Belgique-Luxembourg
France	11.6	10.9	12.3	13.2	France
Germany, Fed. Rep. of	5.6	8.1	15.9	18.5	Allemagne, Rép. féd. d'
Netherlands	5.3	10.5	35.7	57.5	Pays-Bas
Yugoslavia	40.4	43.8	38.2	45.7	Yougoslavie
Romania	31.3	29.5	22.4	17.5	Roumanie
		IM/CO			
Belgium-Luxembourg	33.5	48.4	54.4	71.4	Belgique-Luxembourg
France	5.8	7.5	8.7	19.2	France
Germany, Fed. Rep. of	12.3	18.2	24.1	37.8	Allemagne, Rép. féd. d'
Italy	24.7	31.4	43.7	48.6	Italie
Netherlands	43.8	63.5	83.0	86.7	Pays-Bas
United Kingdom	58.7	52.5	46.6	47.9	Royaume-Uni
Spain	12.7	22.2	34.9	51.5	Espagne

EX/PR = Exports/Production; Exportations/Production.

IM/CO = Imports/Consumption; Importations/Consommation.

Annex table 16.16/Tableau annexe 16.16

Wood-based panels: ratio of exports to production, and imports
to consumption in selected countries 1965 to 1980

Panneaux dérivés du bois : rapport des exportations à la production et des
importations à la consommation, dans certains pays, 1965 à 1980

	1965	1970	1975	1980	
		(%)			
Plywood		EX/PR			**Contreplaqués**
Finland	81.4	85.3	81.7	83.2	Finlande
France	13.7	15.6	30.2	32.1	France
Germany, Fed. Rep. of	7.6	10.6	11.0	12.1	Allemagne, Rép. féd. d'
Italy	12.8	23.2	21.6	22.0	Italie
Greece	–	4.1	31.7	34.0	Grèce
Spain	3.5	10.2	10.5	22.2	Espagne
Romania	47.5	35.3	37.4	46.4	Roumanie
		IM/CO			
France	6.6	18.5	31.2	48.3	France
Germany, Fed. Rep. of	9.9	29.6	43.4	56.3	Allemagne, Rép. féd. d'
Italy	2.1	3.1	6.2	28.6	Italie
United Kingdom	95.3	98.5	98.4	108.8	Royaume-Uni
Greece	49.8	12.3	8.4	4.7	Grèce
German Dem. Rep.	20.9	52.8	58.4	56.9	Rép. dém. allemande
Particle board		EX/PR			**Panneaux de particules**
Finland	25.1	45.2	41.4	46.2	Finlande
Sweden	23.9	23.7	40.7	39.3	Suède
Belgium-Luxembourg	55.8	55.9	68.1	69.9	Belgique-Luxembourg
Germany, Fed. Rep. of	5.6	6.1	8.6	10.6	Allemagne, Rép. féd. d'
Italy	29.1	7.4	5.3	2.2	Italie
Austria	10.7	26.0	41.6	51.6	Autriche
Switzerland	3.0	8.2	20.7	38.3	Suisse
Portugal	21.1	49.9	15.5	22.8	Portugal
Spain	–	0.1	1.7	21.1	Espagne
		IM/CO			
France	4.6	12.5	12.3	16.8	France
United Kingdom	35.8	60.7	65.5	69.0	Royaume-Uni
Greece	42.1	2.1	2.7	0.7	Grèce
Yugoslavia	–	31.7	14.7	1.5	Yougoslavie
Czechoslovakia	4.1	10.2	25.8	11.0	Tchécoslovaquie
Fibreboard		EX/PR			**Panneaux de fibres**
Sweden	52.6	55.5	39.4	44.7	Suède
Poland	24.7	27.7	17.2	30.0	Pologne
Romania	21.4	14.9	21.1	21.5	Roumanie
		IM/CO			
Sweden	0.9	0.8	1.1	2.9	Suède
Germany, Fed. Rep. of	30.4	59.7	41.8	56.3	Allemagne, Rép. féd. d'
United Kingdom	77.3	92.4	93.3	90.0	Royaume-Uni
Spain	0.5	1.2	2.3	0.6	Espagne

EX/PR, IM/CO: see annex table 16.15/Voir tableau annexe 16.15.

Annex table 16.17/Tableau annexe 16.17

Wood pulp: ratio of exports to production, and imports
to consumption in selected countries 1965 to 1980

Pâtes de bois : rapport des exportations à la production et des importations
à la consommation, dans certains pays, 1965 à 1980

	1965	1970	1975	1980	
		(%)			
		EX/PR			
Finland	39.6	33.0	18.2	26.8	Finlande
Norway	48.0	45.1	36.5	35.4	Norvège
Sweden	50.1	46.2	38.5	35.1	Suède
Portugal	72.9	79.6	48.4	67.7	Portugal
		IM/CO			
France	37.8	44.8	39.8	51.1	France
Germany, Fed. Rep. of	44.9	50.4	51.5	56.4	Allemagne, Rép. féd. d'
Italy	58.5	60.9	57.4	71.9	Italie
United Kingdom	91.2	88.0	87.0	88.1	Royaume-Uni
Spain	37.8	34.2	26.9	21.9	Espagne

EX/PR, IM/CO: See annex table 16.15/Voir tableau annexe 16.15.

Annex table 16.18/Tableau annexe 16.18

Paper and paperboard: ratio of exports to production,
and imports to consumption in selected countries, 1965 to 1980

Papiers et cartons : rapport des exportations à la production et des
importations à la consommation, dans certains pays, 1965 à 1980

	1965	1970	1975	1980	
		(%)			
Newsprint		EX/PR			**Papier journal**
Finland	90.7	91.0	78.2	91.3	Finlande
Norway	80.3	85.8	87.4	88.7	Norvège
Sweden	69.9	66.7	75.9	80.7	Suède
Germany, Fed. Rep. of	1.7	6.4	13.2	13.4	Allemagne, Rép. féd. d'
United Kingdom	0.3	0.2	1.5	16.1	Royaume-Uni
		IM/CO			
France	17.8	29.2	52.3	60.4	France
Germany, Fed. Rep. of	71.0	64.5	60.8	62.3	Allemagne, Rép. féd. d'
Italy	3.9	4.9	4.7	19.4	Italie
United Kingdom	43.4	51.1	75.0	77.9	Royaume-Uni
Spain	42.4	40.7	49.9	40.4	Espagne
Printing and writing paper		EX/PR			**Papiers impression et écriture**
Finland	83.2	79.4	87.8	86.3	Finlande
Norway	72.1	70.5	56.3	68.6	Norvège
Sweden	46.3	52.8	61.9	51.1	Suède
Austria	64.3	73.3	78.4	90.9	Autriche
France	8.2	13.4	20.9	26.3	France
Germany, Fed. Rep. of	7.4	20.3	16.2	24.4	Allemagne, Rép. féd. d'
Netherlands	39.4	51.0	47.8	49.7	Pays-Bas
Portugal	0.8	1.5	26.7	19.5	Portugal
Romania	10.8	34.6	31.4	40.4	Roumanie
		IM/CO			
France	9.3	16.2	26.9	28.7	France
Germany, Fed. Rep. of	25.4	24.5	29.6	33.7	Allemagne, Rép. féd. d'
Netherlands	38.8	38.7	54.1	57.8	Pays-Bas
United Kingdom	9.7	20.3	37.3	49.7	Royaume-Uni
Austria	3.0	7.1	9.7	50.2	Autriche
Greece	6.3	43.5	47.6	45.1	Grèce
Czechoslovakia	0.6	15.9	22.5	13.0	Tchécoslovaquie
Other paper and paperboard		EX/PR			**Autres papiers et cartons**
Finland	77.5	80.7	71.2	72.6	Finlande
Sweden	62.6	70.5	60.7	78.8	Suède
France	7.1	9.2	10.0	18.5	France
Germany, Fed. Rep. of	2.9	7.5	14.8	22.6	Allemagne, Rép. féd. d'
Portugal	6.8	6.4	29.6	42.8	Portugal
		IM/CO			
France	19.2	23.1	17.7	32.7	France
Germany, Fed. Rep. of	27.0	32.6	31.2	36.1	Allemagne, Rép. féd. d'
Italy	15.7	16.6	12.4	19.4	Italie
United Kingdom	27.9	33.7	35.3	42.5	Royaume-Uni
Spain	11.1	11.7	7.6	8.6	Espagne

EX/PR, IM/CO: see annex table 16.15/Voir tableau annexe 16.15.

Annex 16.19

TIMTRADE data base

Products: Chips and particles, wood residues, coniferous pulpwood, round and split, non-coniferous pulpwood, round and split, coniferous logs, non-coniferous logs, sawn softwood, sawn hardwood, plywood, particle board, fibreboard, mechanical pulp, chemical woodpulp (including semi-chemical and dissolving), newsprint, paper and paperboard other than newsprint.

Years 1970, 1975, 1980 and 1984.

Method for each product and year a complete global trade matrix by country of origin and destination was constructed. Data were inserted according to the following procedure (in order of priority)

(a) exporters' data from the FAO/ECE questionnaires

(b) where exporters' data were not avilable, importers' data were used (also from the FAO/ECE questionnaires)

(c) data were also taken from the FAO Yearbook of Forest Products and OECD Pulp and Paper statistics

Accuracy

International trade data suffer from numerous problems of coverage and consistency. At the global level, the total for imports, exports and the sum of specified trade flows do not coincide. In order to measure these discrepancies, total imports and total exports, also from the FAO/ECE data base, were entered into the TIMTRADE data base. It was therefore possible to compare the sum of specified trade flows with reported total exports or imports. For many countries, the difference was nil or insignificant. In all cases, it is possible to quantify the difference at the global level between the sum of specified trade flows (SUMSPEC) and total imports or exports. (The difference may be negative or positive). The last table shows SUMSPEC as a percentage of world imports and exports, for 1980.

It will be seen that although the situation varies between products and is far from wholly satisfactory, the broad patterns of world trade are covered and no significant flows, at least involving Europe, are omitted. For some products (e.g. residues and veneer sheets), problems arise from the large discrepancy between total imports and total exports. In the matrices the figure for "world trade" is the sum of specified trade flows (SUMSPEC).

Summary tables for 1980

A world summary with all flows between all continents for all products in the data base, for 1980 is presented below. In view of the importance of this country group, data are also presented for the EEC(9). The six continents or large areas shown (Europe, USSR, North America, Latin America, Africa, Asia/Pacific) theoretically cover all trade. The final column "other" refers to flows whose origin or destination was not specified on the original questionnaire.

Finally summaries of world trade in forest products, in m3EQ, for 1970 and 1980 are also shown.

Further information on the structure of the TIMTRADE data base or data for other years are available on request from the secretariat.

TOTAL 16 PRODUCTS IN 1000 M3 EQ (1970)

FROM

TO	*WORLD*	*EUROPE*	NORDIC	OTHER	*USSR*	*N AMER*	USA	CANADA
WORLD	3569042	132330	84312	48017	32307	133133	46408	86725
EUROPE	181509	115748	73555	42193	22438	29799	15335	14464
NORDIC	10092	7374	6757	617	2397	143	122	21
OTHER	171416	108374	66798	41576	20041	29656	15213	14443
USSR	42240	3521	2255	1266	-	452	311	141
N AMER	71749	2095	1895	200	28	63522	4122	59399
USA	67441	2038	1866	172	9	59399	-	59399
CANADA	4608	57	29	29	19	4122	4122	-
L AMER	11751	2187	1790	396	685	6940	4480	2460
AFRICA	7762	3775	1725	2050	1102	1669	768	900
AS/PAC	79493	4721	2993	1728	7950	30720	21360	9361
JAPAN	54379	178	150	28	6970	23962	17369	6593
OTHER	25113	4543	2843	1700	980	6758	3990	2767
OTHER	439	283	100	183	104	32	31	1

TO	*L AMER*	*AFRICA*	*AS/PAC*	JAPAN	OTHER	*OTHER*
WORLD	3831	10066	45256	2771	42485	19
EUROPE	1022	8612	3890	379	3512	0
NORDIC	9	55	114	46	68	0
OTHER	1012	3557	3776	332	3444	0
USSR	0	58	210	0	210	0
N AMER	822	491	4792	612	4179	0
USA	723	467	4506	558	3948	0
CANADA	99	24	296	55	231	0
L AMER	1907	11	21	13	8	0
AFRICA	46	535	636	121	514	0
AS/PAC	36	359	35706	1646	34060	0
JAPAN	10	91	23167	-	23167	0
OTHER	26	268	12539	1646	10893	0
OTHER	0	0	1	0	1	19

Annex 16.19 (cont.)

TOTAL 16 PRODUCTS IN 1000 M3 EQ (1980)

Annex 16.19 (cont.)

FROM

TO	*WORLD*	*EUROPE*	NORDIC	OTHER	*USSR*	*N AMER*	USA	CANADA
WORLD	473947	167317	90106	77212	35593	181704	62735	118970
EUROPE	224768	140903	71764	69139	24273	38332	17378	20953
NORDIC	15208	9163	7129	2034	4329	832	565	267
OTHER	209560	131740	64634	67105	19944	37500	16813	20686
USSR	5029	4450	3676	775	-	98	98	0
N AMER	90308	1251	997	254	72	84782	5656	79126
USA	84359	1196	966	230	36	79126	-	79126
CANADA	5950	55	31	24	36	5656	5656	-
L AMER	13508	1840	1506	334	552	8113	4992	3121
AFRICA	8835	5287	3018	2270	772	2526	1313	1212
AS/PAC	111952	6796	5288	1508	7534	45107	32441	12666
JAPAN	71272	1082	809	273	6515	33637	24376	9262
OTHER	40679	5714	4478	1235	1018	11469	8065	3404
OTHER	19547	6790	3858	2932	2390	2747	856	1892

TO	*L AMER*	*AFRICA*	*AS/PAC*	JAPAN	OTHER	*OTHER*
WORLD	12177	10199	63213	2611	60603	3744
EUROPE	4088	7392	6577	263	6314	3204
NORDIC	772	23	80	8	72	9
OTHER	3316	7369	6497	255	6242	3195
USSR	0	141	0	0	0	339
N AMER	999	405	2597	213	2385	201
USA	909	405	2486	210	2277	201
CANADA	91	0	111	3	108	0
L AMER	2871	0	133	61	71	0
AFRICA	29	0	221	7	214	0
AS/PAC	2831	139	49545	1347	48198	0
JAPAN	1521	126	28390	-	28390	0
OTHER	1310	13	21155	1347	19808	0
OTHER	1358	2121	4140	720	3420	0

Annex 16.19 (cont.)

WORLD TRADE OF CHIPS AND PARTICLES IN 1980 (1000 M3)

TO	FROM								
	WORLD	EUROPE	.(EEC 9)	USSR	N AMERIC	L AMERIC	AFRICA	ASIA/PAC	OTHER
WORLD	17162	1195	321	547	7988	0	0	7431	0
EUROPE	1658	1195	321	0	463	0	0	0	0
.(EEC 9)	280	230	212	0	0	0	0	0	0
USSR	0	0	0	-	0	0	0	0	0
N AMERIC	1015	0	0	0	1015	0	0	0	0
L AMERIC	1	0	0	0	1	0	0	0	0
AFRICA	0	0	0	0	0	0	0	0	0
ASIA/PAC	14487	0	0	547	6509	0	0	7431	0
OTHER	0	0	0	0	0	0	0	0	0

WORLD TRADE OF WOOD RESIDUES IN 1980 (1000 M3)

TO	FROM								
	WORLD	EUROPE	.(EEC 9)	USSR	N AMERIC	L AMERIC	AFRICA	ASIA/PAC	OTHER
WORLD	3022	1522	1033	1500	0	0	0	0	0
EUROPE	3020	1520	1033	1500	0	0	0	0	0
.(EEC 9)	870	870	641	0	0	0	0	0	0
USSR	0	0	0	-	0	0	0	0	0
N AMERIC	0	0	0	0	0	0	0	0	0
L AMERIC	0	0	0	0	0	0	0	0	0
AFRICA	2	2	0	0	0	0	0	0	0
ASIA/PAC	0	0	0	0	0	0	0	0	0
OTHER	0	0	0	0	0	0	0	0	0

WORLD TRADE OF SOFTWOOD PULPWOOD IN 1980 (1000 M3)

TO	FROM								
	WORLD	EUROPE	.(EEC 9)	USSR	N AMERIC	L AMERIC	AFRICA	ASIA/PAC	OTHER
WORLD	13616	6111	2194	3143	1361	0	0	0	0
EUROPE	9069	6095	2194	2482	493	0	0	0	0
.(EEC 9)	2325	1501	855	575	249	0	0	0	0
USSR	1	1	0	-	0	0	0	0	0
N AMERIC	853	0	0	0	853	0	0	0	0
L AMERIC	0	0	0	0	0	0	0	0	0
AFRICA	0	0	0	0	0	0	0	0	0
ASIA/PAC	677	0	0	662	15	0	0	0	0
OTHER	16	16	0	0	0	0	0	0	0

Annex 16.19 (cont.)

WORLD TRADE OF HARDWOOD PULPWOOD IN 1980 (1000 M3)

TO	FROM								
	WORLD	EUROPE	(EEC 9)	USSR	N AMERIC	L AMERIC	AFRICA	ASIA/PAC	OTHER
WORLD	8053	4760	2228	2593	100	45	270	24	262
EUROPE	7619	4760	2228	2163	96	45	269	23	262
·(EEC 9)	3633	3121	2045	281	46	0	0	0	185
USSR	0	0	0	-	0	0	0	0	0
N AMERIC	4	0	0	0	4	0	0	0	0
L AMERIC	0	0	0	0	0	0	0	0	0
AFRICA	0	0	0	0	0	0	0	0	0
ASIA/PAC	430	0	0	430	0	0	0	0	0
OTHER	0	0	0	0	0	0	0	0	0

WORLD TRADE OF SOFTWOOD LOGS IN 1980 (1000 M3)

TO	FROM								
	WORLD	EUROPE	(EEC 9)	USSR	N AMERIC	L AMERIC	AFRICA	ASIA/PAC	OTHER
WORLD	27887	3967	1060	6514	15134	1004	0	1268	0
EUROPE	5194	3785	1057	1375	22	12	0	0	0
·(EEC 9)	2003	1981	753	10	9	0	0	0	0
USSR	0	0	0	-	0	0	0	0	0
N AMERIC	1638	0	0	0	1638	0	0	0	0
L AMERIC	14	0	0	0	14	0	0	0	0
AFRICA	145	145	0	0	0	0	0	0	0
ASIA/PAC	20688	33	0	5043	13435	988	0	1189	0
OTHER	207	4	2	96	25	4	0	79	0

WORLD TRADE OF HARDWOOD LOGS IN 1980 (1000 M3)

TO	FROM								
	WORLD	EUROPE	(EEC 9)	USSR	N AMERIC	L AMERIC	AFRICA	ASIA/PAC	OTHER
WORLD	42513	3218	1380	115	792	1	6049	31883	453
EUROPE	8831	3115	1279	109	401	1	4637	174	393
·(EEC 9)	6075	1978	1005	29	355	1	3367	139	206
USSR	199	4	4	-	0	0	141	0	54
N AMERIC	247	5	4	0	230	0	5	2	5
L AMERIC	0	0	0	0	0	0	0	0	0
AFRICA	39	38	37	0	0	0	0	0	0
ASIA/PAC	31463	4	4	0	113	0	72	31274	0
OTHER	1734	52	51	6	48	0	1194	433	0

Annex 16.19 (cont.)

WORLD TRADE OF SAWN SOFTWOOD IN 1980 (1000 M3)

TO	FROM WORLD	EUROPE	.(EEC 9)	USSR	N AMERIC	L AMERIC	AFRICA	ASIA/PAC	OTHER
WORLD	64044	24515	844	7173	33616	1397	0	0	342
EUROPE	28522	18207	819	5584	4056	355	0	0	319
.(EEC 9)	22651	15584	548	2884	3854	282	0	0	247
USSR	0	0	0	-	0	0	0	0	0
N AMERIC	22857	11	1	0	22793	31	0	0	22
L AMERIC	1578	20	11	0	1036	522	0	0	0
AFRICA	2675	1597	4	368	722	0	0	0	0
ASIA/PAC	6602	1272	7	366	4880	84	0	0	0
OTHER	1803	419	1	855	130	405	0	0	0

WORLD TRADE OF SAWN HARDWOOD IN 1980 (1000 M3)

TO	FROM WORLD	EUROPE	.(EEC 9)	USSR	N AMERIC	L AMERIC	AFRICA	ASIA/PAC	OTHER
WORLD	12309	3101	1243	5	1196	909	506	6378	213
EUROPE	6267	2554	1212	5	502	298	462	2325	119
.(EEC 9)	4983	1807	756	5	455	167	379	2110	64
USSR	239	239	0	-	0	0	0	0	0
N AMERIC	1115	1	1	0	557	261	12	190	94
L AMERIC	295	2	2	0	61	231	0	0	0
AFRICA	165	160	4	0	5	0	0	0	0
ASIA/PAC	3246	18	4	0	51	8	0	3169	0
OTHER	982	125	20	0	20	111	32	694	0

WORLD TRADE OF VENEER SHEETS IN 1980 (1000 M3)

TO	FROM WORLD	EUROPE	.(EEC 9)	USSR	N AMERIC	L AMERIC	AFRICA	ASIA/PAC	OTHER
WORLD	1308	370	203	7	457	103	144	227	0
EUROPE	747	321	170	0	240	12	139	35	0
.(EEC 9)	619	241	119	0	200	10	135	33	0
USSR	23	23	12	-	0	0	0	0	0
N AMERIC	221	3	3	0	174	16	5	23	0
L AMERIC	86	1	1	0	29	56	0	0	0
AFRICA	16	10	7	0	0	0	0	0	0
ASIA/PAC	108	1	1	0	4	0	0	103	0
OTHER	113	11	10	7	9	19	0	67	0

Annex 16.19 (cont.)

WORLD TRADE OF PLYWOOD IN 1980 (1000 M3)

TO	WORLD	EUROPE	(EEC 9)	USSR	N AMERIC	L AMERIC	AFRICA	ASIA/PAC	OTHER
					FROM				
WORLD	6599	1547	593	335	802	56	50	3668	140
EUROPE	3431	1298	528	259	651	45	50	988	140
.(EEC 9)	2642	1009	467	114	643	44	50	643	140
USSR	11	11	0	-	0	0	0	0	0
N AMERIC	1086	29	1	10	66	12	0	970	0
L AMERIC	71	8	6	30	32	0	0	0	0
AFRICA	151	47	22	10	0	0	0	94	0
ASIA/PAC	1064	34	7	26	38	0	0	965	0
OTHER	785	120	28	0	14	0	0	651	0

WORLD TRADE OF PARTICLE BOARD IN 1980 (1000 M3)

TO	WORLD	EUROPE	(EEC 9)	USSR	N AMERIC	L AMERIC	AFRICA	ASIA/PAC	OTHER
					FROM				
WORLD	5482	4845	2382	332	236	0	0	68	0
EUROPE	5027	4702	2316	303	17	0	0	5	0
.(EEC 9)	4122	4074	2267	28	16	0	0	4	0
USSR	3	3	0	-	0	0	0	0	0
N AMERIC	89	1	1	0	88	0	0	0	0
L AMERIC	146	23	5	11	61	0	0	50	0
AFRICA	41	41	32	0	0	0	0	0	0
ASIA/PAC	90	17	3	0	60	0	0	13	0
OTHER	85	57	25	18	10	0	0	0	0

WORLD TRADE OF FIBREBOARD IN 1980 (1000 M3)

TO	WORLD	EUROPE	(EEC 9)	USSR	N AMERIC	L AMERIC	AFRICA	ASIA/PAC	OTHER
					FROM				
WORLD	2171	1392	297	303	293	183	0	0	0
EUROPE	1447	1130	272	236	24	58	0	0	0
.(EEC 9)	1083	945	258	68	16	55	0	0	0
USSR	23	15	0	-	0	0	0	0	0
N AMERIC	332	4	1	27	217	73	0	0	0
L AMERIC	37	69	12	24	1	8	0	0	0
AFRICA	86	10	1	0	1	16	0	0	0
ASIA/PAC	27	10	1	11	11	5	0	0	0
OTHER	217	140	11	16	38	23	0	0	0

Annex 16.19 (cont.)

WORLD TRADE OF MECHANICAL PULP IN 1980 (1000 MT)

TO	FROM								
	WORLD	EUROPE	(EEC 9)	USSR	N AMERIC	L AMERIC	AFRICA	ASIA/PAC	OTHER
WORLD	1029	669	38	0	343	0	0	0	16
EUROPE	665	582	38	0	67	0	0	0	16
(EEC 9)	594	532	37	0	60	0	0	0	2
USSR	0	0	0	-	0	0	0	0	0
N AMERIC	162	0	0	0	162	0	0	0	0
L AMERIC	22	3	0	0	19	0	0	0	0
AFRICA	7	6	0	0	2	0	0	0	0
ASIA/PAC	146	55	0	0	92	0	0	0	0
OTHER	27	24	0	0	3	0	0	0	0

WORLD TRADE OF CHEMICAL PULP IN 1980 (1000 MT)

TO	FROM								
	WORLD	EUROPE	(EEC 9)	USSR	N AMERIC	L AMERIC	AFRICA	ASIA/PAC	OTHER
WORLD	19583	6468	530	839	10283	1252	541	0	198
EUROPE	11562	5640	492	698	4195	563	268	0	198
(EEC 9)	8047	4236	400	155	3437	395	268	0	155
USSR	220	216	0	-	4	0	0	0	0
N AMERIC	3530	32	0	0	3354	67	77	0	1
L AMERIC	649	43	1	33	297	276	0	0	0
AFRICA	155	101	12	17	37	0	14	0	0
ASIA/PAC	2716	189	4	0	2166	347	0	0	0
OTHER	750	243	21	91	229	0	182	0	0

WORLD TRADE OF NEWSPRINT IN 1980 (1000 MT)

TO	FROM								
	WORLD	EUROPE	(EEC 9)	USSR	N AMERIC	L AMERIC	AFRICA	ASIA/PAC	OTHER
WORLD	12242	3583	208	336	7868	192	0	241	23
EUROPE	3592	2729	181	212	623	5	0	0	23
(EEC 9)	3158	2516	177	8	607	5	0	0	22
USSR	34	34	0	-	0	0	0	0	0
N AMERIC	6303	90	13	0	6212	1	0	0	0
L AMERIC	793	149	3	33	556	55	0	0	0
AFRICA	74	67	3	4	2	0	0	0	0
ASIA/PAC	874	242	0	56	392	5	0	179	0
OTHER	572	272	8	30	82	126	0	62	0

Annex 16.19 (cont.)

WORLD TRADE OF OTHER PAPER AND PAPERBOARD IN 1980 (1000 MT)

TO					FROM				
	WORLD	EUROPE	(EEC 9)	USSR	N AMERIC	L AMERIC	AFRICA	ASIA/PAC	OTHER
WORLD	21168	14019	5053	718	5536	0	16	666	213
EUROPE	12909	10674	4428	576	1462	0	16	51	129
(EEC 9)	10604	9072	4037	71	1314	0	16	46	85
USSR	927	819	5	-	24	0	0	0	84
N AMERIC	1434	197	30	0	1230	0	0	7	0
L AMERIC	1167	302	48	45	802	0	0	18	0
AFRICA	706	353	105	14	336	0	0	2	0
ASIA/PAC	2591	782	114	0	1413	0	0	396	0
OTHER	1434	891	322	83	268	0	0	192	0

PERCENTAGE OF SPECIFIED IMPORTS AND EXPORTS IN TRADE-FLOW TABLES

PRODUCT		IMPORTS	EXPORTS	SUMSPEC	% IMPORT	% EXPORT
				1000 M3 / 1000 MT		
CH	1980	15313	17162	17162	112	100
RE	1980	3006	4398	3022	101	69
PC	1980	11447	10773	10616	93	99
PB	1980	7655	7445	8053	105	108
SL	1980	27906	27806	27887	100	100
HL	1980	43379	42138	42513	98	101
SS	1980	62846	66117	64044	102	97
SH	1980	12644	12685	12309	97	97
VE	1980	1811	1414	1308	72	92
PW	1980	5908	6563	6599	112	101
PA	1980	5754	5951	5482	95	92
FI	1980	2066	2295	2171	105	95
MP	1980	1308	1248	1029	79	82
CP	1980	18942	20264	19583	103	97
NP	1980	12748	12434	12242	96	98
PO	1980	20978	22425	21168	101	94

CH: CHIPS AND PARTICLES
RE: WOOD RESIDUES
PC: SOFTWOOD PULPWOOD
PB: HARDWOOD PULPWOOD
SL: SOFTWOOD LOGS
HL: HARDWOOD LOGS
SS: SAWN SOFTWOOD
SH: SAWN HARDWOOD
VE: VENEER SHEETS
PW: PLYWOOD
PA: PARTICLE BOARD
FI: FIBREBOARD
MP: MECHANICAL PULP
CP: CHEMICAL PULP
NP: NEWSPRINT
PO: OTHER PAPER AND PAPERBOARD (EXCL. NEWSPRINT).

Annex table 19.1
Outlook for consumption of fuelwood

Tableau annexe 19.1
Perspectives pour la consommation de bois de chauffage

| Country | ECE/FAO CEE/FAO 1979-81 a/ | National estimates/Estimations nationales Hypotheses/Hypothèses (1 000 000 m3) | | | | | | Av. annual change Taux de changement annuel moyen 1980-2000 (%) | | Pays |
		1970	1980	A 1990	A 2000	B 1990	B 2000	A	B	
Finland	3.55	9.00	3.70	4.00	4.30	4.20	4.70	+ 0.8	+ 1.2	Finlande
Norway	0.64	1.15	1.75	2.00	2.20	2.20	2.40	+ 1.1	+ 1.6	Norvège
Sweden	1.93	1.00	4.50	7.30	8.90	11.60	13.00	+ 3.5	+ 5.5	Suède
Nordic countries	6.12	11.15	9.95	13.30	15.40	18.00	20.10	+ 2.2	+ 3.6	Pays nordiques
Denmark	0.23	0.10	0.25	0.25	0.20	0.35	0.45	− 1.1	+ 3.0	Danemark
France	3.02	7.60	10.00	10.50	11.20	13.50	18.50	+ 0.6	+ 3.1	France
Germany, Fed. Rep. of	3.60	3.80	3.55	4.50	5.00	..	..	+ 1.7	+ 1.1	Allemagne, Rép. féd. d'
Ireland	0.04	0.10	0.24	0.25	0.28	0.28	0.30	+ 0.8	+ 1.3	Irlande
Italy	4.31	2.20	3.46	4.50	5.53	3.97	4.50	+ 2.4	+ 1.3	Italie
United Kingdom	0.11	0.06	0.15	0.38	0.72	0.16	0.30	+ 8.2	+ 3.6	Royaume-Uni
Six EEC countries	11.31	13.86	17.65	20.38	22.93	22.76	29.05	+ 1.3	+ 2.5	Pays de la CEE (6 pays)
Switzerland	0.91	0.72	0.81	1.28*	1.75	0.86	0.90	+ 3.9	+ 0.5	Suisse
Cyprus	0.02	0.01	0.02	0.03	0.03	..	..	+ 2.1	..	Chypre
Greece	1.84	0.83	0.72	0.75	0.75	0.79	0.79	+ 0.2	+ 0.5	Grèce
Israel	0.01	0.01	0.01	0.03	0.04	0.04	0.05	+ 6.7	+ 7.3	Israël
Portugal	0.50	4.60	5.45	7.20	9.00	6.77	8.56	+ 2.5	+ 2.3	Portugal
Spain	1.32	1.80	2.00	2.60	3.60	2.30	3.10	+ 3.0	+ 2.2	Espagne
Yugoslavia	3.10	6.30	6.10	7.40	7.60	7.80	8.00	+ 1.1	+ 1.4	Yougoslavie
Six southern European countries	6.79	13.55	14.30	18.01	21.02	17.73	20.53	+ 1.9	+ 1.8	Europe méridionale (6 pays)
Bulgaria	1.13	2.26	1.93	1.68	1.61	..	..	− 0.9	..	Bulgarie
Czechoslovakia	1.37	..	1.36	1.00	1.90	..	..	+ 1.7	..	Tchécoslovaquie
Hungary	2.52	2.49	2.38	2.85	2.89	2.90	2.96	+ 1.0	+ 1.1	Hongrie
Poland	2.03	1.89	1.92	2.16	2.16	1.94	1.97	+ 0.6	+ 0.1	Pologne
Four eastern European countries	7.05	8.00	7.59	7.69	8.56	7.52	8.44	+ 0.6	+ 0.5	Europe orientale (4 pays)
Total (20 countries)	32.18	47.28	50.30	60.66	69.66	66.87	79.02	+ 1.6	+ 2.3	Total (20 pays)
Canada	4.83	7.60	11.40	15.00	20.00	18.00	28.00	+ 2.8	+ 4.6	Canada
USA	23.98	..	101.00	229.00	381.00	..	..	+ 6.9	..	USA
USSR	79.30	100.00	89.50	..	..	..	..	..	..	URSS

Source: See text, chapter 19 / Voir texte, chapitre 19.

a/ Recorded data / Données officielles.

Annex table 19.2
Outlook for use of primary processing residues (wood and bark) for energy

Tableau annexe 19.2
Perspectives pour l'utilisation énergétique des déchets
de la première transformation du bois (bois et écorce)

| Country | 1970 | 1980 | National estimates/Estimations nationales — Hypotheses/Hypothèses (1000 m3) | | | | Av. annual change / Taux de changement annuel moyen 1980 – 2000 (%) | | Pays |
			A 1990	A 2000	B 1990	B 2000	A	B	
Finland	1000	1600	1700	1800	2200	2300	+ 0.6	+ 1.8	Finlande
Norway	250	550	550	550	550	550	−	−	Norvège
Sweden	1600	2200	2530	2730	2420	2530	+ 1.1	+ 0.7	Suède
Nordic countries	2850	4350	4780	5080	5170	5380	+ 0.8	+ 1.1	Pays nordiques
Denmark	200	300	200	200	400	500	− 2.0	+ 2.6	Danemark
France	1000	1500	2000	2500	..	..	+ 2.6	..	France
Germany, Fed. Rep. of	2600	2000	2800	3400	..	..	+ 2.7	..	Allemagne, Rép. féd. d'
Ireland	10	75	205	285	215	384	+ 6.9	+ 8.5	Irlande
United Kingdom	110	525	590	640	525	580	+ 1.0	+ 0.5	Royaume-Uni
Five EEC countries	3920	4400	5795	7095	5940	7364	+ 2.4	+ 2.6	Pays de la CEE (5 pays)
Switzerland	100	151	258*	365	180*	210	+ 4.5	+ 1.7	Suisse
Cyprus	4	2	2	2	2	2	−	−	Chypre
Greece	15	50	60	65	70	75	+ 1.3	+ 2.0	Grèce
Portugal	1197	727	300	−	300	−	..	..	Portugal
Spain	1000	1800	2900	3500	2600	3300	+ 3.4	+ 3.1	Espagne
Yugoslavia	2000	2350	2300	2400	2100	2220	+ 0.1	− 0.3	Yougoslavie
Five southern European countries	4216	4929	5562	5967	5072	5597	+ 1.0	+ 0.6	Europe méridionale (5 pays)
Bulgaria	413	332	272	272	..	..	− 1.0	..	Bulgarie
Czechoslovakia	640	757	800	808	..	..	+ 0.3	..	Tchécoslovaquie
Hungary	372	715	800	800	880	880	+ 0.6	+ 1.0	Hongrie
Poland	1390	1527	1567	1812	1410	1631	+ 0.9	+ 0.3	Pologne
Four eastern European countries	2815	3331	3439	3692	3362	3591	+ 0.5	+ 0.6	Europe orientale (4 pays)
Total (18 countries)	13901	17161	19834	22129	19724	22142	+ 1.2	+ 1.3	Total (18 pays)
USSR	12700	14240	17310	21350	18820	28000	..	..	URSS
Canada	4300	9670	..	..	..	..	+ 4.0	+ 5.5	Canada
USA	37000	51000	79000	..	..	..	+ 2.2	..	USA

Source: See chapter 19 / Voir chapitre 19.

Annex table 19.3
Use of primary processing residues for energy:
share of solid wood and of bark

Tableau annexe 19.3
Utilisation énergétique des déchets de la
première transformation du bois : part du bois
massif et de l'écorce

Country	Solid wood Bois massif (1000 m3)			Bark Ecorce (1000 m3)			Other Autres (1000 m3)			Share of solid wood Part du bois massif (%)			Pays
	1980	2000 A	2000 B	1980	2000 A	2000 B	1980	2000 A	2000 B	1980	2000 A	2000 B	
Finland	400	500	700	1200	1300	1600	..	..	..	25	28	30	Finlande
Norway	25	10	15	275	350	350	250	190	185	5	2	3	Norvège
Sweden	700	875	805	1500	1875	1725	..	..	..	32	32	32	Suède
Denmark	270	180	460	30	20	40	-	-	-	90	90	92	Danemark
Germany, Fed. Rep. of	600	600	..	600	1200	..	800	1600	..	30	18	..	Allemagne, Rép. féd. d'
Ireland	72	175	254	1	90	100	2	20	30	96	61	66	Irlande
United Kingdom	490	575	515	15	25	25	20	40	40	93	90	89	Royaume-Uni
Switzerland	67	140	90	22	105	40	62	120	80	44	38	43	Suisse
Greece	50	60	65	-	5	10	..	..	..	100	92	87	Grèce
Israel	26	..	..	9	..	..	..	..	..	74	..	..	Israël
Portugal	727	-	-	-	-	-	-	-	-	100	-	-	Portugal
Spain	1100	2100	2000	700	1200	1100	-	200	200	61	60	61	Espagne
Bulgaria	127	123	123	30	80	80	175	69	69	38	45	45	Bulgarie
Hungary	450	510	550	..	..	..	265	290	330	63	64	63	Hongrie
Poland	625	677	668	59	80	63	843	1055	900	41	37	41	Pologne
Total (15 countries)	5729	6551	6871	4441	6339	6342	2417	3584	3434	46	40	41	Total (15 pays)
USSR	5180	..	..	1230	..	..	7830	..	..	36	..	..	URSS
Canada	4835	10675	14000	4835	10675	4835	..	..	..	50	50	50	Canada

Source: See chapter 19 / Voir chapitre 19.

Annex table 19.4
Outlook for the use of secondary processing residues for energy

Tableau annexe 19.4
Perspectives pour l'utilisation énergétique des déchets de la deuxième transformation du bois

Country	National estimates/Estimations nationales (1000 m3)						Av. annual change Taux de changement annuel moyen 1980 – 2000 (%)		Pays
	1970	1980	Hypotheses/Hypothèses A		B		A	B	
			1990	2000	1990	2000			
Finland	250	400	500	500	600	700	+ 1.1	+ 2.8	Finlande
Norway	80	110	130	150	140	160	+ 1.6	+ 1.9	Norvège
Sweden	..	500	..	..	..	..	..	..	Suède
Denmark	300	400	300	400	300	400	–	–	Danemark
France	500	800	1200	1700	1400	1900	+ 3.9	+ 4.4	France
Germany, Fed. Rep. of	4000	5000	6500	7000	7000	8000	+ 1.7	+ 2.4	Allemagne, Rép. féd. d'
Ireland	–	1	1	1	1	2	–	+ 3.5	Irlande
United Kingdom	40	95	155	230	135	165	+ 4.5	+ 2.8	Royaume-Uni
Switzerland	200	243	372*	500	272*	300	+ 3.7	+ 1.1	Suisse
Greece	25	35	25	29	30	35	– 1.0	–	Grèce
Spain	100	200	300	400	200	300	+ 3.5	+ 2.1	Espagne
Yugoslavia	80	70	80	80	100	100	+ 0.7	+ 1.8	Yougoslavie
Bulgaria	185	145	139	139	139	139	– 0.2	– 0.2	Bulgarie
Czechoslovakia	..	250	295	300	..	..	+ 0.9	..	Tchécoslovaquie
Hungary	100*	159	175	190	170	180	+ 0.9	+ 0.6	Hongrie
Poland	682	763	1050	1256	878	1005	+ 2.5	+ 1.4	Pologne
Total (15 countries)	6792	8671	11222	12875	11660	13686	+ 2.0	+ 2.3	Total (15 pays)
USSR	3650	3330	..	..	..	..	..	..	URSS
Canada	20	250	1110	3100	1600	4600	+ 13.4	+ 15.7	Canada
USA	1300	1800	..	2800	..	..	+ 2.2	..	USA

Source: See chapter 19 / Voir chapitre 19.

203

Annex table 19.5
Outlook for the use for energy of wood and bark
residues by the pulp and paper industry

Tableau annexe 19.5
Perspectives pour l'utilisation énergétique des déchets
(bois et écorce) de l'industrie des pâtes et papiers

Country	National estimates/Estimations nationales						Av. annual change Taux de changement annuel moyen 1980 – 2000 (%)		Pays
	1970	1980	Hypotheses/Hypothèses A		B		A	B	
			1990	2000	1990	2000			
			(1000 m3)						
Finland	750	1200	1200	1300	1600	1600	− 0.4	+ 1.4	Finlande
Norway	50	175	235	300	225	275	+ 2.7	+ 2.3	Norvège
Sweden	1600	2000	2500	2500	2500	2500	+ 1.1	+ 1.1	Suède
France	500	600	720	900	720	900	+ 2.1	+ 2.1	France
Germany, Fed. Rep. of	250	350	350	300	..	..	− 0.8	..	Allemagne, Rép. féd. d'
United Kingdom	5	10	35	35	35	35	+ 6.4	+ 6.4	Royaume-Uni
Switzerland	10	40	70	100	45	50	+ 4.7	+ 1.1	Suisse
Portugal	–	298	872	2184	872	2184	+ 10.5	+ 10.5	Portugal
Spain	200	400	500	600	400	500	+ 2.1	+ 1.1	Espagne
Bulgaria	5	25	50	75	50	75	+ 5.7	+ 5.7	Bulgarie
Poland	5	87	280	372	252	335	+ 7.5	+ 7.0	Pologne
Total (11 countries)	4015	5185	6812	8666	7049	8754	+ 2.6	+ 2.7	Total (11 pays)
USSR	1210	1850	..	..	..	..	..	..	URSS
Canada	4200	8580	9000	10000	9000	10000	+ 0.7	+ 0.7	Canada
USA	14900	24700	..	78800	..	..	+ 6.0	..	USA

Source: See chapter 19 / Voir chapitre 19.

Annex table 19.6
Outlook for the use of recycled wood products for energy

Tableau annexe 19.6
Perspectives pour l'utilisation énergétique des produits recyclés

Country	National estimates/Estimations nationales						Av. annual change / Taux de changement annuel moyen 1980 – 2000 (%)		Pays
			Hypotheses/Hypothèses						
			A		B		A	B	
	1970	1980	1990	2000	1990	2000			
	(1000 m3)								
Finland	-	-	-	-	-	-	-	-	Finlande
Norway	90	225	250	300	300	350	+ 1.5	+ 2.2	Norvège
Sweden	..	1500	..	..	..	..	..	..	Suède
Denmark	10	50	50	50	100	150	-	+ 5.6	Danemark
France	2000	3000	3500	4000	4000	5200	+ 1.5	+ 2.8	France
Germany, Fed. Rep. of	100	500	3000	5000	3000	5000	+ 12.2	+ 12.2	Allemagne, Rép. féd. d'
Ireland	2	4	5	6	7	7	+ 2.1	+ 2.8	Irlande
United Kingdom	5	10	10	10	10	10	-	-	Royaume-Uni
Switzerland	1000	1000	1500	2000	1100	1200	+ 3.6	+ 0.9	Suisse
Cyprus	-	1	1	1	1	1	-	-	Chypre
Spain	-	100	300	500	200	300	+ 8.4	+ 5.7	Espagne
Bulgaria	-	17	15	15	15	15	- 0.6	- 0.6	Bulgarie
Poland	198	264	300	320	280	300	+ 1.0	+ 0.6	Pologne
Total (12 countries) a/	3405	5171	8931	12202	9013	12533	+ 4.4	+ 4.5	Total (12 pays) a/
Canada	-	-	2	5	4	10	b/	b/	Canada
USA	1500	..	..	..	..	..	..	..	USA

Source: See text, chapter 19 / Voir texte, chapitre 19.

a/ Excluding Sweden / Suède exclue.
b/ Infinity / Infini.

Annex table 19.7
Outlook for wood combustion in households

Tableau annexe 19.7
Perspectives pour la combustion du bois par les particuliers

Country	National estimates/Estimations nationales						Av. annual change Taux de changement annuel moyen 1980 – 2000 (%)		Pays
	1970	1980	Hypotheses/Hypothèses				A	B	
			A		B				
			1990	2000	1990	2000			
	(1000 m3)								
Finland	9000	4000	4200	4400	5300	5500	+ 0.5	+ 1.6	Finlande
Norway	1160	2100	2250	2450	2500	2700	+ 0.8	+ 1.3	Norvège
Sweden	1000	3800	4600	5600	6600	7000	+ 2.6	+ 3.1	Suède
Denmark	190	300	250	200	200	250	– 2.0	– 0.9	Danemark
Germany, Fed. Rep. of	4100	4150	5300	6100	5300	6100	+ 1.9	+ 1.9	Allemagne, Rép. féd. d'
Ireland	110	280	280	300	310	400	+ 0.3	+ 1.8	Irlande
United Kingdom	105	470	555	606	410	445	+ 1.3	– 0.3	Royaume-Uni
Switzerland	1712	1777	2706*	3635	1938*	2100	+ 3.7	+ 0.8	Suisse
Cyprus	6	7	10	10	10	10	+ 1.8	+ 1.8	Chypre
Greece	700	600	650	650	680	680	+ 0.4	+ 0.6	Grèce
Portugal	3450	3779	3225	2634	3166	2836	– 1.8	– 1.4	Portugal
Spain	800	1000	1700	2300	1500	2000	+ 4.3	+ 3.5	Espagne
Yugoslavia	4600	4600	5900	5900	6350	6550	+ 1.2	+ 1.8	Yougoslavie
Bulgaria	1963	1680	1370	1312	1370	1312	– 1.2	– 1.2	Bulgarie
Czechoslovakia a/	..	1987	1480	1300	1480	1300	– 2.1	– 2.1	Tchécoslovaquie a/
Hungary	2170	2680	3130	3150	3255	3305	+ 0.8	+ 1.1	Hongrie
Poland	1880	2059	2301	2507	2070	2267	+ 1.0	+ 0.5	Pologne
Total (17 countries)	34933	35269	39907	43054	42439	44755	+ 1.0	+ 1.2	Total (17 pays)
USSR	73740	65790	..	..	..	..	..	..	URSS
Canada	7600	11400	13900	16400	17400	22000	+ 1.8	+ 3.4	Canada
USA	25700	90300	147200	164200	..	..	+ 3.0	..	USA

Source: See text, chapter 19 / Voir texte, chapitre 19.

a/ Fuelwood removals plus residues burned in households / Quantités enlevées de bois de chauffage plus déchets brûlés par les particuliers.

Annex table 19.8
Outlook for wood combustion by
the forest industries

Tableau annexe 19.8
Perspectives pour la combustion du bois par
les industries forestières

Country	National estimates/Estimations nationales — Hypotheses/Hypothèses						Av. annual change / Taux de changement annuel moyen 1980 – 2000 (%)		Pays
			A		B		A	B	
	1970	1980	1990	2000	1990	2000			
	(1000 m3)								
Finland	3000	3200	3400	3700	4300	4400	+ 0.7	+ 1.6	Finlande
Norway	355	550	700	800	650	705	+ 1.9	+ 1.2	Norvège
Sweden	1725	2275	4500	4000	5000	4500	+ 2.9	+ 3.5	Suède
Denmark	410	510	450	550	690	830	+ 0.4	+ 2.5	Danemark
Germany, Fed. Rep. of	6450	6950	9450	10800	9450	10800	+ 2.2	+ 2.2	Allemagne, Rép. féd. d'
Ireland	2	10	70	120	80	180	+ 13.2	+ 15.6	Irlande
United Kingdom	100	240	325	460	285	355	+ 3.3	+ 2.0	Royaume-Uni
Switzerland	310	434	650*	865	470*	510	+ 3.5	+ 0.8	Suisse
Cyprus	–	–	–	–	–	–	–	–	Chypre
Greece	20	40	120	120	100	100	+ 5.7	+ 5.7	Grèce
Portugal	175	607	1492	3147	1428	3033	+ 8.5	+ 8.4	Portugal
Spain	–	100	200	400	200	300	+ 7.2	+ 5.7	Espagne
Yugoslavia	100	100	100	100	100	100	–	–	Yougoslavie
Bulgaria	380	365	420	480	420	480	+ 1.4	+ 1.4	Bulgarie
Czechoslovakia	..	377	615	708	615	708	+ 3.2	+ 3.2	Tchécoslovaquie
Hungary	..	199	340	360	340	360	+ 3.0	+ 3.0	Hongrie
Poland	1614	1776	2076	2295	1847	2031	+ 1.3	+ 0.7	Pologne
Total (17 countries)	15217	17783	24908	28905	25975	29392	+ 2.5	+ 2.6	Total (17 pays)
USSR	30150	31600	..	..	..	..	..	..	URSS
Canada	7500	18300	25000	26000	25000	26000	+ 1.8	+ 1.8	Canada
USA	51900	75700	..	317800	..	..	+ 7.4	..	USA

Source: See chapter 19 / Voir chapitre 19.

Annex table 19.9
Outlook for wood combustion in intermediate-size units (outside the forest industries)

Tableau annexe 19.9
Perspectives pour la combustion du bois par les unités de taille intermédiaire (industries forestières exclues)

Country	National estimates/Estimations nationales						Av. annual change Taux de changement annuel moyen 1980 – 2000 (%)		Pays
			Hypotheses/Hypothèses						
			A		B				
	1970	1980	1990	2000	1990	2000	A	B	
	(1000 m3)								
Finland	-	40	100	160	120	200	+ 7.2	+ 8.4	Finlande
Norway	100	130	150	170	160	180	+ 1.3	+ 1.6	Norvège
Sweden	-	500	1300	2100	3000	3400	+ 7.5	+ 10.1	Suède
Denmark	10	50	100	100	250	400	+ 3.6	+ 11.0	Danemark
Germany, Fed. Rep. of	-	100	2200	3600	..	..	+ 19.7	..	Allemagne, Rép. féd. d'
Ireland	-	-	1	2	2	3	+ 3.6	+ 5.7	Irlande
United Kingdom	-	30	160	340	60	120	+ 12.9	+ 7.2	Royaume-Uni
Switzerland	5	32	116*	200	41*	50	+ 9.6	+ 2.3	Suisse
Greece	220	280	300	300	300	300	+ 0.3	+ 0.3	Grèce
Portugal	1767	1447	2904	4508	2601	3975	+ 5.9	+ 5.2	Portugal
Spain	2200	2900	3600	3900	3200	3800	+ 1.5	+ 1.4	Espagne
Yugoslavia	1000	800	700	700	600	600	- 0.7	- 1.4	Yougoslavie
Bulgaria	410	384	350	300	350	300	- 1.2	- 1.2	Bulgarie
Hungary	750	325	300	300	300	300	- 0.4	- 0.4	Hongrie
Poland	540	591	695	768	618	680	+ 1.3	+ 0.7	Pologne
Total (15 countries)	7002	7609	12976	17448	13802	17908	+ 4.2	+ 4.4	Total (15 pays)
USSR	13010	11610	..	..	..	..	..	..	URSS
Canada	-	5	2000	6000	2500	12000	a/	a/	Canada
USA	850	1760	..	..	..	..	..	..	USA

Source: See text, chapter 19 / Voir texte, chapitre 19.

a/ Growth over 40% p.a. because of low start point / Croissance de plus de 40% p.a. à cause du point de départ bas.

Annex table 19.10
Outlook for use of wood for charcoal

Tableau annexe 19.10
Perspectives pour l'utilisation du bois
pour fabriquer le charbon de bois

Country	National estimates/Estimations nationales						Av. annual change Taux de changement annuel moyen 1980 - 2000 (%)		Pays
	Hypotheses/Hypothèses (1000 m3)								
			A		B		A	B	
	1970	1980	1990	2000	1990	2000			
Norway	3	3	5	10	10	20	+ 6.2	+ 9.9	Norvège
Denmark	-	-	-	-	-	-	-	-	Danemark
Germany, Fed. Rep. of	200	200	200	200	..	..	-	..	Allemagne, Rép. féd. d'
Ireland	-	-	-	-	-	-	-	-	Irlande
United Kingdom	5	10	15	20	15	20	+ 3.6	+ 3.6	Royaume-Uni
Switzerland	-	-	-	-	-	-	-	-	Suisse
Cyprus	12	14	20	20	20	20	+ 1.8	+ 1.8	Chypre
Greece	ro	12	20	20	20	20	+ 2.6	+ 2.6	Grèce
Portugal	400	643	750	900	750	900	+ 1.7	+ 1.7	Portugal
Spain	100	300	600	900	500	700	+ 5.7	+ 4.3	Espagne
Yugoslavia	600	500	500	500	500	500	-	-	Yougoslavie
Bulgaria	110	19	14	14	14	14	- 1.5	- 1.5	Bulgarie
Hungary	39	48	50	50	50	50	+ 0.2	+ 0.2	Hongrie
Poland	127	132	181	209	154	165	+ 2.3	+ 1.1	Pologne
Europe	1606	1881	2355	2843	2233	2609	+ 2.1	+ 1.6	Europe
Canada	-	-	10	50	10	50	a/	a/	Canada
USA	2000	1700	..	2800	..	..	+ 2.5	..	USA

Source: See text, chapter 19 / Voir texte, chapitre 19.

a/ Infinity / Infinité.

Annex table 19.11
Outlook for use of wood for electricity, solid, liquid and gaseous fuels

Tableau annexe 19.11
Perspective pour l'utilisation du bois pour l'électricité et les combustibles solides, liquides ou gaseux

(1000 m3)

Country	Electricity / Electricité			Solid fuels / Combustibles solides			Liquid or gaseous fuels / Combustibles liquides ou gaseux			Pays
	1980	2000 A	2000 B	1980	2000 A	2000 B	1980	2000 A	2000 B	
Finland	-	-	-	-	-	-	-	-	-	Finlande
Norway	25	50	100	2	20	30	-	..	..	Norvège
Denmark	-	-	-	1	1	20	-	-	-	Danemark
Germany, Fed. Rep. of	-	-	-	-	-	-	..	..	..	Allemagne, Rép. féd. d'
Ireland	30	-	-	-	150	170	-	-	-	Irlande
United Kingdom	-	-	-	40	130	130	-	80	80	Royaume-Uni
Switzerland	-	-	-	-	10	-	-	5	-	Suisse
Greece	-	-	-	-	-	-	-	-	-	Grèce
Portugal	-	-	-	-	-	-	-	-	-	Portugal
Yugoslavia	-	-	-	100	200	250	-	-	-	Yougoslavie
Bulgaria	-	-	-	-	-	-	-	-	-	Bulgarie
Hungary	-	-	-	..	20	5	-	-	-	Hongrie
Poland	-	80	55	-	30	20	-	30	20	Pologne
Total (13 countries)	55	130	155	143	561	625	690	115	40	Total (13 pays)
Canada	-	-	50	200	3000	4500	-	3000	6000	Canada
USA	180	..	1200	580	..	580	690	..	1400	USA

Source: See chapter 19 / Voir chapitre 19.

Annex table 19.12
Outlook for energy from pulping liquors

Tableau annexe 19.12
Perspectives pour l'énergie tirée
des lessives résiduaires

| Country | National estimates/Estimations nationales | | | | | | Av. annual change Taux de changement annuel moyen 1980 - 2000 (%) | | Pays |
| | 1970 | 1980 | Hypotheses/Hypothèses A | | B | | A | B | |
	(1000 TJ)		1990	2000	1990	2000			
Finland	70.0	100.0	110.0	120.0	130.0	140.0	+ 0.9	+ 1.7	Finlande
Norway	2.0	3.2	3.7	3.7	3.6	3.6	+ 0.7	+ 0.6	Norvège
Sweden	a/95.5	b/97.1	105.0	105.0	105.0	105.0	+ 0.4	+ 0.4	Suède
Germany, Fed. Rep. of	5.0	22.5	22.5	22.5	..	..	-	..	Allemagne, Rép. féd. d'
Switzerland	0.5	1.1	..	1.2	..	1.2	+ 0.5	+ 0.5	Suisse
Portugal	7.5	14.7	22.4	27.6	22.4	27.6	+ 3.2	+ 3.2	Portugal
Spain	5.0	10.0	15.0	20.0	10.0	15.0	+ 3.6	+ 2.1	Espagne
Bulgaria	0.7	1.3	2.3	3.0	2.3	3.0	+ 4.3	+ 4.3	Bulgarie
Poland	4.5	7.2	10.0	14.5	9.0	13.0	+ 3.6	+ 3.0	Pologne
Total (9 countries)	190.7	257.1	292.0	317.5	305.9	330.9	+ 1.1	+ 1.3	Total (9 pays)
USSR	110.0	130.1	..	..	..	..	..	..	URSS
Canada	158.0	244.8	260.0	270.0	260.0	270.0	+ 0.5	+ 0.5	Canada
USA	750.0	856.0	..	2800.0	..	..	+ 6.1	..	USA

Source: See text, chapter 19 / Voir texte, chapitre 19.

a/ 1973.
b/ 1979.

Note: Not specified whether these data are for gross or net energy input. See text chapt. 19. / Non précisé si ces données se réfèrent à la production brute ou nette d'énergie. Voir texte, chap. 19.

Annex 19.13

Replies to the enquiry on the outlook for non conventional uses of wood for energy

- ### Generation of electricity for the public grid

Sweden   Generation of electricity, including cogeneration, will be technically and economically possible in the middle or the end of this decade. However, due to the elctricity supply situation, with a current surplus of cheap electricity, the demand for electricity from forest fuel is expected to be low over the next twenty years. Furthermore, peat will probably be more interesting as a fuel for electricity generation than forest fuel.

Federal Republic of Germany   Generation of electricity in special power plants on the basis of wood is not to be expected. The timber industry, however, already today sells part of the electric energy produced to the public grid. This share could increase further. Data are not available. This will, however, depend less on the general energy price situation than on the prices to be settled between the public grid and the supplying industry which will be rather decisive (at present 0.02 - 0.06 DM/kWh).

Ireland   While extensive trials on the use of sawmill residues for the generation of electricity have been carried out in recent years in Ireland, there are no firm plans at present for the generation of electricity on a commercial basis using wood as a fuel.

United Kingdom   We can see little or no interest in this particular outlet for wood energy as we have sufficient coal and are building a number of nuclear power stations.

Spain   Wood is hardly used in Spain to generate electricity other than a certain amount for cogeneration. Some increase is expected in the use of fuelwood to produce electricity.

Canada   Electricity generation should increase moderately, mostly in the pulp and paper sector and in remote locations using diesel-fired generators. Small-scale cogeneration will depend on economic return and legislation developments. Electricity from surplus mill residues would be structurally feasible in many areas in Canada.

USA   Incentives for cogeneration are just now being classified and defined in many States. Large generation facilities (up to 55 MW) with wood are beginning to be demonstrated successfully. Because generation and cogeneration appear to be catching on, a steady growth up to a maximum of the equipment of fifty 50 MW plants may be anticipated.

USSR   The generation of electricity from forest biomass (including cogeneration) will remain stable.

- ### Wood-based solid fuels

Norway   (This) has been tried by several mills, as an outlet for wood residues, but so far without any economic success. There ought to be a good basis for such production as far as raw material is concerned. The problem is profitability.

<u>Sweden</u> The manufacture of wood-based solid fuels in the form of pellets
and briquettes has not yet reached the stage of commercial viability.
The recent decision by the Government to propose the removal of the
value-added tax upon domestic fuels may have a stimulating effect
regarding demand for these products. However, it is likely that
different materials like peat, industrial by-products and straw will be
used together with forest products as raw material when possible. The
commercial large-scale introduction may come in the middle of this decade.

<u>The Federal Republic of Germany</u> Manufactured solid fuels are at present
being produced in small quantities within the timber industry
(briquetting of shavings) and are also used there. The market (trade)
for such products has been small so far. The possibilities of
development are generally assessed favourably, although no forecasts are
available. Possible consumers: households, intermediate units and
forest industry itself. The quantities of residues to be briquetted are
already included in the other data. Manufactured solid fuels are, like
charcoal, only another form of the energy source wood.

<u>Ireland</u> The manufacture of wood-based briquettes started in this country
in 1980 and there are plans to increase the number of briquette producing
plants.

<u>United Kingdom</u> There are likely to be small increases in the quantities
of pellets and briquettes which are manufactured if only because they are
a convenient form of wood fuel for the average user. At present they are
certainly not economical when compared with coal or smokeless fuel.

<u>Spain</u> Pellet and briquette manufacture was begun some years ago but has
not achieved any degree of importance. There is a factory in Soria where
fuelwood is mixed with urban refuse and pig farm waste. Partial
oxidation by aerobic micro-organisms is achieved by open air
fermentation. The fermented mixture is then further homogenized and
dried in an anaerobic fermentation reactor. Particles are then sorted by
size, drying is completed and finally the briquettes are made. At
present another briquette plant is being set up in Bailén (Jaén) where
fuelwood will be mixed with urban refuse and solid waste from olive oil
manufacture.

<u>Canada</u> Future growth depends on fossil fuel price escalation rate(s) to
make densification plants and chipping operations economically viable.
It is likely to become an important fuel supply in remote residential
areas and small industries as the supply infrastructure becomes better
established.

<u>USA</u> Despite the advantages of wood-based solid fuels, no clearcut
success in producing and using them economically has been demonstrated.
Pellet manufacturing plants, characteristically, have opened with wide
publicity, operations proceeded for months or, at most, a year or two,
then the plants closed. Briquette plants have operated more successfully
over the long term. But they have catered to the smaller, fire log
aesthetic markets rather than to the industrial or residential
competitive fuel market. No significant increase in the production of
these fuels is foreseen.

<u>USSR</u> The manufacture of solid fuels from wood residues will increase as
these fuels will replace in part round and split wood, which will be
redirected to wood processing uses.

- Synthetic liquid or gaseous fuels from wood

Norway (This) seems to be rather far off in Norway. Developments in other countries, e.g. Sweden, are followed with interest.

Sweden The manufacture of synthetic liquid and gaseous fuels may not be commercially viable until after the turn of the century. One crucial point is that the volumes of raw material needed for economical production of i.e. methanol are so huge that it is not likely that there will be any entirely wood-based liquefaction plants in Sweden unless economical techniques for production on a smaller scale are established. The Swedish experiments are mainly directed towards the use of energy crops and not towards woody biomass.

The Federal Republic of Germany At present, no liquid fuels are being manufactured from wood and it cannot be estimated what will be the development by 1990 and/or 2000. Wood gas is being produced in small quantities in some timber industries to produce heat and power for their own requirements. This could, provided there was some progress in technology, increase further.

Ireland Synthetic liquid and gaseous fuels from wood are not produced on a commercial basis in Ireland nor are there at present any plans for development in this field.

United Kingdom We can see little or no interest in this particular outlet for wood energy due largely to our own supplies of oil and also to the more productive route using coal if this became necessary.

Spain The wood distillation plants that used to exist have been disappearing over the last few years, and today only small quantities of liquid and gas derivatives are produced. Production is expected to increase in the future.

Poland Nowadays, synthetic liquid or gaseous fuels from wood are not produced on a large scale in Poland. The designation of a limited quantity of wood for gaseous fuel in gas generators installed in cars for wood transport, and for stationary gas generators in particle board plants, has been estimated.

Canada The penetration of alcohol fuels in the transport sector will remain slow. Gasification systems will become very popular once the technology is perfected and small-scale wood gas to electricity systems become economically viable.

USA Use of wood for liquid fuels has remained constant since the last wood distillation plant was closed in the early 60s. The use of corn as a source of fermentation alcohol, however, has grown significantly in the last few years. Wood has some advantages over corn as a feedstock, and it could demonstrate significant growth as a source of liquid fuels if it displaced some of the corn. However, in the past, wood has only been able to compete with other sources of alcohol, during normal times, in the Soviet Union. Therefore, we are predicting only a doubling by 2000. However, because of the potential for research breakthroughs, this could be considered a very conservative prediction.

Gas from wood shows even more potential for successful implementation with research breakthroughs. Although significant growth is possible we have not guessed what consumption may be in 2000.

<table>
<tr><td colspan="2">

Annex 20.1
Structure of consistency
analysis model used in ETTS IV
</td><td>

Annexe 20.1
Structure du modèle d'analyse
de la cohérence des prévisions
utilisé dans ETTS IV
</td></tr>
</table>

Set out below are the equations used to construct the model presented in chapter 20. The structure is presented in graphic form in fig. 20.1.

Le modèle du chapitre 20 a été construit à partir des équations présentées ci-dessous. La structure est presentée, en forme de graphique à la fig. 20.1.

A. Codes

Each code is in two parts:
- an indication of the "transaction"
- an indication of the product (or raw material) to which the transaction refers e.g. P = production
SS = sawn softwood
PSS = production of sawn softwood.

Chaque code a deux parties :
- une indication de la "transaction"
- une indication du produit (ou matière première) auquel se réfère la transaction p. ex. P = production
SS = sciages résineux
PSS = production des sciages résineux

A.1 **Transactions**

P..	Production	Production
I..	Imports	Importations
E..	Exports	Exportations
C..	Apparent consumption (P+I-E)	Consommation apparente (P+I-E)
REC..	Recovery (of waste paper)	Récupération (vieux papiers)
DS..	Domestic supply (of residues)	Approvisionnement domestique (de déchets de bois)
RESGEN	Residues generated	Déchets "produits"
CFRES	DSRES RESGEN	DSRES RESGEN
CF..	Conversion factor raw material/ product	
CLOG..	Consumption of logs for (a product) e.g. CLOGSS = consumption of logs for sawn softwood	Consommation de grumes pour la fabrication (d'un produit) p. ex. CLOGSS = consommation de grumes pour la fabrication de sciages résineux

CPW..	Consumption of pulpwood for (a product)	Consommation de bois de trituration pour la fabrication (d'un produit)
CFFIBPAP	Conversion factor: fibres used/ paper produced: $\dfrac{\text{CFIB}}{\text{PPAP}}$	Secteur de conversion : fibres utilisées/papier produit : $\dfrac{\text{CFIB}}{\text{PPAP}}$
RECRATE	Recovery rate of waste paper	Taux de récupération des vieux papiers
RC..	Derived removals	Quantités enlevées dérivées
REM..	Removals, recorded or forecast (chapt. 5)	Quantitées enlevées, enregistrées ou prévues (chap. 5)
LOGDIFF	Difference, in 1979–81, between derived and recorded removals of logs (constant)	Différence, en 1979–81, entre les quantités enlevées dérivées et réelles (constant)
SMALLDIFF	Difference, in 1979–81, between derived and recorded removals of smallwood (constant)	Différence, en 1979–81, entre les quantités enlevées dérivées et réelles de petits bois ronds (constant)
DIS..	Difference between derived and recorded removals (in 1979–81, zero or insignifiant)	Différence entre les quantités enlevées dérivées et réelles (en 1979–81, nul ou insignifiant)

A.2   Products/produits

..SS	Sawn softwood	Sciages résineux
..SH	Sawn hardwood	Sciages feuillus
..SL	Sleepers	Traverses
..PY	Plywood	Contreplaqués
..VEN	Veneer sheets	Feuilles de placage
..PB	Particle board	Panneaux de particules
..FB	Fibreboard	Panneaux de fibres
..PULP	Pulp (for paper making)	Pâtes (pour la fabrication des papiers et cartons)
..DISS	Dissolving pulp	Pâte à dissoudre
..PAP	Paper and paperboard	Papiers et cartons
..LOG	Sawlogs and veneer logs	Grumes de sciage et de placage
..PWRS	Pulpwood, round and split	Bois de trituration, rondins et quartiers

..PIT	Pitprops	Bois de mine
..FUEL	Fuelwood	Bois de chauffage
..OIW	Other industrial wood	Autres bois industriels
..SMALL	Smallwood (i.e. PWRS+PIT+FUEL + OIW)	Petits bois ronds (c'est-à-dire PWRS+PIT+FUEL+OIW)
..RES	Residues, chips and particles	Déchets, plaquettes et particules
..PWTOT	Pulpwood total (i.e. PWRS+RES)	Bois de trituration, total (c'est-à-dire PWRS+RES)
..WP	Waste paper	Vieux papiers
..FIB	Fibres used for paper making (i.e. PULP+WP)	Fibres utilisées pour la fabrication de papiers et cartons (c'est-à-dire PULP+WP)
..TOT	Total removals (i.e. LOG+SMALL)	Quantités enlevées, total (LOG+SMALL)

B. Equations

Log sector/Secteur grumes

1. PSS   = CSS + ESS − ISS
2. CLOGSS   = PSS x CFSS
3. PSH   = CSH + ESH − ISH
4. CLOGSH   = PSH x CFSH
5. PSL   = CSL + ESL − ISL
6. CLOGSL   = PSL x CFSL
7. PPLY   = CPLY + EPLY − IPLY
8. CLOGPLY   = PPLY x CFPLY
9. PVEN   = CVEN + EVEN − IVEN
10. CLOGVEN   = PVEN x CFVEN
11. CLOG   = CLOGSS + CLOGSH + CLOGLY + CLOGSL + CLOGVEN
12. RCLOG   = CLOG + ELOG − ILOG + LOGDIFF

Smallwood sector/Secteur petit bois

13. PPB   = CPB + EPB − IPB
14. CPWPB   = PPB x CFPB
15. PFB   = CFB + EFB − IFB
16. CPWFB   = PFB x CFFB
17. PPAP   = CPAP + EPAP − IPAP
18. CFIB   = $\dfrac{PPAP}{CFFIBPAP}$

19.	PWP	= CPAP x RECRATE
20.	CWP	= PWP + IWP - EWP
21.	CPULP	= CFIB - CWP
22.	PPULP	= CPULP + EPULP - IPULP
23.	CPWPULP	= PPULP x CFPULP
24.	PDISS	= CDISS + EDISS - IDISS
25	CPWDISS	= PDISS x CFDISS
26.	CPWTOT	= CPWPB + CPWFB + CPWPULP + CPWDISS
27.	RESGEN	= CLOG - PSS - PSH - PPLY - PVEN
28.	DSRES	= RESGEN x CFRES
29.	CRES	= DSRES + IRES - ERES
30.	CPWRS	= CPWTOT - CRES
31.	RCPWRS	= CPWRS + EPWRS - IPWRS
32.	RCOIW	= COIW + EOIW - IOIW
33.	RCFUEL	= CFUEL + EFUEL - IFUEL
34.	RCPIT	= CPIT + EPIT- IPIT
35.	RCSMALL	= RCPWRS + RCOIW + RCFUEL + RCPIT
		+ SMALLDIFF

Comparison derived and forecast removals/Comparaison des quantités enlevées dérivées et réelles

36.	RCTOT	= RCLOG + RCSMALL
37.	DISTOT	= REMTOT - RCTOT
38.	DISLOG	= REMLOG - RCLOG
39.	DISSMALL	= REMSMALL - RCSMALL

C.   Sources of exogenous variables/Sources des variables exogènes

C..	Chapter 12	Chapitre 12
REM..	Chapter 5	Chapitre 5
CFRES	Chapter 15	Chapitre 15
RECRATE	Chapter 15	Chapitre 15
I..))) E..)	All trade is unchanged, 1979-81 level	Importations et exportations inchangées au niveau de 1979-81
CF..	Calculated from 1979-81 data base and checked with correspondents. It is assumed that these factors do not change between 1979-81 and 2000.	Calculés à partir des données de 1979-81 et vérifiés par les correspondants. On a supposé que ces facteurs ne vont pas changer entre 1979-81 et 2000.

Annex 20.2
__Consistency analysis, by country__

Annexe 20.2
__Analyse de cohérence, par pays__

Annex 20.2. presents a summary of the consistency analysis by country. The concepts and structure of the material balance are presented in chapter 20 and annex 20.1.

L'annexe 20.2 présente un résumé de l'analyse par pays. Les idées sous-jacentes et la structure du bilan matière sont présentées au chapitre 20 et à l'annexe 20.1.

AUSTRIA – AUTRICHE

	Unit / Unité 1000	1979-81 (average) (moyenne)	Low-bas 1990	Low-bas 2000	High-haut 1990	High-haut 2000	
FORECAST CONSUMPTION :							**PREVISIONS DE CONSOMMATION :**
Sawnwood	m3(s)	2940	3080	3390	3230	3900	Sciages
Wood-based panels	m3	695	785	971	845	1191	Panneaux a base de bois
Paper and paperboard	m.t.	916	1010	1190	1150	1580	Papiers et cartons
Fuelwood	m3	1410	1500	1570	2230	3050	Bois de chauffage
DERIVED PRODUCTION :							**PRODUCTION CALCULEE :**
Sawnwood	m3(s)	6587	6727	7037	6877	7547	Sciages
Wood-based panels	m3	1353	1442	1629	1502	1849	Panneaux a base de bois
Woodpulp	m.t.	1098	1128	1217	1193	1390	Pates de bois
Paper and paperboard	m.t.	1617	1711	1891	1851	2281	Papiers et cartons
RECOVERY :							**RECUPERATION :**
Waste paper	m.t.	273	332	415	401	614	Vieux papiers
DOMESTIC SUPPLY :							**APPROVISIONNEMENT NATIONAL :**
Residues and chips	m3	2812	2896	3035	3083	3526	Dechets et plaquettes
DERIVED CONSUMPTION :							**CONSOMMATION CALCULEE :**
Sawlogs and veneer logs	m3 u.b.	9748	9952	10376	10165	11122	Grumes de sciage et placage
Total pulpwood	m3	7963	8258	9003	8654	10150	Total bois de trituration
Fibres for paper	m.t.	1544	1633	1804	1766	2177	Fibres pour papier
DERIVED REMOVALS :							**QUANTITES ENLEVEES CALCULEES :**
Sawlogs and veneer logs	m3 u.b.	7470	7673	8098	7887	8844	Grumes de sciage et placage
Smallwood	m3 u.b.	4696	4997	5643	5935	7780	Petits bois
Total removals	m3 u.b.	12166	12671	13742	13822	16624	Total quantites enlevees
FORECAST REMOVALS :							**PREVISIONS DE QUANTITES ENLEVEES :**
Sawlogs and veneer logs	m3 u.b.	7470	8550	8835	8835	9415	Grumes de sciage et placage
Smallwood	m3 u.b.	4700	6350	6515	6565	6935	Petits bois
Total removals	m3 u.b.	12170	14900	15350	15400	16350	Total quantites enlevees
DIFFERENCE :							**DIFFERENCE :**
Sawlogs and veneer logs	m3 u.b.	0	876	736	947	570	Grumes de sciage et placage
Smallwood	m3 u.b.	3	1352	871	629	846-	Petits bois
Total removals	m3 u.b.	4	2228	1607	1577	275-	Total quantites enlevees

Trade is assumed to remain unchanged, at 1979-81 level.
Le commerce est suppose rester inchange a son niveau de 1979-81.

BELGIUM-LUXEMBOURG - BELGIQUE-LUXEMBOURG

	Unit / Unite 1000	1979-81 (average) (moyenne)	Low-bas 1990	Low-bas 2000	High-haut 1990	High-haut 2000	
FORECAST CONSUMPTION							**PREVISIONS DE CONSOMMATION :**
Sawnwood	m3(s)	2053	2050	2330	2150	2800	Sciages
Wood-based panels	m3	859	933	1116	1003	1416	Panneaux à base de bois
Paper and paperboard	m.t.	1390	1490	1850	1710	2500	Papiers et cartons
Fuelwood	m3	370	430	480	480	610	Bois de chauffage
DERIVED PRODUCTION							**PRODUCTION CALCULEE :**
Sawnwood	m3(s)	669	666	946	766	1416	Sciages
Wood-based panels	m3	1750	1824	2006	1894	2306	Panneaux à base de bois
Woodpulp	m.t.	344	358	527	450	785	Pâtes de bois
Paper and paperboard	m.t.	871	971	1331	1191	1981	Papiers et cartons
RECOVERY							**RECUPERATION :**
Waste paper	m.t.	379	451	597	552	907	Vieux papiers
DOMESTIC SUPPLY							**APPROVISIONNEMENT NATIONAL :**
Residues and chips	m3	273	333	568	458	975	Dechets et plaquettes
DERIVED CONSUMPTION							**CONSOMMATION CALCULEE :**
Sawlogs and veneer logs	m3 u.b.	1369	1363	1953	1582	2909	Grumes de sciage et placage
Total pulpwood	m3	2990	3105	3879	3508	5116	Total bois de trituration
Fibres for paper	m.t.	761	848	1162	1040	1730	Fibres pour papier
DERIVED REMOVALS							**QUANTITES ENLEVEES CALCULEES :**
Sawlogs and veneer logs	m3 u.b.	1350	1344	1934	1563	2890	Grumes de sciage et placage
Smallwood	m3 u.b.	1379	1473	2032	1801	2992	Petits bois
Total removals	m3 u.b.	2730	2817	3966	3364	5882	Total quantites enlevees
FORECAST REMOVALS							**PREVISIONS DE QUANTITES ENLEVEES :**
Sawlogs and veneer logs	m3 u.b.	1350	1632	1749	1726	1951	Grumes de sciage et placage
Smallwood	m3 u.b.	1380	1561	1658	1639	1823	Petits bois
Total removals	m3 u.b.	2730	3193	3407	3365	3774	Total quantites enlevees
DIFFERENCE							**DIFFERENCE :**
Sawlogs and veneer logs	m3 u.b.	0	287	186-	162	940-	Grumes de sciage et placage
Smallwood	m3 u.b.	0	87	375-	163-	1170-	Petits bois
Total removals	m3 u.b.	0	375	560-	0	2109-	Total quantites enlevees

Trade is assumed to remain unchanged, at 1979-81 level.
Le commerce est suppose rester inchange a son niveau de 1979-81.

BULGARIA - BULGARIE

	Unit Unite 1000	1979-81 (average) (moyenne)	Low-bas 1990	Low-bas 2000	High-haut 1990	High-haut 2000
FORECAST CONSUMPTION :						PREVISIONS DE CONSOMMATION :
Sawnwood	m3(s)	1492	1660	1660	1780	1920	Sciages
Wood-based panels	m3	486	691	904	731	1004	Panneaux a base de bois
Paper and paperboard	m.t.	485	520	560	570	680	Papiers et cartons
Fuelwood	m3	1930	1680	1610	1680	1610	Bois de chauffage
DERIVED PRODUCTION :						PRODUCTION CALCULEE :
Sawnwood	m3(s)	1368	1535	1535	1655	1795	Sciages
Wood-based panels	m3	547	752	965	792	1065	Panneaux a base de bois
Woodpulp	m.t.	216	237	261	270	339	Pates de bois
Paper and paperboard	m.t.	385	420	460	470	580	Papiers et cartons
RECOVERY :						RECUPERATION :
Waste paper	m.t.	154	175	200	204	270	Vieux papiers
DOMESTIC SUPPLY :						APPROVISIONNEMENT NATIONAL :
Residues and chips	m3	342	491	605	882	1084	Dechets et plaquettes
DERIVED CONSUMPTION :						CONSOMMATION CALCULEE :
Sawlogs and veneer logs	m3 u.b.	2455	2730	2737	2929	3169	Grumes de sciage et placage
Total pulpwood	m3	1820	2230	2656	2451	3182	Total bois de trituration
Fibres for paper	m.t.	476	518	567	580	715	Fibres pour papier
DERIVED REMOVALS :						QUANTITES ENLEVEES CALCULEES :
Sawlogs and veneer logs	m3 u.b.	1479	1754	1761	1953	2193	Grumes de sciage et placage
Smallwood	m3 u.b.	2960	2928	3140	2758	3188	Petits bois
Total removals	m3 u.b.	4439	4682	4902	4711	5382	Total quantites enlevees
FORECAST REMOVALS :						PREVISIONS DE QUANTITES ENLEVEES :
Sawlogs and veneer logs	m3 u.b.	1479	1660	1705	1890	2175	Grumes de sciage et placage
Smallwood	m3 u.b.	2960	2740	2805	2850	2965	Petits bois
Total removals	m3 u.b.	4439	4400	4510	4740	5140	Total quantites enlevees
DIFFERENCE :						DIFFERENCE :
Sawlogs and veneer logs	m3 u.b.	0	95-	57-	64-	19-	Grumes de sciage et placage
Smallwood	m3 u.b.	0	189-	336-	91-	224-	Petits bois
Total removals	m3 u.b.	0	283-	393-	28-	243-	Total quantites enlevees

Trade is assumed to remain unchanged, at 1979-81 level.

Le commerce est suppose rester inchange a son niveau de 1979-81.

CYPRUS - CHYPRE

	Unit Unite 1000	1979-81 (average) (moyenne)	Low-bas 1990	Low-bas 2000	High-haut 1990	High-haut 2000	
FORECAST CONSUMPTION :							**PREVISIONS DE CONSOMMATION :**
Sawnwood	m3(s)	133	120	140	120	140	Sciages
Wood-based panels	m3	25	47	65	47	65	Panneaux a base de bois
Paper and paperboard	m.t.	48	50	60	60	90	Papiers et cartons
Fuelwood	m3	20	30	30	30	30	Bois de chauffage
DERIVED PRODUCTION :							**PRODUCTION CALCULEE :**
Sawnwood	m3(s)	74	60	80	60	80	Sciages
Wood-based panels	m3	0	21	39	21	39	Panneaux a base de bois
Woodpulp	m.t.	0	2-	5-	4-	11-	Pates de bois
Paper and paperboard	m.t.	0	1	11	11	41	Papiers et cartons
RECOVERY :							**RECUPERATION :**
Waste paper	m.t.	0	2	4	3	10	Vieux papiers
DOMESTIC SUPPLY :							**APPROVISIONNEMENT NATIONAL :**
Residues and chips	m3	0	0	0	0	0	Dechets et plaquettes
DERIVED CONSUMPTION :							**CONSOMMATION CALCULEE :**
Sawlogs and veneer logs	m3 u.b.	123	124	176	124	176	Grumes de sciage et placage
Total pulpwood	m3	0	17	33	17	33	Total bois de trituration
Fibres for paper	m.t.	0	0	0	0	0	Fibres pour papier
DERIVED REMOVALS :							**QUANTITES ENLEVEES CALCULEES :**
Sawlogs and veneer logs	m3 u.b.	40	41	93	41	93	Grumes de sciage et placage
Smallwood	m3 u.b.	35	58	74	58	74	Petits bois
Total removals	m3 u.b.	76	100	168	100	168	Total quantites enlevees
FORECAST REMOVALS :							**PREVISIONS DE QUANTITES ENLEVEES :**
Sawlogs and veneer logs	m3 u.b.	40	27	27	27	27	Grumes de sciage et placage
Smallwood	m3 u.b.	35	28	28	28	28	Petits bois
Total removals	m3 u.b.	75	55	55	55	55	Total quantites enlevees
DIFFERENCE :							**DIFFERENCE :**
Sawlogs and veneer logs	m3 u.b.	0	15-	67-	15-	67-	Grumes de sciage et placage
Smallwood	m3 u.b.	0	31-	47-	31-	47-	Petits bois
Total removals	m3 u.b.	0	46-	114-	46-	114-	Total quantites enlevees

Trade is assumed to remain unchanged, at 1979-81 level.
Le commerce est suppose rester inchange a son niveau de 1979-81.

CZECHOSLOVAKIA - TCHECOSLOVAQUIE

	Unit Unité 1000	1979-81 (average) (moyenne)	Low-bas 1990	Low-bas 2000	High-haut 1990	High-haut 2000	
FORECAST CONSUMPTION :							**PREVISIONS DE CONSOMMATION :**
Sawnwood	m3(s)	3873	3870	3870	4160	4470	Sciages
Wood-based panels	m3	1214	1625	2046	1675	2146	Panneaux à base de bois
Paper and paperboard	m.t.	1069	1160	1250	1300	1510	Papiers et cartons
Fuelwood	m3	1360	1000	1900	1000	1900	Bois de chauffage
DERIVED PRODUCTION :							**PRODUCTION CALCULEE :**
Sawnwood	m3(s)	4880	4877	4877	5167	5477	Sciages
Wood-based panels	m3	1168	1579	2000	1629	2100	Panneaux à base de bois
Woodpulp	m.t.	815	850	882	912	979	Pâtes de bois
Paper and paperboard	m.t.	1177	1268	1358	1408	1618	Papiers et cartons
RECOVERY :							**RECUPERATION :**
Waste paper	m.t.	377	432	491	510	653	Vieux papiers
DOMESTIC SUPPLY :							**APPROVISIONNEMENT NATIONAL :**
Residues and chips	m3	1036	1323	1611	1998	2422	Déchets et plaquettes
DERIVED CONSUMPTION :							**CONSOMMATION CALCULEE :**
Sawlogs and veneer logs	m3 u.b.	8104	8098	8100	8538	9008	Grumes de sciage et placage
Total pulpwood	m3	5168	5937	6667	6276	7232	Total bois de trituration
Fibres for paper	m.t.	1177	1268	1359	1409	1619	Fibres pour papier
DERIVED REMOVALS :							**QUANTITES ENLEVEES CALCULEES :**
Sawlogs and veneer logs	m3 u.b.	10574	10568	10569	11008	11478	Grumes de sciage et placage
Smallwood	m3 u.b.	8747	8760	10022	8425	9777	Petits bois
Total removals	m3 u.b.	19321	19328	20592	19433	21256	Total quantités enlevées
FORECAST REMOVALS :							**PREVISIONS DE QUANTITES ENLEVEES :**
Sawlogs and veneer logs	m3 u.b.	10574	9270	9140	9910	9990	Grumes de sciage et placage
Smallwood	m3 u.b.	8743	8280	8410	8540	8910	Petits bois
Total removals	m3 u.b.	19317	17550	17550	18450	18900	Total quantités enlevées
DIFFERENCE :							**DIFFERENCE :**
Sawlogs and veneer logs	m3 u.b.	0	1299-	1430-	1099-	1489-	Grumes de sciage et placage
Smallwood	m3 u.b.	0	481-	1613-	114-	868-	Petits bois
Total removals	m3 u.b.	0	1779-	3043-	984-	2357-	Total quantités enlevées

Trade is assumed to remain unchanged, at 1979-81 level.
Le commerce est supposé rester inchangé à son niveau de 1979-81.

DENMARK - DANEMARK

	Unit Unité 1000	1979-81 (average) (moyenne)	Low-bas 1990	Low-bas 2000	High-haut 1990	High-haut 2000
FORECAST CONSUMPTION : PREVISIONS DE CONSOMMATION						
Sawnwood / Sciages	m3(s)	1921	1970	2220	2070	2530
Wood-based panels / Panneaux a base de bois	m3	680	692	829	742	969
Paper and paperboard / Papiers et cartons	m.t.	761	850	1010	970	1350
Fuelwood / Bois de chauffage	m3	250	250	200	350	450
DERIVED PRODUCTION : PRODUCTION CALCULEE						
Sawnwood / Sciages	m3(s)	818	866	1116	966	1426
Wood-based panels / Panneaux a base de bois	m3	379	391	528	441	668
Woodpulp / Pates de bois	m.t.	62	99	195	167	381
Paper and paperboard / Papiers et cartons	m.t.	245	333	493	453	833
RECOVERY : RECUPERATION						
Waste paper / Vieux papiers	m.t.	214	264	334	321	500
DOMESTIC SUPPLY : APPROVISIONNEMENT NATIONAL						
Residues and chips / Dechets et plaquettes	m3	193	295	516	554	1016
DERIVED CONSUMPTION : CONSOMMATION CALCULEE						
Sawlogs and veneer logs / Grumes de sciage et placage	m3 u.b.	1793	1860	2433	2102	3179
Total pulpwood / Total bois de trituration	m3	974	1128	1642	1380	2357
Fibres for paper / Fibres pour papier	m.t.	258	345	511	470	864
DERIVED REMOVALS : QUANTITES ENLEVEES CALCULEES						
Sawlogs and veneer logs / Grumes de sciage et placage	m3 u.b.	1050	1117	1690	1359	2436
Smallwood / Petits bois	m3 u.b.	839	893	1136	985	1601
Total removals / Total quantites enlevees	m3 u.b.	1890	2011	2827	2345	4038
FORECAST REMOVALS : PREVISIONS DE QUANTITES ENLEVEES						
Sawlogs and veneer logs / Grumes de sciage et placage	m3 u.b.	1050	1100	1050	1090	1120
Smallwood / Petits bois	m3 u.b.	840	1150	1290	1250	1310
Total removals / Total quantites enlevees	m3 u.b.	1890	2250	2340	2340	2430
DIFFERENCE : DIFFERENCE						
Sawlogs and veneer logs / Grumes de sciage et placage	m3 u.b.	0	18-	641-	270-	1317-
Smallwood / Petits bois	m3 u.b.	0	256	153	264	292-
Total removals / Total quantites enlevees	m3 u.b.	0	238	488-	6-	1609-

Trade is assumed to remain unchanged at 1979-81 level.
Le commerce est suppose rester inchange a son niveau de 1979-81.

FINLAND - FINLANDE

	Unit Unite 1000	1979-81 (average) (moyenne)	Low-bas 1990	Low-bas 2000	High-haut 1990	High-haut 2000	
FORECAST CONSUMPTION :							**PREVISIONS DE CONSOMMATION :**
Sawnwood	m3(s)	3103	3190	3450	3310	4020	Sciages
Wood-based panels	m3	697	750	901	830	1131	Panneaux à base de bois
Paper and paperboard	m.t.	1138	1340	1680	1520	2260	Papiers et cartons
Fuelwood	m3	3700	4000	4300	4200	4700	Bois de chauffage
DERIVED PRODUCTION :							**PRODUCTION CALCULEE :**
Sawnwood	m3(s)	9405	9492	9752	9612	10322	Sciages
Wood-based panels	m3	1681	1734	1885	1814	2115	Panneaux à base de bois
Woodpulp	m.t.	7010	7102	7271	7197	7583	Pates de bois
Paper and paperboard	m.t.	5930	6132	6472	6312	7052	Papiers et cartons
RECOVERY :							**RECUPERATION :**
Waste paper	m.t.	234	329	480	404	714	Vieux papiers
DOMESTIC SUPPLY :							**APPROVISIONNEMENT NATIONAL :**
Residues and chips	m3	7829	8017	8336	8491	9585	Dechets et plaquettes
DERIVED CONSUMPTION :							**CONSOMMATION CALCULEE :**
Sawlogs and veneer logs	m3 u.b.	19101	19273	19811	19522	20916	Grumes de sciage et placage
Total pulpwood	m3	29294	29748	30605	30219	32119	Total bois de trituration
Fibres for paper	m.t.	5577	5765	6084	5934	6630	Fibres pour papier
DERIVED REMOVALS :							**QUANTITES ENLEVEES CALCULEES :**
Sawlogs and veneer logs	m3 u.b.	20456	20629	21167	20878	22271	Grumes de sciage et placage
Smallwood	m3 u.b.	25319	25888	26725	26084	27390	Petits bois
Total removals	m3 u.b.	45776	46517	47892	46962	49662	Total quantites enlevees
FORECAST REMOVALS :							**PREVISIONS DE QUANTITES ENLEVEES :**
Sawlogs and veneer logs	m3 u.b.	20456	20875	20885	21600	22525	Grumes de sciage et placage
Smallwood	m3 u.b.	25320	25055	27715	28610	30555	Petits bois
Total removals	m3 u.b.	45776	45930	48600	50210	53080	Total quantites enlevees
DIFFERENCE :							**DIFFERENCE :**
Sawlogs and veneer logs	m3 u.b.	0	245	283-	721	253	Grumes de sciage et placage
Smallwood	m3 u.b.	0	834-	989	2525	3164	Petits bois
Total removals	m3 u.b.	0	588-	707	3247	3417	Total quantites enlevees

Trade is assumed to remain unchanged, at 1979-81 level.
Le commerce est suppose rester inchange a son niveau de 1979-81.

FRANCE - FRANCE

	Unit / Unité 1000	1979-81 (average) (moyenne)	Low-bas 1990	Low-bas 2000	High-haut 1990	High-haut 2000	
FORECAST CONSUMPTION :							**PREVISIONS DE CONSOMMATION :**
Sawnwood	m3(s)	11750	12230	14550	12930	17550	Sciages
Wood-based panels	m3	3315	3455	4181	3705	5351	Panneaux à base de bois
Paper and paperboard	m.t.	6203	7030	9100	8000	12250	Papiers et cartons
Fuelwood	m3	10000	13500	18500	10500	11200	Bois de chauffage
DERIVED PRODUCTION :							**PRODUCTION CALCULEE :**
Sawnwood	m3(s)	9467	9946	12266	10646	15266	Sciages
Wood-based panels	m3	3071	3211	3937	3461	5107	Panneaux à base de bois
Woodpulp	m.t.	1737	2143	3296	2609	4773	Pates de bois
Paper and paperboard	m.t.	5189	6015	8085	6985	11235	Papiers et cartons
RECOVERY :							**RECUPERATION :**
Waste paper	m.t.	1983	2390	3276	2880	4900	Vieux papiers
DOMESTIC SUPPLY :							**APPROVISIONNEMENT NATIONAL :**
Residues and chips	m3	4077	5079	6750	6286	10266	Dechets et plaquettes
DERIVED CONSUMPTION :							**CONSOMMATION CALCULEE :**
Sawlogs and veneer logs	m3 u.b.	17548	18430	22724	19754	28432	Grumes de sciage et placage
Total pulpwood	m3	10697	12392	17780	14462	24762	Total bois de trituration
Fibres for paper	m.t.	5109	5923	7961	6878	11062	Fibres pour papier
DERIVED REMOVALS :							**QUANTITES ENLEVEES CALCULEES :**
Sawlogs and veneer logs	m3 u.b.	18490	19372	23666	20695	29373	Grumes de sciage et placage
Smallwood	m3 u.b.	19990	24103	32792	21965	28957	Petits bois
Total removals	m3 u.b.	38480	43475	56458	42661	58331	Total quantites enlevees
FORECAST REMOVALS :							**PREVISIONS DE QUANTITES ENLEVEES :**
Sawlogs and veneer logs	m3 u.b.	18490	21495	23870	22700	26890	Grumes de sciage et placage
Smallwood	m3 u.b.	19990	20355	20120	22740	23990	Petits bois
Total removals	m3 u.b.	38480	41850	43990	45440	50880	Total quantites enlevees
DIFFERENCE :							**DIFFERENCE :**
Sawlogs and veneer logs	m3 u.b.	0	2122	203	2004	2484-	Grumes de sciage et placage
Smallwood	m3 u.b.	0	3749-	12673-	774	4968-	Petits bois
Total removals	m3 u.b.	0	1626-	12469-	2778	7452-	Total quantites enlevees

Trade is assumed to remain unchanged, at 1979-81 level.
Le commerce est suppose rester inchange a son niveau de 1979-81.

GERMAN DEM. REP. - REP. DEM. ALLEMANDE

	Unit Unite 1000	1979-81 (average) (moyenne)	Low-bas 1990	Low-bas 2000	High-haut 1990	High-haut 2000	
FORECAST CONSUMPTION :							**PREVISIONS DE CONSOMMATION :**
Sawnwood	m3(s)	3717	3720	3720	4000	4300	Sciages
Wood-based panels	m3	1427	2122	2915	2222	3105	Panneaux a base de bois
Paper and paperboard	m.t.	1383	1500	1610	1650	1920	Papiers et cartons
Fuelwood	m3	680	670	760	690	700	Bois de chauffage
DERIVED PRODUCTION :							**PRODUCTION CALCULEE :**
Sawnwood	m3(s)	2393	2396	2396	2676	2976	Sciages
Wood-based panels	m3	1077	1772	2566	1872	2756	Panneaux a base de bois
Woodpulp	m.t.	640	704	761	771	889	Pates de bois
Paper and paperboard	m.t.	1240	1356	1466	1506	1776	Papiers et cartons
RECOVERY :							**RECUPERATION :**
Waste paper	m.t.	596	661	726	760	942	Vieux papiers
DOMESTIC SUPPLY :							**APPROVISIONNEMENT NATIONAL :**
Residues and chips	m3	1313	1385	1527	1635	1976	Dechets et plaquettes
DERIVED CONSUMPTION :							**CONSOMMATION CALCULEE :**
Sawlogs and veneer logs	m3 u.b.	4226	4180	4319	4653	5300	Grumes de sciage et placage
Total pulpwood	m3	5309	6747	8032	7165	8824	Total bois de trituration
Fibres for paper	m.t.	1377	1506	1628	1673	1972	Fibres pour papier
DERIVED REMOVALS :							**QUANTITES ENLEVEES CALCULEES :**
Sawlogs and veneer logs	m3 u.b.	4046	3999	4139	4473	5119	Grumes de sciage et placage
Smallwood	m3 u.b.	5939	7251	8452	7439	8736	Petits bois
Total removals	m3 u.b.	9985	11251	12592	11912	13856	Total quantites enlevees
FORECAST REMOVALS :							**PREVISIONS DE QUANTITES ENLEVEES :**
Sawlogs and veneer logs	m3 u.b.	4046	4420	4400	4240	4470	Grumes de sciage et placage
Smallwood	m3 u.b.	5943	6625	6595	6360	6705	Petits bois
Total removals	m3 u.b.	9989	11045	10995	10600	11175	Total quantites enlevees
DIFFERENCE :							**DIFFERENCE :**
Sawlogs and veneer logs	m3 u.b.	0	420	260	234-	650-	Grumes de sciage et placage
Smallwood	m3 u.b.	4	627-	1858-	1080-	2032-	Petits bois
Total removals	m3 u.b.	4	207-	1598-	1313-	2682-	Total quantites enlevees

Trade is assumed to remain unchanged, at 1979-81 level.
Le commerce est suppose rester inchange a son niveau de 1979-81.

GERMANY, FED. REP. OF - REP. FED. D'ALLEMAGNE

	Unit Unite 1000	1979-81 (average) (moyenne)	Low-bas 1990	Low-bas 2000	High-haut 1990	High-haut 2000
FORECAST CONSUMPTION : / PREVISIONS DE CONSOMMATION :						
Sawnwood / Sciages	m3(s)	14445	14510	16120	15310	18990
Wood-based panels / Panneaux à base de bois	m3	8003	8425	10053	8995	12303
Paper and paperboard / Papiers et cartons	m.t.	9599	10710	13030	12190	17530
Fuelwood / Bois de chauffage	m3	3550	4500	5000	4500	5000
DERIVED PRODUCTION : / PRODUCTION CALCULEE :						
Sawnwood / Sciages	m3(s)	10118	10183	11793	10983	14663
Wood-based panels / Panneaux à base de bois	m3	7030	7452	9081	8022	11331
Woodpulp / Pâtes de bois	m.t.	1830	2286	3381	2907	5219
Paper and paperboard / Papiers et cartons	m.t.	7617	8727	11047	10207	15547
RECOVERY : / RECUPERATION :						
Waste paper / Vieux papiers	m.t.	3339	3941	5055	4729	7502
DOMESTIC SUPPLY : / APPROVISIONNEMENT NATIONAL :						
Residues and chips / Déchets et plaquettes	m3	5019	5077	5855	5468	7522
DERIVED CONSUMPTION : / CONSOMMATION CALCULEE :						
Sawlogs and veneer logs / Grumes de sciage et placage	m3 u.b.	15594	15747	18154	17028	22816
Total pulpwood / Total bois de trituration	m3	16226	18293	24264	21116	33317
Fibres for paper / Fibres pour papier	m.t.	7253	8311	10520	9720	14805
DERIVED REMOVALS : / QUANTITES ENLEVEES CALCULEES :						
Sawlogs and veneer logs / Grumes de sciage et placage	m3 u.b.	17699	17853	20259	19134	24921
Smallwood / Petits bois	m3 u.b.	15329	18201	23840	20632	31226
Total removals / Total quantites enlevees	m3 u.b.	33029	36054	44099	39766	56147
FORECAST REMOVALS : / PREVISIONS DE QUANTITES ENLEVEES :						
Sawlogs and veneer logs / Grumes de sciage et placage	m3 u.b.	17699	17000	17000	19000	19400
Smallwood / Petits bois	m3 u.b.	15330	13150	13150	14150	14250
Total removals / Total quantites enlevees	m3 u.b.	33030	30150	30150	33150	33650
DIFFERENCE : / DIFFERENCE :						
Sawlogs and veneer logs / Grumes de sciage et placage	m3 u.b.	0	854-	3260-	135-	5522-
Smallwood / Petits bois	m3 u.b.	0	5052-	10691-	6483-	16977-
Total removals / Total quantites enlevees	m3 u.b.	0	5905-	13950-	6617-	22498-

Trade is assumed to remain unchanged, at 1979-81 level.
Le commerce est supposé rester inchangé a son niveau de 1979-81.

GREECE - GRECE

	Unit Unite 1000	1979-81 (average) (moyenne)	Low-bas 1990	Low-bas 2000	High-haut 1990	High-haut 2000	
FORECAST CONSUMPTION :							PREVISIONS DE CONSOMMATION :
Sawnwood	m3(s)	751	1050	1340	1140	1730	Sciages
Wood-based panels	m3	451	422	591	452	741	Panneaux à base de bois
Paper and paperboard	m.t.	442	500	700	600	1020	Papiers et cartons
Fuelwood	m3	720	750	750	790	790	Bois de chauffage
DERIVED PRODUCTION :							PRODUCTION CALCULEE :
Sawnwood	m3(s)	407	705	995	795	1385	Sciages
Wood-based panels	m3	480	450	619	480	769	Panneaux à base de bois
Woodpulp	m.t.	25	31	86	60	165	Pâtes de bois
Paper and paperboard	m.t.	302	359	559	459	879	Papiers et cartons
RECOVERY :							RECUPERATION :
Waste paper	m.t.	40	65	119	90	214	Vieux papiers
DOMESTIC SUPPLY :							APPROVISIONNEMENT NATIONAL :
Residues and chips	m3	150	283	481	452	897	Déchets et plaquettes
DERIVED CONSUMPTION :							CONSOMMATION CALCULEE :
Sawlogs and veneer logs	m3 u.b.	895	1376	1933	1551	2684	Grumes de sciage et placage
Total pulpwood	m3	572	573	899	673	1245	Total bois de trituration
Fibres for paper	m.t.	165	195	304	250	478	Fibres pour papier
DERIVED REMOVALS :							QUANTITES ENLEVEES CALCULEES :
Sawlogs and veneer logs	m3 u.b.	486	967	1524	1142	2275	Grumes de sciage et placage
Smallwood	m3 u.b.	2086	1978	2105	1948	2075	Petits bois
Total removals	m3 u.b.	2572	2945	3630	3091	4351	Total quantités enlevées
FORECAST REMOVALS :							PREVISIONS DE QUANTITES ENLEVEES :
Sawlogs and veneer logs	m3 u.b.	486	550	650	580	730	Grumes de sciage et placage
Smallwood	m3 u.b.	2086	2460	2700	2560	2925	Petits bois
Total removals	m3 u.b.	2572	3010	3350	3140	3655	Total quantités enlevées
DIFFERENCE :							DIFFERENCE :
Sawlogs and veneer logs	m3 u.b.	0	418-	875-	563-	1546-	Grumes de sciage et placage
Smallwood	m3 u.b.	0	481	594	611	849	Petits bois
Total removals	m3 u.b.	0	64	281-	48	697-	Total quantités enlevées

Trade is assumed to remain unchanged, at 1979-81 level.
Le commerce est supposé rester inchangé à son niveau de 1979-81.

HUNGARY - HONGRIE

	Unit Unite 1000	1979-81 (average) (moyenne)	Low-bas 1990	Low-bas 2000	High-haut 1990	High-haut 2000	
FORECAST CONSUMPTION :							**PREVISIONS DE CONSOMMATION :**
Sawnwood	m3(s)	2057	2070	2070	2230	2400	Sciages
Wood-based panels	m3	470	717	967	747	1027	Panneaux a base de bois
Paper and paperboard	m.t.	657	710	760	770	890	Papiers et cartons
Fuelwood	m3	2380	2850	2890	2900	2960	Bois de chauffage
DERIVED PRODUCTION :							**PRODUCTION CALCULEE :**
Sawnwood	m3(s)	1295	1308	1308	1468	1638	Sciages
Wood-based panels	m3	400	647	897	677	957	Panneaux a base de bois
Woodpulp	m.t.	96	118	145	144	196	Pates de bois
Paper and paperboard	m.t.	444	497	547	557	677	Papiers et cartons
RECOVERY :							**RECUPERATION :**
Waste paper	m.t.	250	285	313	324	402	Vieux papiers
DOMESTIC SUPPLY :							**APPROVISIONNEMENT NATIONAL :**
Residues and chips	m3	128	245	362	530	870	Dechets et plaquettes
DERIVED CONSUMPTION :							**CONSOMMATION CALCULEE :**
Sawlogs and veneer logs	m3 u.b.	2481	2505	2507	2791	3096	Grumes de sciage et placage
Total pulpwood	m3	883	1259	1682	1368	1901	Total bois de trituration
Fibres for paper	m.t.	480	538	592	603	732	Fibres pour papier
DERIVED REMOVALS :							**QUANTITES ENLEVEES CALCULEES :**
Sawlogs and veneer logs	m3 u.b.	1839	1864	1865	2150	2455	Grumes de sciage et placage
Smallwood	m3 u.b.	4326	4934	5179	4808	4960	Petits bois
Total removals	m3 u.b.	6166	6798	7045	6958	7415	Total quantites enlevees
FORECAST REMOVALS :							**PREVISIONS DE QUANTITES ENLEVEES :**
Sawlogs and veneer logs	m3 u.b.	1839	1930	1940	2160	2310	Grumes de sciage et placage
Smallwood	m3 u.b.	4326	4830	5020	5400	5950	Petits bois
Total removals	m3 u.b.	6166	6760	6960	7560	8260	Total quantites enlevees
DIFFERENCE :							**DIFFERENCE :**
Sawlogs and veneer logs	m3 u.b.	0	65	74	9	146-	Grumes de sciage et placage
Smallwood	m3 u.b.	0	105-	160-	591	989	Petits bois
Total removals	m3 u.b.	0	39-	86-	601	844	Total quantites enlevees

Trade is assumed to remain unchanged, at 1979-81 level.
Le commerce est suppose rester inchange a son niveau de 1979-81.

ICELAND – ISLANDE

	Unit / Unité 1000	1979-81 (average / moyenne)	Low-bas 1990	Low-bas 2000	High-haut 1990	High-haut 2000	
FORECAST CONSUMPTION :							**PREVISIONS DE CONSOMMATION :**
Sawnwood	m3(s)	67	70	80	70	80	Sciages
Wood-based panels	m3	28	30	50	30	50	Panneaux à base de bois
Paper and paperboard	m.t.	19	10	30	20	30	Papiers et cartons
Fuelwood	m3	0	0	0	0	0	Bois de chauffage
DERIVED PRODUCTION :							**PRODUCTION CALCULEE :**
Sawnwood	m3(s)	0	2	12	2	12	Sciages
Wood-based panels	m3	0	1	21	1	21	Panneaux à base de bois
Woodpulp	m.t.	0	0	0	0	0	Pâtes de bois
Paper and paperboard	m.t.	0	9-	11	1	11	Papiers et cartons
RECOVERY :							**RECUPERATION :**
Waste Paper	m.t.	0	0	0	0	0	Vieux papiers
DOMESTIC SUPPLY :							**APPROVISIONNEMENT NATIONAL :**
Residues and chips	m3	0	0	0	0	0	Dechets et plaquettes
DERIVED CONSUMPTION :							**CONSOMMATION CALCULEE :**
Sawlogs and veneer logs	m3 u.b.	0	7-	11	7-	11	Grumes de sciage et placage
Total pulpwood	m3	0	10	42	10	42	Total bois de trituration
Fibres for paper	m.t.	0	0	0	0	0	Fibres pour papier
DERIVED REMOVALS :							**QUANTITES ENLEVEES CALCULEES :**
Sawlogs and veneer logs	m3 u.b.	0	7-	11	7-	11	Grumes de sciage et placage
Smallwood	m3 u.b.	0	11	43	11	43	Petits bois
Total removals	m3 u.b.	0	4	54	4	54	Total quantites enlevees
FORECAST REMOVALS :							**PREVISIONS DE QUANTITES ENLEVEES :**
Sawlogs and veneer logs	m3 u.b.	0	0	0	0	0	Grumes de sciage et placage
Smallwood	m3 u.b.	0	0	0	0	0	Petits bois
Total removals	m3 u.b.	0	0	0	0	0	Total quantites enlevees
DIFFERENCE :							**DIFFERENCE :**
Sawlogs and veneer logs	m3 u.b.	0	6	12-	6	12-	Grumes de sciage et placage
Smallwood	m3 u.b.	0	12-	44-	12-	44-	Petits bois
Total removals	m3 u.b.	0	5-	55-	5-	55-	Total quantites enlevees

Trade is assumed to remain unchanged, at 1979-81 level.
Le commerce est suppose rester inchange a son niveau de 1979-81.

IRELAND - IRLANDE

	Unit Unité 1000	1979-81 (average) (moyenne)	Low-bas 1990	Low-bas 2000	High-haut 1990	High-haut 2000	
FORECAST CONSUMPTION :							**PREVISIONS DE CONSOMMATION :**
Sawnwood	m3(s)	635	640	720	700	970	Sciages
Wood-based panels	m3	174	203	255	223	385	Panneaux à base de bois
Paper and paperboard	m.t.	273	270	360	330	520	Papiers et cartons
Fuelwood	m3	240	250	280	280	300	Bois de chauffage
DERIVED PRODUCTION :							**PRODUCTION CALCULEE :**
Sawnwood	m3(s)	157	162	242	222	492	Sciages
Wood-based panels	m3	49	79	131	99	261	Panneaux à base de bois
Woodpulp	m.t.	15	0	53	38	146	Pates de bois
Paper and paperboard	m.t.	67	63	153	123	313	Papiers et cartons
RECOVERY :							**RECUPERATION :**
Waste paper	m.t.	47	58	91	77	153	Vieux papiers
DOMESTIC SUPPLY :							**APPROVISIONNEMENT NATIONAL :**
Residues and chips	m3	71	85	145	134	352	Dechets et plaquettes
DERIVED CONSUMPTION :							**CONSOMMATION CALCULEE :**
Sawlogs and veneer logs	m3 u.b.	287	294	460	424	970	Grumes de sciage et placage
Total pulpwood	m3	114	121	274	201	590	Total bois de trituration
Fibres for paper	m.t.	66	61	148	119	302	Fibres pour papier
DERIVED REMOVALS :							**QUANTITES ENLEVEES CALCULEES :**
Sawlogs and veneer logs	m3 u.b.	340	347	513	477	1023	Grumes de sciage et placage
Smallwood	m3 u.b.	190	225	349	286	477	Petits bois
Total removals	m3 u.b.	530	572	862	763	1501	Total quantites enlevees
FORECAST REMOVALS :							**PREVISIONS DE QUANTITES ENLEVEES :**
Sawlogs and veneer logs	m3 u.b.	340	865	2045	865	2045	Grumes de sciage et placage
Smallwood	m3 u.b.	190	735	995	735	995	Petits bois
Total removals	m3 u.b.	530	1600	3040	1600	3040	Total quantites enlevees
DIFFERENCE :							**DIFFERENCE :**
Sawlogs and veneer logs	m3 u.b.	0	517	1531	387	1021	Grumes de sciage et placage
Smallwood	m3 u.b.	0	509	645	448	517	Petits bois
Total removals	m3 u.b.	0	1027	2177	836	1538	Total quantites enlevees

Trade is assumed to remain unchanged, at 1979-81 level.
Le commerce est suppose rester inchange a son niveau de 1979-81.

ITALY – ITALIE

	Unit Unite 1000	1979-81 (average) (moyenne)	Low-bas 1990	Low-bas 2000	High-haut 1990	High-haut 2000	
FORECAST CONSUMPTION :							**PREVISIONS DE CONSOMMATION :**
Sawnwood	m3(s)	7954	8140	9500	8610	11100	Sciages
Wood-based panels	m3	3425	3531	4276	3661	4766	Panneaux a base de bois
Paper and paperboard	m.t.	5194	5650	7340	6700	10750	Papiers et cartons
Fuelwood	m3	3460	3970	4500	4500	5530	Bois de chauffage
DERIVED PRODUCTION :							**PRODUCTION CALCULEE :**
Sawnwood	m3(s)	2555	2740	4100	3210	5700	Sciages
Wood-based panels	m3	2890	2997	3741	3127	4231	Panneaux a base de bois
Woodpulp	m.t.	659	774	1635	1266	3169	Pates de bois
Paper and paperboard	m.t.	4963	5418	7108	6468	10518	Papiers et cartons
RECOVERY :							**RECUPERATION :**
Waste paper	m.t.	1491	1791	2473	2257	4052	Vieux papiers
DOMESTIC SUPPLY :							**APPROVISIONNEMENT NATIONAL :**
Residues and chips	m3	1405	1802	2974	2379	4492	Dechets et plaquettes
DERIVED CONSUMPTION :							**CONSOMMATION CALCULEE :**
Sawlogs and veneer logs	m3 u.b.	6284	6714	9502	7619	12624	Grumes de sciage et placage
Total pulpwood	m3	4746	5174	8461	6750	13457	Total bois de trituration
Fibres for paper	m.t.	4533	4947	6491	5906	9604	Fibres pour papier
DERIVED REMOVALS :							**QUANTITES ENLEVEES CALCULEES :**
Sawlogs and veneer logs	m3 u.b.	2309	2739	5527	3644	8649	Grumes de sciage et placage
Smallwood	m3 u.b.	6649	7175	9820	8704	14328	Petits bois
Total Removals	m3 u.b.	8958	9915	15347	12349	22978	Total quantites enlevees
FORECAST REMOVALS :							**PREVISIONS DE QUANTITES ENLEVEES :**
Sawlogs and veneer logs	m3 u.b.	2309	2610	2760	2715	3080	Grumes de sciage et placage
Smallwood	m3 u.b.	6649	6925	7010	7080	7490	Petits bois
Total removals	m3 u.b.	8959	9535	9770	9795	10570	Total quantites enlevees
DIFFERENCE :							**DIFFERENCE :**
Sawlogs and veneer logs	m3 u.b.	0	130-	2768-	930-	5570-	Grumes de sciage et placage
Smallwood	m3 u.b.	0	251-	2811-	1625-	6839-	Petits bois
Total removals	m3 u.b.	0	381-	5578-	2555-	12409-	Total quantites enlevees

Trade is assumed to remain unchanged, at 1979-81 level.
Le commerce est suppose rester inchange a son niveau de 1979-81.

MALTA - MALTE

	Unit Unite 1000	1979-81 (average) (moyenne)	Low-bas 1990	Low-bas 2000	High-haut 1990	High-haut 2000
FORECAST CONSUMPTION : / PREVISIONS DE CONSOMMATION :						
Sawnwood / Sciages	m3(s)	24	30	40	30	40
Wood-based panels / Panneaux a base de bois	m3	5	10	20	10	20
Paper and paperboard / Papiers et cartons	m.t.	6	20	40	30	60
Fuelwood / Bois de chauffage	m3	0	0	0	0	0
DERIVED PRODUCTION : / PRODUCTION CALCULEE :						
Sawnwood / Sciages	m3(s)	0	5	15	5	15
Wood-based panels / Panneaux a base de bois	m3	0	4	14	4	14
Woodpulp / Pates de bois	m.t.	0	1-	4-	2-	8-
Paper and paperboard / Papiers et cartons	m.t.	0	13	33	23	53
RECOVERY : / RECUPERATION :						
Waste paper / Vieux papiers	m.t.	0	0	3	1	7
DOMESTIC SUPPLY : / APPROVISIONNEMENT NATIONAL :						
Residues and chips / Dechets et plaquettes	m3	0	0	0	0	0
DERIVED CONSUMPTION : / CONSOMMATION CALCULEE :						
Sawlogs and veneer logs / Grumes de sciage et placage	m3 u.b.	0	19	58	19	58
Total pulpwood / Total bois de trituration	m3	0	0	0	0	0
Fibres for paper / Fibres pour papier	m.t.	0	0	0	0	0
DERIVED REMOVALS : / QUANTITES ENLEVEES CALCULEES :						
Sawlogs and veneer logs / Grumes de sciage et placage	m3 u.b.	0	19	58	19	58
Smallwood / Petits bois	m3 u.b.	0	0	0	0	0
Total removals / Total quantites enlevees	m3 u.b.	0	19	58	19	58
FORECAST REMOVALS : / PREVISIONS DE QUANTITES ENLEVEES :						
Sawlogs and veneer logs / Grumes de sciage et placage	m3 u.b.	0	0	0	0	0
Smallwood / Petits bois	m3 u.b.	0	0	0	0	0
Total removals / Total quantites enlevees	m3 u.b.	0	0	0	0	0
DIFFERENCE : / DIFFERENCE :						
Sawlogs and veneer logs / Grumes de sciage et placage	m3 u.b.	0	20-	59-	20-	59-
Smallwood / Petits bois	m3 u.b.	0	0	0	0	0
Total removals / Total quantites enlevees	m3 u.b.	0	20-	59-	20-	59-

Trade is assumed to remain unchanged, at 1979-81 level.
Le commerce est suppose rester inchange a son niveau de 1979-81.

NETHERLANDS - PAYS-BAS

	Unit Unite 1000	1979-81 (average) (moyenne)	Low-bas 1990	Low-bas 2000	High-haut 1990	High-haut 2000	
FORECAST CONSUMPTION :							**PREVISIONS DE CONSOMMATION :**
Sawnwood	m3(s)	3146	3120	3480	3240	4150	Sciages
Wood-based panels	m3	1231	1202	1321	1252	1671	Panneaux à base de bois
Paper and paperboard	m.t.	2194	2280	2710	2580	3610	Papiers et cartons
Fuelwood	m3	80	90	100	100	130	Bois de chauffage
DERIVED PRODUCTION :							**PRODUCTION CALCULEE :**
Sawnwood	m3(s)	294	267	627	387	1297	Sciages
Wood-based panels	m3	170	141	260	191	610	Panneaux à base de bois
Woodpulp	m.t.	169	223	430	334	765	Pates de bois
Paper and paperboard	m.t.	1697	1783	2213	2083	3113	Papiers et cartons
RECOVERY :							**RECUPERATION :**
Waste paper	m.t.	966	996	1211	1179	1758	Vieux papiers
DOMESTIC SUPPLY :							**APPROVISIONNEMENT NATIONAL :**
Residues and chips	m3	166	157	365	276	1030	Dechets et plaquettes
DERIVED CONSUMPTION :							**CONSOMMATION CALCULEE :**
Sawlogs and veneer logs	m3 u.b.	642	531	1166	806	2670	Grumes de sciage et placage
Total pulpwood	m3 u.b.	537	674	1258	951	2279	Total bois de trituration
Fibres for paper	m.t.	1665	1748	2170	2042	3052	Fibres pour papier
DERIVED REMOVALS :							**QUANTITES ENLEVEES CALCULEES :**
Sawlogs and veneer logs	m3 u.b.	140	28	664	304	2168	Grumes de sciage et placage
Smallwood	m3 u.b.	875	1026	1421	1194	1806	Petits bois
Total removals	m3 u.b.	1016	1055	2085	1498	3975	Total quantites enlevees
FORECAST REMOVALS :							**PREVISIONS DE QUANTITES ENLEVEES :**
Sawlogs and veneer logs	m3 u.b.	140	230	270	245	325	Grumes de sciage et placage
Smallwood	m3 u.b.	870	790	860	815	935	Petits bois
Total removals	m3 u.b.	1011	1020	1130	1060	1260	Total quantites enlevees
DIFFERENCE :							**DIFFERENCE :**
Sawlogs and veneer logs	m3 u.b.	0	201	395-	60-	1844-	Grumes de sciage et placage
Smallwood	m3 u.b.	0	237-	562-	380-	872-	Petits bois
Total removals	m3 u.b.	0	36-	956-	439-	2716-	Total quantites enlevees

Trade is assumed to remain unchanged, at 1979-81 level.
Le commerce est suppose rester inchange a son niveau de 1979-81.

NORWAY – NORVEGE

	Unit Unite 1000	1979-81 (average) (moyenne)	Low-bas 1990	Low-bas 2000	High-haut 1990	High-haut 2000	
FORECAST CONSUMPTION :							PREVISIONS DE CONSOMMATION :
Sawnwood	m3(s)	2468	2470	2530	2580	2740	Sciages
Wood-based panels	m3	637	663	733	723	843	Panneaux a base de bois
Paper and paperboard	m.t.	519	640	770	640	990	Papiers et cartons
Fuelwood	m3	1750	2000	2200	2200	2400	Bois de chauffage
DERIVED PRODUCTION :							PRODUCTION CALCULEE :
Sawnwood	m3(s)	2427	2428	2488	2538	2698	Sciages
Wood-based panels	m3	588	614	684	674	794	Panneaux a base de bois
Woodpulp	m.t.	1419	1490	1566	1477	1686	Pates de bois
Paper and paperboard	m.t.	1382	1502	1632	1502	1852	Papiers et cartons
RECOVERY :							RECUPERATION :
Waste paper	m.t.	118	170	228	183	333	Vieux papiers
DOMESTIC SUPPLY :							APPROVISIONNEMENT NATIONAL :
Residues and chips	m3	1866	1896	1973	2068	2324	Dechets et plaquettes
DERIVED CONSUMPTION :							CONSOMMATION CALCULEE :
Sawlogs and veneer logs	m3 u.b.	4533	4551	4682	4754	5071	Grumes de sciage et placage
Total pulpwood	m3	6160	6410	6711	6465	7274	Total bois de trituration
Fibres for paper	m.t.	1418	1542	1675	1542	1901	Fibres pour papier
DERIVED REMOVALS :							QUANTITES ENLEVEES CALCULEES :
Sawlogs and veneer logs	m3 u.b.	4770	4788	4919	4991	5308	Grumes de sciage et placage
Smallwood	m3 u.b.	4749	5224	5647	5307	6060	Petits bois
Total removals	m3 u.b.	9519	10012	10567	10298	11369	Total quantites enlevees
FORECAST REMOVALS :							PREVISIONS DE QUANTITES ENLEVEES :
Sawlogs and veneer logs	m3 u.b.	4770	4389	4550	5145	5115	Grumes de sciage et placage
Smallwood	m3 u.b.	4749	4811	4950	5555	6185	Petits bois
Total removals	m3 u.b.	9519	9200	9500	10700	11300	Total quantites enlevees
DIFFERENCE :							DIFFERENCE :
Sawlogs and veneer logs	m3 u.b.	0	400-	370-	153	194-	Grumes de sciage et placage
Smallwood	m3 u.b.	0	414-	698-	247	124-	Petits bois
Total removals	m3 u.b.	0	813-	1068-	401	70-	Total quantites enlevees

Trade is assumed to remain unchanged, at 1979-81 level.
Le commerce est suppose rester inchange a son niveau de 1979-81.

POLAND - POLOGNE

	Unit Unite 1000	1979-81 (average) (moyenne)	Low-bas 1990	Low-bas 2000	High-haut 1990	High-haut 2000	
FORECAST CONSUMPTION :							PREVISIONS DE CONSOMMATION :
Sawnwood	m3(s)	6767	7980	7980	7980	8560	Sciages
Wood-based panels	m3	2081	2726	3359	2786	3479	Panneaux à base de bois
Paper and paperboard	m.t.	1373	1550	1670	1560	1830	Papiers et cartons
Fuelwood	m3	1920	1940	1970	2160	2160	Bois de chauffage
DERIVED PRODUCTION :							PRODUCTION CALCULEE :
Sawnwood	m3(s)	7263	8475	8475	8475	9055	Sciages
Wood-based panels	m3	1932	2577	3209	2637	3329	Panneaux à base de bois
Woodpulp	m.t.	605	698	747	674	779	Pâtes de bois
Paper and paperboard	m.t.	1216	1392	1512	1402	1672	Papiers et cartons
RECOVERY :							RECUPERATION :
Waste paper	m.t.	443	531	606	566	737	Vieux papiers
DOMESTIC SUPPLY :							APPROVISIONNEMENT NATIONAL :
Residues and chips	m3	1683	2353	2768	2772	3385	Déchets et plaquettes
DERIVED CONSUMPTION :							CONSOMMATION CALCULEE :
Sawlogs and veneer logs	m3 u.b.	11156	12921	12907	12921	13748	Grumes de sciage et placage
Total pulpwood	m3	6001	7541	8434	7509	8764	Total bois de trituration
Fibres for paper	m.t.	1245	1426	1549	1437	1713	Fibres pour papier
DERIVED REMOVALS :							QUANTITES ENLEVEES CALCULEES :
Sawlogs and veneer logs	m3 u.b.	10787	12552	12537	12552	13379	Grumes de sciage et placage
Smallwood	m3 u.b.	10410	11012	11297	10782	11200	Petits bois
Total removals	m3 u.b.	21197	23564	23835	23334	24580	Total quantités enlevées
FORECAST REMOVALS :							PREVISIONS DE QUANTITES ENLEVEES :
Sawlogs and veneer logs	m3 u.b.	10787	10340	11030	10710	11370	Grumes de sciage et placage
Smallwood	m3 u.b.	10410	11110	12170	11740	13590	Petits bois
Total removals	m3 u.b.	21197	21450	23200	22450	24960	Total quantités enlevées
DIFFERENCE :							DIFFERENCE :
Sawlogs and veneer logs	m3 u.b.	0	2213-	1508-	1843-	2010-	Grumes de sciage et placage
Smallwood	m3 u.b.	0	97	872	957	2389	Petits bois
Total removals	m3 u.b.	0	2115-	636-	885-	379	Total quantités enlevées

Trade is assumed to remain unchanged, at 1979-81 level.

Le commerce est supposé rester inchangé à son niveau de 1979-81.

PORTUGAL - PORTUGAL

	Unit Unite 1000	1979-81 (average) (moyenne)	Low-bas 1990	Low-bas 2000	High-haut 1990	High-haut 2000	
FORECAST CONSUMPTION :							**PREVISIONS DE CONSOMMATION :**
Sawnwood	m3(s)	1199	1550	1960	1700	2530	Sciages
Wood-based panels	m3	347	440	596	470	736	Panneaux a base de bois
Paper and paperboard	m.t.	432	490	680	580	1000	Papiers et cartons
Fuelwood	m3	5450	6770	8560	7200	9000	Bois de chauffage
DERIVED PRODUCTION :							**PRODUCTION CALCULEE :**
Sawnwood	m3(s)	2156	2506	2916	2656	3486	Sciages
Wood-based panels	m3	460	552	708	582	848	Panneaux a base de bois
Woodpulp	m.t.	777	806	927	855	1099	Pates de bois
Paper and paperboard	m.t.	519	576	766	666	1086	Papiers et cartons
RECOVERY :							**RECUPERATION :**
Waste paper	m.t.	171	204	290	253	467	Vieux papiers
DOMESTIC SUPPLY :							**APPROVISIONNEMENT NATIONAL :**
Residues and chips	m3	765	1135	1570	1439	2189	Dechets et plaquettes
DERIVED CONSUMPTION :							**CONSOMMATION CALCULEE :**
Sawlogs and veneer logs	m3 u.b.	4079	4784	5558	5066	6658	Grumes de sciage et placage
Total pulpwood	m3	3641	3922	4642	4161	5507	Total bois de trituration
Fibres for paper	m.t.	566	628	835	726	1184	Fibres pour papier
DERIVED REMOVALS :							**QUANTITES ENLEVEES CALCULEES :**
Sawlogs and veneer logs	m3 u.b.	4549	5254	6028	5536	7128	Grumes de sciage et placage
Smallwood	m3 u.b.	3870	5118	7187	5483	7873	Petits bois
Total removals	m3 u.b.	8419	10373	13216	11019	15001	Total quantites enlevees
FORECAST REMOVALS :							**PREVISIONS DE QUANTITES ENLEVEES :**
Sawlogs and veneer logs	m3 u.b.	4549	4970	5560	5180	5760	Grumes de sciage et placage
Smallwood	m3 u.b.	3869	3790	4420	4180	4980	Petits bois
Total removals	m3 u.b.	8419	8760	9980	9360	10740	Total quantites enlevees
DIFFERENCE :							**DIFFERENCE :**
Sawlogs and veneer logs	m3 u.b.	0	285-	469-	357-	1369-	Grumes de sciage et placage
Smallwood	m3 u.b.	0	1329-	2768-	1304-	2894-	Petits bois
Total removals	m3 u.b.	0	1614-	3237-	1660-	4262-	Total quantites enlevees

Trade is assumed to remain unchanged, at 1979-81 level.
Le commerce est suppose rester inchange a son niveau de 1979-81.

ROMANIA - ROUMANIE

		Unit Unite 1000	1979-81 (average) (moyenne)	Low-bas 1990	Low-bas 2000	High-haut 1990	High-haut 2000
FORECAST CONSUMPTION :	**PREVISIONS DE CONSOMMATION :**						
Sawnwood	Sciages	m3(s)	3648	3640	3640	3920	4220
Wood-based panels	Panneaux a base de bois	m3	1201	1698	2199	1818	2439
Paper and paperboard	Papiers et cartons	m.t.	698	760	820	840	980
Fuelwood	Bois de chauffage	m3	3400	3370	3780	3440	3780
DERIVED PRODUCTION :	**PRODUCTION CALCULEE :**						
Sawnwood	Sciages	m3(s)	4638	4629	4629	4909	5209
Wood-based panels	Panneaux a base de bois	m3	1619	2116	2617	2236	2857
Woodpulp	Pates de bois	m.t.	651	680	704	728	799
Paper and paperboard	Papiers et cartons	m.t.	824	686	946	966	1106
RECOVERY :	**RECUPERATION :**						
Waste paper	Vieux papiers	m.t.	187	227	269	267	351
DOMESTIC SUPPLY :	**APPROVISIONNEMENT NATIONAL :**						
Residues and chips	Dechets et plaquettes	m3	1383	1723	2066	2171	2665
DERIVED CONSUMPTION :	**CONSOMMATION CALCULEE :**						
Sawlogs and veneer logs	Grumes de sciage et placage	m3 u.b.	8415	8398	8402	8854	9348
Total pulpwood	Total bois de trituration	m3 u.b.	5968	6954	7842	7389	8710
Fibres for paper	Fibres pour papier	m.t.	911	979	1046	1068	1222
DERIVED REMOVALS :	**QUANTITES ENLEVEES CALCULEES :**						
Sawlogs and veneer logs	Grumes de sciage et placage	m3 u.b.	8576	8559	8563	9014	9509
Smallwood	Petits bois	m3 u.b.	9839	10330	11186	10388	11454
Total removals	Total quantites enlevees	m3 u.b.	18416	18890	19749	19403	20963
FORECAST REMOVALS :	**PREVISIONS DE QUANTITES ENLEVEES :**						
Sawlogs and veneer logs	Grumes de sciage et placage	m3 u.b.	8577	9090	9570	9370	10080
Smallwood	Petits bois	m3 u.b.	9840	10330	10920	10680	11480
Total removals	Total quantites enlevees	m3 u.b.	18417	19420	20490	20050	21560
DIFFERENCE :	**DIFFERENCE :**						
Sawlogs and veneer logs	Grumes de sciage et placage	m3 u.b.	0	530	1006	355	570
Smallwood	Petits bois	m3 u.b.	0	1-	267-	291	25
Total removals	Total quantites enlevees	m3 u.b.	0	529	740	646	596

Trade is assumed to remain unchanged, at 1979-81 level.
Le commerce est suppose rester inchange a son niveau de 1979-81.

SPAIN - ESPAGNE

	Unit Unite 1000	1979-81 (average) (moyenne)	High-haut 1990	High-haut 2000	Low-bas 1990	Low-bas 2000	
FORECAST CONSUMPTION :							**PREVISIONS DE CONSOMMATION :**
Sawnwood	m3(s)	3437	3940	5260	3680	4340	Sciages
Wood-based panels	m3	1415	1631	2236	1541	1626	Panneaux a base de bois
Paper and paperboard	m.t.	2673	3660	5840	3060	3950	Papiers et cartons
Fuelwood	m3	2000	2600	3600	2300	3100	Bois de chauffage
DERIVED PRODUCTION :							**PRODUCTION CALCULEE :**
Sawnwood	m3(s)	2453	2955	4275	2695	3355	Sciages
Wood-based panels	m3	1816	2031	2637	1941	2027	Panneaux a base de bois
Woodpulp	m.t.	1166	1595	2610	1329	1754	Pates de bois
Paper and paperboard	m.t.	2563	3549	5729	2949	3839	Papiers et cartons
RECOVERY :							**RECUPERATION :**
Waste paper	m.t.	991	1504	2575	1196	1623	Vieux papiers
DOMESTIC SUPPLY :							**APPROVISIONNEMENT NATIONAL :**
Residues and chips	m3	220	1014	2221	474	888	Dechets et plaquettes
DERIVED CONSUMPTION :							**CONSOMMATION CALCULEE :**
Sawlogs and veneer logs	m3 u.b.	4665	5606	8179	5124	6434	Grumes de sciage et placage
Total pulpwood	m3	7566	9641	14692	8397	10116	Total bois de trituration
Fibres for paper	m.t.	2454	3396	5482	2822	3673	Fibres pour papier
DERIVED REMOVALS :							**QUANTITES ENLEVEES CALCULEES :**
Sawlogs and veneer logs	m3 u.b.	3552	4493	7067	4011	5321	Grumes de sciage et placage
Smallwood	m3 u.b.	8620	10365	15106	9361	11363	Petits bois
Total removals	m3 u.b.	12172	14859	22173	13372	16685	Total quantites enlevees
FORECAST REMOVALS :							**PREVISIONS DE QUANTITES ENLEVEES :**
Sawlogs and veneer logs	m3 u.b.	3552	4620	5970	4350	5610	Grumes de sciage et placage
Smallwood	m3 u.b.	8620	13020	16600	11790	14920	Petits bois
Total removals	m3 u.b.	12172	17640	22570	16140	20530	Total quantites enlevees
DIFFERENCE :							**DIFFERENCE :**
Sawlogs and veneer logs	m3 u.b.	0	126	1098-	338	288	Grumes de sciage et placage
Smallwood	m3 u.b.	0	2654	1493	2428	3556	Petits bois
Total removals	m3 u.b.	0	2780	396	2767	3844	Total quantites enlevees

Trade is assumed to remain unchanged, at 1979-81 level.
Le commerce est suppose rester inchange a son niveau de 1979-81.

SWEDEN - SUEDE

	Unit / Unité 1000	1979-81 (average) (moyenne)	Low-bas 1990	Low-bas 2000	High-haut 1990	High-haut 2000
FORECAST CONSUMPTION : / PREVISIONS DE CONSOMMATION :						
Sawnwood / Sciages	m3(s)	5112	4660	4910	4840	5520
Wood-based panels / Panneaux à base de bois	m3	1228	1240	1395	1340	1755
Paper and paperboard / Papiers et cartons	m.t.	1661	1870	2340	2130	3130
Fuelwood / Bois de chauffage	m3	4500	7300	8900	11600	13000
DERIVED PRODUCTION : / PRODUCTION CALCULEE :						
Sawnwood / Sciages	m3(s)	11007	10555	10805	10735	11415
Wood-based panels / Panneaux à base de bois	m3	1885	1897	2052	1997	2412
Woodpulp / Pates de bois	m.t.	8632	8736	9003	8867	9389
Paper and paperboard / Papiers et cartons	m.t.	6198	6406	6876	6666	7666
RECOVERY : / RECUPERATION :						
Waste paper / Vieux papiers	m.t.	546	652	863	785	1280
DOMESTIC SUPPLY : / APPROVISIONNEMENT NATIONAL :						
Residues and chips / Dechets et plaquettes	m3	9967	9627	9960	10230	11474
DERIVED CONSUMPTION : / CONSOMMATION CALCULEE :						
Sawlogs and veneer logs / Grumes de sciage et placage	m3 u.b.	22147	21195	21697	21576	22996
Total pulpwood / Total bois de trituration	m3 u.b.	38196	38708	39955	39383	42031
Fibres for paper / Fibres pour papier	m.t.	6301	6511	6988	6775	7791
DERIVED REMOVALS : / QUANTITES ENLEVEES CALCULEES :						
Sawlogs and veneer logs / Grumes de sciage et placage	m3 u.b.	22058	21106	21608	21487	22907
Smallwood / Petits bois	m3 u.b.	26459	26506	28220	27470	29482
Total removals / Total quantites enlevees	m3 u.b.	48517	47612	49828	48957	52389
FORECAST REMOVALS : / PREVISIONS DE QUANTITES ENLEVEES :						
Sawlogs and veneer logs / Grumes de sciage et placage	m3 u.b.	22058	27940	28240	29910	30840
Smallwood / Petits bois	m3 u.b.	26460	21260	23460	36090	34860
Total removals / Total quantites enlevees	m3 u.b.	48518	49200	51700	66000	65700
DIFFERENCE : / DIFFERENCE :						
Sawlogs and veneer logs / Grumes de sciage et placage	m3 u.b.	0	6833	6631	8422	7932
Smallwood / Petits bois	m3 u.b.	0	-5246	-4760	8620	5378
Total removals / Total quantites enlevees	m3 u.b.	0	1588	1872	17043	13311

Trade is assumed to remain unchanged, at 1979-81 level.
Le commerce est suppose rester inchange a son niveau de 1979-81.

SWITZERLAND - SUISSE

	Unit Unité 1000	1979-81 (average) (moyenne)	Low-bas 1990	Low-bas 2000	High-haut 1990	High-haut 2000	
FORECAST CONSUMPTION :							**PREVISIONS DE CONSOMMATION :**
Sawnwood	m3(s)	2118	2230	2380	2340	2800	Sciages
Wood-based panels	m3	656	673	774	713	964	Panneaux a base de bois
Paper and paperboard	m.t.	1020	1080	1270	1240	1700	Papiers et cartons
Fuelwood	m3	810	860	900	1280	1750	Bois de chauffage
DERIVED PRODUCTION :							**PRODUCTION CALCULEE :**
Sawnwood	m3(s)	1710	1821	1971	1931	2391	Sciages
Wood-based panels	m3	736	752	853	792	1043	Panneaux a base de bois
Woodpulp	m.t.	284	295	371	356	543	Pates de bois
Paper and paperboard	m.t.	907	966	1156	1126	1586	Papiers et cartons
RECOVERY :							**RECUPERATION :**
Waste paper	m.t.	385	429	530	518	761	Vieux papiers
DOMESTIC SUPPLY :							**APPROVISIONNEMENT NATIONAL :**
Residues and chips	m3	568	642	748	732	987	Dechets et plaquettes
DERIVED CONSUMPTION :							**CONSOMMATION CALCULEE :**
Sawlogs and veneer logs	m3 u.b.	2510	2667	2906	2846	3572	Grumes de sciage et placage
Total pulpwood	m3	1908	1972	2340	2214	3128	Total bois de trituration
Fibres for paper	m.t.	848	903	1081	1053	1483	Fibres pour papier
DERIVED REMOVALS :							**QUANTITES ENLEVEES CALCULEES :**
Sawlogs and veneer logs	m3 u.b.	2613	2770	3009	2949	3675	Grumes de sciage et placage
Smallwood	m3 u.b.	1781	1822	2123	2394	3523	Petits bois
Total removals	m3 u.b.	4394	4593	5133	5344	7198	Total quantites enlevees
FORECAST REMOVALS :							**PREVISIONS DE QUANTITES ENLEVEES :**
Sawlogs and veneer logs	m3 u.b.	2613	2340	2480	2670	2910	Grumes de sciage et placage
Smallwood	m3 u.b.	1780	1950	2110	2320	2580	Petits bois
Total removals	m3 u.b.	4393	4290	4590	4990	5490	Total quantites enlevees
DIFFERENCE :							**DIFFERENCE :**
Sawlogs and veneer logs	m3 u.b.	0	431-	530-	280-	766-	Grumes de sciage et placage
Smallwood	m3 u.b.	0	127-	14-	75-	944-	Petits bois
Total removals	m3 u.b.	0	304-	544-	355-	1709-	Total quantites enlevees

Trade is assumed to remain unchanged, at 1979-81 level.
Le commerce est suppose rester inchange a son niveau de 1979-81.

TURKEY - TURQUIE

	Unit Unité 1000	1979-81 (average) (moyenne)	Low-bas 1990	Low-bas 2000	High-haut 1990	High-haut 2000	
FORECAST CONSUMPTION :							**PREVISIONS DE CONSOMMATION :**
Sawnwood	m3(s)	4641	5360	6900	6140	9670	Sciages
Wood-based panels	m3	481	580	810	640	1080	Panneaux à base de bois
Paper and paperboard	m.t.	554	760	1190	900	1740	Papiers et cartons
Fuelwood	m3	15700	14500	13000	17000	18000	Bois de chauffage
DERIVED PRODUCTION :							**PRODUCTION CALCULEE :**
Sawnwood	m3(s)	4650	5368	6908	6148	9678	Sciages
Wood-based panels	m3	480	580	809	640	1079	Panneaux à base de bois
Woodpulp	m.t.	270	387	645	457	898	Pâtes de bois
Paper and paperboard	m.t.	477	683	1113	823	1663	Papiers et cartons
RECOVERY :							**RECUPERATION :**
Waste paper	m.t.	218	314	504	390	824	Vieux papiers
DOMESTIC SUPPLY :							**APPROVISIONNEMENT NATIONAL :**
Residues and chips	m3	453	913	1694	1929	4489	Déchets et plaquettes
DERIVED CONSUMPTION :							**CONSOMMATION CALCULEE :**
Sawlogs and veneer logs	m3 u.b.	8022	9297	12050	10643	16887	Grumes de sciage et placage
Total pulpwood	m3	1877	2543	3882	2917	5282	Total bois de trituration
Fibres for paper	m.t.	499	712	1160	858	1733	Fibres pour papier
DERIVED REMOVALS :							**QUANTITES ENLEVEES CALCULEES :**
Sawlogs and veneer logs	m3 u.b.	4943	6217	8970	7564	13807	Grumes de sciage et placage
Smallwood	m3 u.b.	17442	16325	15283	18182	18887	Petits bois
Total removals	m3 u.b.	22385	22543	24254	25746	32695	Total quantités enlevées
FORECAST REMOVALS :							**PREVISIONS DE QUANTITES ENLEVEES :**
Sawlogs and veneer logs	m3 u.b.	4943	5300	5600	5570	6140	Grumes de sciage et placage
Smallwood	m3 u.b.	17439	17535	18125	18275	19595	Petits bois
Total removals	m3 u.b.	22383	22835	23725	23845	25735	Total quantités enlevées
DIFFERENCE :							**DIFFERENCE :**
Sawlogs and veneer logs	m3 u.b.	0	918-	3371-	1995-	7668-	Grumes de sciage et placage
Smallwood	m3 u.b.	0	1209	2841	92	707	Petits bois
Total removals	m3 u.b.	0	291-	530-	1902-	6961-	Total quantités enlevées

Trade is assumed to remain unchanged, at 1979-81 level.
Le commerce est supposé rester inchangé à son niveau de 1979-81.

UNITED KINGDOM — ROYAUME-UNI

	Unit Unite 1000	1979-81 (average) (moyenne)	Low-bas 1990	Low-bas 2000	High-haut 1990	High-haut 2000	
FORECAST CONSUMPTION :							**PREVISIONS DE CONSOMMATION :**
Sawnwood	m3(s)	8469	9040	9680	9590	11200	Sciages
Wood-based panels	m3	3379	3600	4098	3880	4938	Panneaux a base de bois
Paper and paperboard	m.t.	7161	7520	8630	8830	12310	Papiers et cartons
Fuelwood	m3	150	160	300	380	720	Bois de chauffage
DERIVED PRODUCTION :							**PRODUCTION CALCULEE :**
Sawnwood	m3(s)	1592	2162	2802	2712	4322	Sciages
Wood-based panels	m3	679	900	1398	1180	2238	Panneaux a base de bois
Woodpulp	m.t.	251	353	955	1091	2957	Pates de bois
Paper and paperboard	m.t.	3795	4154	5264	5464	8944	Papiers et cartons
RECOVERY :							**RECUPERATION :**
Waste paper	m.t.	2165	2436	2968	3037	4727	Vieux papiers
DOMESTIC SUPPLY :							**APPROVISIONNEMENT NATIONAL :**
Residues and chips	m3	567	978	1568	1486	2901	Dechets et plaquettes
DERIVED CONSUMPTION :							**CONSOMMATION CALCULEE :**
Sawlogs and veneer logs	m3 u.b.	2783	3933	5313	5068	8499	Grumes de sciage et placage
Total pulpwood	m3	1371	1793	3480	3968	9442	Total bois de trituration
Fibres for paper	m.t.	3871	4244	5378	5583	9138	Fibres pour papier
DERIVED REMOVALS :							**QUANTITES ENLEVEES CALCULEES :**
Sawlogs and veneer logs	m3 u.b.	2649	3799	5179	4934	8365	Grumes de sciage et placage
Smallwood	m3 u.b.	1670	1640	2836	3527	7885	Petits bois
Total removals	m3 u.b.	4320	5439	8016	8461	16251	Total quantites enlevees
FORECAST REMOVALS :							**PREVISIONS DE QUANTITES ENLEVEES :**
Sawlogs and veneer logs	m3 u.b.	2650	3300	4060	3680	5155	Grumes de sciage et placage
Smallwood	m3 u.b.	1670	2350	3190	2470	3895	Petits bois
Total removals	m3 u.b.	4320	5650	7250	6150	9050	Total quantites enlevees
DIFFERENCE :							**DIFFERENCE :**
Sawlogs and veneer logs	m3 u.b.	0	500-	1120-	1255-	3211-	Grumes de sciage et placage
Smallwood	m3 u.b.	0	709	353	1058-	3991-	Petits bois
Total removals	m3 u.b.	0	210	767-	2312-	7202-	Total quantites enlevees

Trade is assumed to remain unchanged, at 1979-81 level.

Le commerce est suppose rester inchange a son niveau de 1979-81.

YUGOSLAVIA - YOUGOSLAVIE

	Unit Unite 1000	1979-81 (average) (moyenne)	Low-bas 1990	Low-bas 2000	High-haut 1990	High-haut 2000	
FORECAST CONSUMPTION :							**PREVISIONS DE CONSOMMATION :**
Sawnwood	m3(s)	3499	4210	5730	4540	7120	Sciages
Wood-based panels	m3	1179	1371	1876	1451	2216	Panneaux a base de bois
Paper and paperboard	m.t.	1069	1350	2100	3150	2740	Papiers et cartons
Fuelwood	m3	6100	7400	7600	7800	8000	Bois de chauffage
DERIVED PRODUCTION :							**PRODUCTION CALCULEE :**
Sawnwood	m3(s)	4246	4957	6477	5287	7867	Sciages
Wood-based panels	m3	1291	1482	1988	1562	2328	Panneaux a base de bois
Woodpulp	m.t.	613	804	1327	2098	1687	Pates de bois
Paper and paperboard	m.t.	1076	1357	2107	3157	2747	Papiers et cartons
RECOVERY :							**RECUPERATION :**
Waste paper	m.t.	352	472	777	1165	1123	Vieux papiers
DOMESTIC SUPPLY :							**APPROVISIONNEMENT NATIONAL :**
Residues and chips	m3	846	1395	2370	2351	4793	Dechets et plaquettes
DERIVED CONSUMPTION :							**CONSOMMATION CALCULEE :**
Sawlogs and veneer logs	m3 u.b.	8199	9493	12383	10089	14949	Grumes de sciage et placage
Total pulpwood	m3	4622	5615	8766	11298	10757	Total bois de trituration
Fibres for paper	m.t.	1188	1498	2326	3486	3033	Fibres pour papier
DERIVED REMOVALS :							**QUANTITES ENLEVEES CALCULEES :**
Sawlogs and veneer logs	m3 u.b.	7626	8920	11810	9517	14376	Grumes de sciage et placage
Smallwood	m3 u.b.	6160	7801	10106	12928	10075	Petits bois
Total removals	m3 u.b.	13786	16721	21917	22445	24452	Total quantites enlevees
FORECAST REMOVALS :							**PREVISIONS DE QUANTITES ENLEVEES :**
Sawlogs and veneer logs	m3 u.b.	7626	9450	10840	9755	11235	Grumes de sciage et placage
Smallwood	m3 u.b.	6160	7850	8460	8045	8765	Petits bois
Total removals	m3 u.b.	13786	17300	19300	17800	20000	Total quantites enlevees
DIFFERENCE :							**DIFFERENCE :**
Sawlogs and veneer logs	m3 u.b.	0	529	971-	237-	3142-	Grumes de sciage et placage
Smallwood	m3 u.b.	0	48	1647-	4884-	1311-	Petits bois
Total removals	m3 u.b.	0	578	2618-	4646-	4453-	Total quantites enlevees

Trade is assumed to remain unchanged, at 1979-81 level.
Le commerce est suppose rester inchange a son niveau de 1979-81.

FUELWOOD SCENARIOS: ALTERNATIVE

FRANCE - FRANCE

	Unit Unité 1000	1979-81 (average) (moyenne)	Low-bas 1990	Low-bas 2000	High-haut 1990	High-haut 2000	
FORECAST CONSUMPTION :							**PREVISIONS DE CONSOMMATION :**
Sawnwood	m3(s)	11750	12230	14550	12930	17550	Sciages
Wood-based panels	m3	3315	3455	4181	3705	5351	Panneaux a base de bois
Paper and paperboard	m.t.	6203	7030	9100	8000	12250	Papiers et cartons
Fuelwood	m3	10000	10500	11200	13500	18500	Bois de chauffage
DERIVED PRODUCTION :							**PRODUCTION CALCULEE :**
Sawnwood	m3(s)	9467	9946	12266	10646	15266	Sciages
Wood-based panels	m3	3071	3211	3937	3461	5107	Panneaux a base de bois
Woodpulp	m.t.	1737	2143	3296	2609	4773	Pates de bois
Paper and paperboard	m.t.	5189	6015	8085	6985	11235	Papiers et cartons
RECOVERY :							**RECUPERATION :**
Waste paper	m.t.	1983	2390	3276	2880	4900	Vieux papiers
DOMESTIC SUPPLY :							**APPROVISIONNEMENT NATIONAL :**
Residues and chips	m3	4077	5079	6750	6286	10266	Dechets et plaquettes
DERIVED CONSUMPTION :							**CONSOMMATION CALCULEE :**
Sawlogs and veneer logs	m3 u.b.	17548	18430	22724	19754	28432	Grumes de sciage et placage
Total pulpwood	m3	10697	12392	17780	14462	24762	Total bois de trituration
Fibres for paper	m.t.	5109	5923	7961	6878	11062	Fibres pour papier
DERIVED REMOVALS :							**QUANTITES ENLEVEES CALCULEES :**
Sawlogs and veneer logs	m3 u.b.	18490	19372	23666	20695	29373	Grumes de sciage et placage
Smallwood	m3 u.b.	19990	21103	25492	24965	36257	Petits bois
Total removals	m3 u.b.	38480	40475	49158	45661	65631	Total quantites enlevees
FORECAST REMOVALS :							**PREVISIONS DE QUANTITES ENLEVEES :**
Sawlogs and veneer logs	m3 u.b.	18490	21495	23870	22700	26890	Grumes de sciage et placage
Smallwood	m3 u.b.	19990	20355	20120	22740	23990	Petits bois
Total removals	m3 u.b.	38480	41850	43990	45440	50880	Total quantites enlevees
DIFFERENCE :							**DIFFERENCE :**
Sawlogs and veneer logs	m3 u.b.	0	2122	203	2004	2484-	Grumes de sciage et placage
Smallwood	m3 u.b.	0	749-	5373-	2226-	12268-	Petits bois
Total removals	m3 u.b.	0	1374-	5169-	222-	14752-	Total quantites enlevees

Trade is assumed to remain unchanged, at 1979-81 level.
Le commerce est supposé rester inchangé a son niveau de 1979-81.

Annex A/Annexe A

National correspondents for ETTS IV
Correspondants nationaux pour ETTS IV

Austria/Autriche   Messrs. J. Foissner, H. Lackner and F.Tersch, Federal Ministry of Agriculture and Forestry

Belgium/Belgique   M. E. Clicheroux, Ministère de la Région Wallone

Bulgaria/Bulgarie   Mr. D. Doitchinov, Ministry of Forests and Forest Industry; Ms. Tosheva, Stara Planina; Ms. Dicheva, Lesoimpex

Cyprus/Chypre   Mr. V. Pantelas, Ministry of Agriculture and Natural Resources

Canada   Messrs. E.L. Kelly, K. Aird, J. Melnyk, Department of Industry, Trade and Commerce; Mr. D.F. Ketcheson, Canadian Forestry Service

Denmark/Danemark   Mr. F. Helles, Faculty of Forestry, Royal Veterinary and Agricultural University

Finland/Finlande   Messrs. K. Kuusela and L. Heikinheimo, Forest Research Institute; Messrs. V. Toppari and H. Valtanen, Central Association of Finnish Forest Industries; Ms. E. Rutanen, Forestry Council of the Central Union of Agricultural Producers

France   MM. P. Cros et P. Bazire, Direction des Forêts; M. J. Guillard, Conseil général du GREF, M. G.-A. Morin, DATAR

Germany, Fed. Rep. of
Allemagne, Rép. féd. d'   Mr. H. Ollman, Federal Research Centre for Forestry and Forest Products

Greece/Grèce   Messrs. G. Kokkinidis and S. Vogiatzis, Directorate of Planning and Studies (Forests), Ministry of Agriculture

Hungary/Hongrie   Mr. A. Madas, Vice-Minister (retired); Mr. A. Halasz, Head of Department, (retired); Mr. T. Bencze and Mr. J.Gemesi, National Planning Committee; Messrs. G. Varadi and A. Szenthe, Office for Forestry and Wood Industry, FAO Hungarian National Committee

Ireland/Irlande   Messrs. T. Rea, C. Tucker, G. Byrne, L. Condon, D. McGuire, L. Quinn, Department of Fisheries and Forestry

Israel/Israël   Mr. M. Kolar, Forest Department, Land Development Authority

Italy/Italie   Mr. A. Froncillo, Fedecomlegno; Messrs. S. Salvatici and G. Zanoni, General Directorate for the Mountain Economy and Forests

<u>Netherlands/Pays-Bas</u>	Mr. H.A. van der Meiden, Stichting Bos en Hout, Mr. T.M. Ritskes, State Forest Service
<u>Norway/Norvège</u>	Mr. K.E. Fjulsrud, Department of Agriculture; Mr.T. Slettemoen, Department of Industry; Mr. K. Morkved, Norsk Treteknisk Institut; Mr. E. Bohmer, Papirindustriens Forskningsinstitutt
<u>Poland/Pologne</u>	Mr. Z. Przyborski, Ministry of Forestry and Forest Industries; Mr. W. Strykowski, Wood Technology Institute; Mr. A. Bosiak, Forest Research Institute
<u>Portugal</u>	MM. M. Ferreirinha et J. Soares, Institut des Produits Forestiers; M. F. Veloso Gaio, Direction Génerale des Forêts; M. P. Mendes Cabrita, Estaçao Florestal Nacional
<u>Romania/Roumanie</u>	M. G. Borhan, Institut de Recherches et Projets pour les pâtes et les papiers; M. P. Necsulescu, Institut de Recherche et Projets pour l'industrie du bois, MM. C. Floricel, V. Moscu, R. Preda, spécialistes
<u>Spain/Espagne</u>	Messrs. J.J. Nicolas Isasa, I. Astorga, F. Barrientos, M.A. Gonzales Alvares, J. Martinez Millan, J. Pelfort Batalla, R. Velez Munnoz, ICONA
<u>Sweden/Suède</u>	Mr. K. Janz, National Board of Forestry, Mr. Z. Tamminen, Department of Forest Products, Swedish University of Agricultural Sciences; Mr. H. von Sydow, Joint Committee of the Swedish Forest Industry, Mr. L. Engstroem, Advisory Council for Forest Industry, Ministry of Industry; Mr. G. Lundbäck, Domänverket
<u>Switzerland/Suisse</u>	Messrs. A. Semadeni, M. Zanetti, Federal Forest Office; Messrs. E.P. Grieder, P. Meyer, Federal Polytechnic Zürich
<u>United Kingdom Royaume Uni</u>	Messrs. J. Oakley and G.M.L. Locke, Forestry Commission; Ms. R. Neale, Department of Industry
<u>United States of America/Etats-Unis d'Amérique</u>	Messrs. D.R. Darr and D. Hair, USDA Forest Service
<u>Yugoslavia/Yougoslavie</u>	Mr. D. Orescanin, consultant; Mr. I. Knezevic, Federal Institute for Social Planning; Mr. M. Petrovic, General Directorate for Forestry, Wood-working Industry, Pulp and Paper; Ms. M. Jakić, Federal Office of Statistics

In addition to the officially nominated correspondents the secretariat was helped by many experts participating in the activities of the Timber Committee, the European Forestry Commission or their subsidiary bodies. In particular the following experts contributed to the study by participation in the review meeting.

Outre l'assistance de correspondants nommés officiellement, le secrétariat a reçu celle de beaucoup d'experts participant aux activités du Comité du bois, de la Commission européenne des forêts ou de leurs organes subsidiaires. En particulier, les experts suivants ont contribué à l'étude par leur participation à la revue du projet de l'étude.

<u>Canada</u>	Mr. L. Reed, University of British Columbia, Mr. D. Luck, Canadian Forestry Service
<u>Czechoslovakia</u> <u>Tchécoslovaquie</u>	Messrs. A. Cerny and M. Zdarek, Ministry of Industry, Mr. J. Orech, Timber Industry Research Project, Mr. M. Kolejak, Lignoprojekt
<u>Netherlands/Pays-Bas</u>	Mr. K. Alkema, Ministry of Economic Affairs
<u>Poland/Pologne</u>	Mr. J. Kozinski, Ministry of Forestry and Forest Industries
<u>Portugal</u>	Mme. M.J. Pinto, Institut des Produits Forestiers
<u>United Kingdom</u> <u>Royaume-Uni</u>	Mr. D. Grundy, Forestry Commission
<u>United States of</u> <u>America/Etats-Unis</u> <u>d'Amérique</u>	Mr. D. Brooks, USDA Forest Service

Annex B

Data for EEC(12)

Throughout the study, one of the country groups has been the EEC(9) i.e. the European Economic Community as it was before its recent enlargements. Data are presented here for the EEC(12) i.e. the Community as it has been since January 1986, with the inclusion of Greece, Portugal and Spain.

Data for the EEC(12) are presented in Forest products statistics: production trade and apparent consumption, 1969-1984. For definitions etc., see source tables.

B.1 Population and GDP (annex tables 1.1 and 1.2)

	Unit	1965	1980	Low		High	
				1990	2000	1990	2000
Population	million	293.7	317.1	..	331.0a/	..	
GDP b/	billion US$ 1980 prices	1785.0	3004.5	3534.7	4840.3	3664.0	5357.5

a/ UN medium variant. 2025:333.3
b/ 1973: 2565.7

Rate of growth of GDP (% per year)

1965-73	4.6%
1973-80	2.3%
1980-2000	
Low	2.4%
High	2.9%

B.2 Forest resource around 1980 (tables 3.2, 3.6, 3.8)

B.2.1 Main land-use categories

	Unit	
Total land (excl. water)	1000 ha	222 112
Forest and other wooded land	1000 ha	66 898
- as % of total land	%	31.0
Exploitable closed forest	1000 ha	38 673
- as % of forest and OWL	%	57.8
Unexploitable closed forest	1000 ha	5 934
Other wooded land	1000 ha	24 291
Non-forest land, total	1000 ha	153 214

B.2.2 Forestry situation on exploitable closed forest

-	Growing stock, total	million m3 o.b.	4 340
	of which:		
	coniferous	"	2 248
	non-coniferous	"	2 092
-	Growing stock per hectare	m3/ha	112
-	Net annual increment (NAI),		
	total	1000 m3 o.b.	170 546
	of which:		
	coniferous	"	103 813
	non-coniferous	"	66 733
-	NAI per hectare	m3/ha	4.41
-	NAI as % of growing stock		
	total		3.93
	coniferous		4.62
	non-coniferous		3.19

B.2.3 Forest area by main ownership groups

Area covered	1000 ha	66 759
Publicly owned, total	1000 ha	26 956
as % of area covered	%	40.4
State owned	1000 ha	11 021
Other publicly owned	1000 ha	15 935
Privately owned, total	1000 ha	39 803
as % of area covered	%	59.6
Forest industries	1000 ha	513
Farm forests	1000 ha	27 734
Other privately owned	1000 ha	10 323

B.3 Forecasts for the forest resource (tables 5.18, 5.19, annex tables 5.2, 5.3)

	Unit	Base period	2000	
			Low	High
Area of exploitable closed forest (ECF)	1000 ha	39 267	41 659	43 798
Growing stock (GS) on ECF	million m3 o.b.	4 805	5 578	5 464
- coniferous	"	2 580	3 090	3 040
- non-coniferous	"	2 225	2 488	2 424
- GS per hectare	m3 o.b./ha	122	134	125
Net annual increment (NAI) on ECF	million m3 o.b.	171.53	187.24	197.27
- coniferous	"	104.21	117.31	125.21
- non-coniferous	"	67.32	69.93	72.06
- NAI per hectare	m3 o.b./ha	4.4	4.5	4.5
Removals, total	million m3 u.b.	114.27	134.94	151.63
- coniferous	"	60.31	73.38	87.91
- non-coniferous	"	53.95	61.56	63.72
- sawlogs etc.	"	52.72	64.63	72.45
- smallwood	"	61.52	70.31	79.18

B.4 <u>ETTS IV consumption scenarios</u> (annex tables 12.13 to 12.26)

	Unit	1979-81	Low		High	
			1990	2000	1990	2000
			(million units)			
Sawn softwood	m3	41.68	42.47	46.50	44.96	55.14
Sawn hardwood	m3	13.66	14.66	18.89	15.57	22.82
Sleepers	m3	0.86	0.86	0.86	0.86	0.86
Sawnwood, total	m3	56.20	57.98	66.24	61.38	78.81
Particle board	m3	15.84	17.20	21.03	18.28	25.45
Fibreboard	m3	1.95	1.97	2.26	2.11	2.79
Plywood	m3	3.90	4.02	4.74	4.37	6.09
Veneer sheets	m3	1.21	1.21	1.21	1.21	1.21
Wood-based panels	m3	22.91	24.40	29.24	25.97	35.54
Newsprint	m.t.	4.85	5.32	6.43	5.92	8.13
Printing and writing paper	m.t.	10.98	14.32	20.74	16.36	28.05
Other paper and paperboard	m.t.	20.41	20.21	22.19	23.87	32.50
Paper and paperboard	m.t.	36.25	39.86	49.38	46.13	77.67
Dissolving pulp	m.t.	0.83	0.74	0.68	0.74	0.68
Pitprops	m3	1.82	1.44	1.16	1.44	1.16
Other industrial wood	m3	4.46	4.46	4.46	4.46	4.46
Fuelwood	m3	26.29	29.97	34.47	34.68	44.63

B.5 <u>Production and self-sufficiency</u> (Timber Bulletin)

	Production		Self-sufficiency	
	1969-71	1979-81	1969-71	1979-81
	(million units)		(%)	
Sawnwood (m3)	29.22	30.71	54.4	55.8
Wood-based panels (m3)	12.76	18.74	83.2	80.5
Wood pulp (m.t.)	6.41	7.31	45.2	47.3
Paper and paperboard (m.t.)	22.06	17.76	77.0	76.3

B.6 Structure and capacity of the industries (annex tables 14.2 to 14.7)

B.6.1 Wood-based panels

	Capacity (1000 m3/yr)		Number of mills (No)	Average size (1000 m3/yr)
	1972	1982	1982	1982
Particle board	12528	17738	192	92
Plywood	2896	2094	271	8
Fibreboard	1436	1646	33	50

B.6.2 Pulp, paper and paperboard, total

	Capacity		Change 1970 to 1983	
	1970	1983	Quantity	%
		(1000 m.t./year)		
Pulp	8416	9360	+ 944	+ 11.2
Paper and paperboard	25760	31922	+ 6162	+ 23.9

B.6.3 Capacity by grade, 1983

Pulp

	1000 m.t./year	Share of total
Mechanical (incl. thermo-mechanical)	3467	36.8
Chemical for paper (incl. semi-chemical)	4986	52.9
Other fibre pulp	675	7.2
Dissolving pulp	292	3.1
Total pulp	9420	100.0

Paper and paperboard

	1000 m.t./year	Share of total
Newsprint	1977	6.2
Printing and writing	11242	35.2
Other paper and paperboard	18703	58.6
of which:		
household and sanitary	2184	6.8
wrapping and packaging	14611	45.8
Total paper and paperboard	31922	100.0

B.7 Residues and waste paper (annex tables 15.1 and 15.2 and ECE/FAO data base)

B.7.1 Residues, chips and particles

	Unit	1969-71	1979-81
Domestic supply	million m3	11.33	12.39
Recovery ratio	%	53.1	54.8
Apparent consumption	million m3	11.74	12.07
Imports	million m3	0.94	1.66
Exports	million m3	0.53	1.98
Net trade	million m3	− 0.41	+ 0.32

B.7.2 Waste paper

	Unit	1969-71	1979-81
Recovery	million m.t.	7.81	11.79
Recovery rate	%	27.3	32.4